D0200455

Into That Silent Sea

Outward Odyssey
A People's History of Spaceflight

Series editor
Colin Burgess

INTO THAT SILENT SEA

Trailblazers of the Space Era, 1961–1965

Francis French and Colin Burgess

With a foreword by Paul Haney

UNIVERSITY OF NEBRASKA PRESS • LINCOLN AND LONDON

Publication of this volume was assisted by The Virginia Faulkner Fund, established in memory of Virginia Faulkner, editor in chief of the University of Nebraska Press. ¶ ¶ Manufactured in the United States of America ¶ ♾ ¶ Set in Adobe Garamond and Futura by Kim Essman. ¶ Designed by R. W. Boeche.

Library of Congress Cataloging-in-Publication Data
French, Francis.
Into that silent sea : trailblazers of the Space Era, 1961–1965 / Francis French and Colin Burgess ; with a foreword by Paul Haney.
p. cm.—(Outward odyssey, a people's history of spaceflight)
Includes bibliographical references.
ISBN-13: 978-0-8032-1146-9
(cl. : alk. paper)
ISBN-10: 0-8032-1146-5
(cl. : alk. paper)
1. Astronautics — History — 20th century. 2. Manned space flight — History — 20th century. 3. Astronauts — Biography.
I. Burgess, Colin, 1947– . II. Title.
TL788.5.F74 2007
629.450092'2—dc22
2006026745

For Scott Carpenter and Gherman Titov

Sometimes being next is harder than being first.

We were the first that ever burst
Into that silent sea.

Samuel Taylor Coleridge,
The Rime of the Ancient Mariner

Contents

Illustrations

Foreword

Reflections of a Golden Era

During the extraordinarily dynamic period covered in this book, Paul Haney became widely known as NASA's "voice of Mission Control," and later its "voice of Apollo." A journalist and news editor from Akron, Ohio, Haney joined the fledgling space agency as an information officer based in Washington DC and later served in NASA's Public Affairs Office as its first news director. In 1963 he moved to Houston's Manned Spacecraft Center (later the Johnson Space Center [JSC]) as NASA's public affairs officer in their Office of Manned Spaceflight.

In October 1957 the Soviet Union launched its *Sputnik* satellite. It shocked, surprised, amazed, and confounded the federal establishment in Washington. At the time I was working for the 140-year-old *Washington Evening Star.*

Following the *Sputnik* announcement the *Star* commissioned me to do something that usually preceded a congressional vote on whether to go to war—a special section called a "man-in-the-street" reaction. The national reaction went deep: a majority decided there was something wrong with America's education system. Teachers began doubting the efficacy of the McGuffey Reader, an educational standard since the 1840s. At the very least, the Soviet space challenge ushered something called New Math into American grade schools.

In December 1958 I quit the *Star*, having accepted an invitation to join America's brand-new space agency, NASA, as their first news division chief. To me, the space challenge seemed capable of becoming at least as big as the opening up of the western United States, 150 years earlier. It was one of the few times in my life when I guessed right.

In early January 1959, NASA inherited eight Redstone boosters from the U.S. Army for use in its Mercury program. We wondered what fools would actually fly on these things. Three or four earlier rocket flights lofted by the U.S. Air Force had barely cleared their launch pads before blowing up in thousands of pieces.

For a while things were slow at NASA headquarters, located in the old Dolley Madison House across the street from the White House on Lafayette Square in Washington DC. There was plenty of time to read the morning *Post* and *New York Times*. The most exciting thing that happened back then was the day a yard of plaster let go from the ceiling of a downstairs parlor—which was serving as my new office. The plaster hit me squarely on the head.

One of the few launches that worked in those early days was the first of two 100-foot, self-inflating Mylar-plastic sphere satellites named Echo. It was sent up in 1959 from NASA's own special launch site on Wallops Island, Virginia. As the name suggests, radio-television signals could be bounced off these satellites. The inventor was a NASA man named William O'Sullivan. Thus the first Echo became known, at least in house, as one of O'Sullivan's Balls. An unexpected dividend was that O'Sullivan's Ball could be seen brilliantly on the horizon with a rising or a setting sun. For millions around the Earth, it was the first satellite ever observed. We got idle engineers using their handy slide rules to plot the times and places where the satellite could be viewed. Then we turned this information over to the wire services, who moved it to newspapers around the world. The *New York Times* went so far as to create a satellite-viewing "ear" at the top of page one, as part of their daily weather information. One day we got a call from a newspaper in India wondering when the satellite could be seen. Suddenly I knew NASA Public Affairs was in business.

Another incident involved a NASA experimental launch from the pad at Wallops Island, which was situated about one hundred miles south of Washington DC. Engineers wanted to see what impact a flash of light one hundred miles or so above the Earth would have while the East Coast was in darkness. I argued that we should warn the public beforehand that the test was coming. The all-knowing engineers running the show overruled me. The test lit up every emergency switchboard on the East Coast from Maine to

Florida. People wanted to know if that flash of light was some sort of atomic bomb test. Even the White House called to find out what was going on.

Looking back, I was privileged to observe the infant NASA in its bureaucratic birthing crib in Washington DC. The Eisenhower administration, like so many real-life fathers, wasn't sure what it had sired, so it was understandably reluctant to take too much credit.

In 1961 when Houston was selected as the home of what is now the Johnson Space Center, it was the last time in the modern era that cities bid for the right to be home to a new national facility. Houston's psychic payoff for volunteering came when *Apollo 11* landed on the moon, and Neil Armstrong's first words back to mother earth were: "Houston . . . the Eagle has landed."

In the course of 1961–63, everyone eventually "transitioned" to Houston and the new home the human spaceflight program was building on a swamp-flood plain twenty miles south of the city. I set up a weekly news conference that identified all coming events in the manned space effort—flight hardware tests in California, geology hikes in Iceland, jungle training in Panama, simulations in Houston. It was like opening my day-to-day office planner to anyone who cared to look.

The news organizations responded overwhelmingly. The wire services sent more hands to cover NASA in Houston full time. So did the major national newspapers and news magazines. We put leftover space hardware in the auditorium, which attracted casual Houston visitors from all over the world. The late Walt Disney and his entire board of directors spent three days at the center looking at high-speed NASA film techniques. The State Department began scheduling heads of state for visits to the center after meetings in Washington or the United Nations.

The associations I developed among that first cut of engineers and pilots who created the NASA manned flight program are inspiring—at least to me. People like Robert Gilruth, who graduated from the University of Minnesota. He stayed in school an extra year and got a master's degree, because in the early 1930s there weren't any jobs anywhere. Then came work at the NACA (National Advisory Committee for Aeronautics) at Langley Center in Virginia, airplane research during World War II, and rocket research. When NASA was chartered in 1958, Gilruth became the director of the Space Task Group at Langley, and three years later was appointed director of NASA's Manned Spacecraft Center in Houston. Gilruth once told me that the first thing he did on NASA's first day in 1958 was to write himself a one-sentence memo assigning himself direction of a manned flight program called Project Mercury. No one else had such authority under the confused hierarchy of the emerging NASA.

My first encounter with a real, live astronaut was on 8 April 1959—the day before the seven original astronauts were introduced to the Washington press corps. I met with the seven pilots in Walt Bonney's office—he was my boss, the director of public information. Bonney had four photos showing a rocket exploding at two-second intervals on the wall above his desk. Under the pictures were the words: "How to get on Page One." My instructions were to brief the pilots on the sort of questions they were likely to get from several hundred reporters. Topics like birthplace. Hometown. Education. Flying experience. Space goals. Married. Kids. I told them they could answer or not answer questions about their religious preference, but they would certainly be asked. These topics provoked a lot of discussion. Several decided to respond: "No religious preference."

We agreed it would be best to go down the line alphabetically so each man could answer, whatever the question. Gus Grissom was selected as spokesman for the pilots where one-answer questions seemed obvious. That was because he would be approximately in the middle of a single line of seated astronauts arrayed across the auditorium stage. There was also the fact that Grissom had a huge booming, broadcaster-type voice and sounded the most confident.

The pilots ranged in age from John Glenn at thirty-seven to Gordon Cooper, thirty-two. Three from the air force, three navy, and one marine—all test pilots. Because NASA was a civilian agency they dressed in civilian attire, but they were introduced according to their military rank. Office types would later call the group CCGGSSS, for that was the alphabetical order in which they would naturally line up—Carpenter, Cooper, Glenn, Grissom, Schirra, Shepard, and Slayton.

NASA Administrator T. Keith Glennan said a few words of introduction, and then the press conference was off and running. I quickly noticed that answers needing elaboration weren't really answered until Glenn and/

or Shepard answered them. Usually it was Glenn *and* Shepard answering. Glenn had by far the most experience as a test pilot and wartime aviator in both World War II and the Korean War. Shepard reasoned well, standing or sitting. It was a question from the late May Craig of the *Portland (Maine) Press-Herald* that broke up the conference. She asked what was the most difficult part of the physical at Lovelace Hospital in Albuquerque, New Mexico. With a straight face, Glenn began his answer. "Think of all the orifices in the human body, and then how far up any one orifice a medic might go . . . "

A roar of laughter swept the auditorium. Everyone stood and smiled at everyone else. The conference was over.

When the original seven astronauts moved in with NASA in the summer of 1959, they were secreted away on the NASA side of Langley Air Force Base, 150 miles south of Washington DC, for about eighteen months. Each pilot was given an assigned specialty to follow—certain kinds of boosters, spacecraft production, recovery, and communication.

There was almost no public contact for the pilots, except for personal stories, which ran in *Life* magazine. Under the *Life* contract, the pilots were paid about $15,000 each per year. That was in addition to their military pay, which ran about $12,000 a year. You can imagine how these "personal stories" went over with competing media. What NASA was trying to do was give the pilots a moat of privacy around their homes. The pilots were fair game for the press before, during, and after flight tests.

The original seven were clearly all for one and one for all. They met and voted frequently on issues, sometimes in unlikely places. I recall that during Al Shepard's parade in Washington, all seven wedged into a men's room inside the front door of the Foggy Bottom State Department. They were deciding how to respond to questions during an ensuing press conference. NASA Administrator James E. Webb viewed this as astronaut arrogance. But there was nothing he could do about it—not as long as John Glenn was giving First Lady Jackie Kennedy weekend water-ski lessons at Cape Cod.

I have been asked whom among the original group of pilots selected to fly on Project Mercury impressed me the most. I would have to say they impressed me about equally. I had no background in flying, so I just viewed them as highly intelligent human beings who enjoyed hard work and, with the exception of Glenn, the pleasures of the opposite sex.

Of all the early astronauts, I found Wally Schirra most interesting and spontaneously funny. He was also spontaneously serious at work. But he could take or deal a joke with the best, which always lightened otherwise heavy loads. We used to trade double and triple entendres to keep from falling asleep in some briefings. Wally was a renowned Mr. Eloquence and always ready with the incisive final word, but in the forty-odd years I've known him there was only one time when he had nothing to say. That was in the JSC Teague Auditorium in 1998, where we had gathered on 1 August for a tribute to Al Shepard, who had died on 21 July. Wally got up on stage but could hardly say a word he was so choked up, and then he broke down. It was the only time I ever knew this wonderful guy to be lost for words.

Schirra showed class, I thought, by resigning immediately after he captained the first postfire Apollo mission. Grissom, who was killed in the *Apollo 1* fire, was Schirra's next-door neighbor in the Clear Lake area. Schirra appeared to have promised Gus's ghost that he would do the mission correctly. And he did, despite lots of disputes with Mission Control about when scheduled tasks were going to start and end.

Scott Carpenter was another of the original seven guys I liked. He was truly a master of his emotions. The only exception to this that I noticed was when the two of us were leaving Baltimore on a 707. We were rolling up to near rotate (takeoff) speed when the plane began slowing dramatically. "Guess we're not going just yet," I said to Scott innocently. "That," said Scott, fists gripping the seat rails, "was an aborted takeoff!" Somehow I got the idea that it was a bad thing.

After Scott Carpenter's flight, there was a huge lunch at the Waldorf in New York to honor Scott for his Mercury mission, with a double tier of guests from the political and business communities. Among the honored guests, Walt Williams, deputy director of the Manned Spacecraft Center for Flight Operations, wound up sitting next to former president Hoover, who was almost deaf. Hoover, I noticed, was dozing while NASA Administrator Jim Webb was at the microphone expounding on NASA's plans. A burst of applause woke Hoover up. At this point, he put in his earpiece and asked Williams who was speaking. Williams told him, "Webb, the head of NASA." Whereupon Hoover took out his ear piece and went back to sleep.

Gordon Cooper—Gordo—was always the "kid" of the group. He was the youngest, and he rarely had much to say except when discussing space hardware, simulators, and such. When Tom Wolfe's book *The Right Stuff* came out, poor Gordo was saddled forever with that pompous line, "Who's the best pilot you ever saw? You're lookin' at him!" This came about when he was discussing his persona with Wolfe, and the author asked him if that was something he was likely to say. He squirmed, shrugged, and said, "I guess so." And that was it—the later movie almost made it the only thing Gordo said. Everywhere he went, people always dropped that corny line on him. The funny thing is, that was not really Gordo stuff—he was far more unassuming and reserved. But at least he was gracious enough to give the expected response. Sometimes.

One of the most beautiful postflight parades was Gordo's. He and his first wife, Trudy, had met at the University of Honolulu, so we had a parade through that city, whose locals have a custom of throwing rose petals. It was beautiful, but he probably didn't make any friends in his native Oklahoma.

Deke Slayton was your basic good ol' midwestern boy from Sparta, Wisconsin. He had done the "milk run" in the late stages of World War II, flying out of Africa and later Italy to bomb the Polesti oil fields. Sadly for Deke he brought along some unknown baggage—he had a slight heart murmur, and concerned doctors grounded him a few weeks before he was due to fly his *Delta 7* mission. Losing his flight status really burned Deke, but he overcame this to become head of the astronaut office and the guy who selected all the early crews. He shouldered a lot of executive responsibility, and the younger pilots would cringe when they were called into his office. No question, he was the boss man in the pilots' building. And of course he finally got to fly on that Apollo-Soyuz test mission in 1975.

One night around midnight late in 1968, Deke and I walked out of Mission Control together after the first shift. We both looked up at the moon because it was an unusually bright night. Frank Borman and his *Apollo 8* crew were on their early orbits around the moon. "Think of it; those three fuckers up there orbiting that thing," Slayton said. Sometime after Deke died of cancer in June 1993 I told Bobbie, his second wife, about that incident. "He really had a way with words, didn't he?" she laughed.

I also recall as we were leaving the center for Washington and a joint session of Congress after *Apollo 8*, Deke and I wound up sitting together. *Apollo 8* had been a trendsetter; flights like *9* and *10* would become much easier. "What kind of date do we have on *11*?" I asked Deke. "If everything goes perfectly, we'll land on the 20th of July," he said. "I'm going to hold you to that," I told him. "That's my birthday—my forty-first birthday." Good old Deke—he didn't let me down.

Senator Stuart Symington (D-Missouri) coined the phrase *missile gap*, which came into play for the first time in the 1960 presidential campaign. John F. Kennedy, who won the presidential sweepstakes that year, gave an interview to one of the space magazines during the race. The story entertained the thought of changing the basic booster of the then evolving Mercury program, something that scared the hell out of the fledgling NASA program engineers. The idea never surfaced again. Yet three weeks after Alan Shepard's first manned ballistic flight of the same Mercury program, in May 1961, Kennedy went before Congress to propose a race with the Soviet Union to see who could put a man on the moon first. It was as if we entered into a space Olympic contest. All the space boosters of war could be modified for use in this program, including some new ones that no one had dreamed of.

A few days before Kennedy delivered his lunar proposal, he sat in the Oval Office and toyed with the seven astronauts. Did we really need a manned space program? Wouldn't instruments or computers do just as well? I was privileged to sit in on that dialogue; the heads of several space committees; Jim Webb, head of NASA; and Vice President Lyndon B. Johnson were also there. JFK was obviously prepping himself for questions he would get should he go ahead with his lunar proposal. As usual, Shepard and John Glenn supplied most of the answers.

Kennedy conceded later that he didn't give a hoot about the science involved in going to the moon. He thought it would be a worthy contest: something like a horse race between two gentlemen owners. This was the same man who, only weeks earlier on 17 April, had pulled the air cover for a loose, unsuccessful 1,300-man hired-patriot invasion of Cuba at Bahia de Cochinos—the Bay of Pigs. The same man who supported a few hundred U.S. troops fighting-training in Vietnam, just as the Eisenhower administration had.

Once we got into the flight phase of Mercury from May 1961 virtually all of the work was done at Cape Canaveral. Flight controllers, simulations, mission control, booster status—all these functions were handled at the Cape. By the end of the Mercury program in the middle of 1963, the space business was booming. The new center was under construction in Houston, the Cape was trebling in size, and centers had sprouted in the Washington area, Alabama, and Mississippi. A company called IBM invited me to join them in a new division building in suburban Washington.

The night I was due to go to New York and give IBM my answer, Jim Webb asked me to take the senior information job in Houston. Webb's question to me was several hundred words and one or two wars long. It covered a much larger U.S. involvement in Southeast Asia, countering a much larger Russian involvement in the same region. "For those and other considerations, how and where would you like to serve your country?" Webb asked. "Houston," I said.

So I made some youthful mistakes. Squandered opportunities. But it was still one helluva ride.

The war in Vietnam escalated beyond belief under President Johnson. Somewhere in there the first manned Gemini rendezvous took place, two hundred miles above Saigon. Even so, more and more were questioning why America was pursuing a civilian-oriented space program and a war simultaneously.

During President Johnson's term, he stopped in Houston frequently on Friday nights on his way to his ranch, two hundred miles west of the space center. When things went bad briefly during *Gemini 8*, NASA Administrator Webb called me in Mission Control from the White House. "Paul, the president wants you to know that you're doing a great job in handling the situation." You can imagine how all this sat with my tiers of public information bosses in Washington. If they insisted on "coordinating" whatever news we released at weekly Houston briefings (as in ensuring they released the news first in Washington), they came close to violating the golden "no-scoop" rule. I had them exactly where I wanted them.

Because of the media contract mentioned earlier, it was my responsibility to review each astronaut story before it went to *Life* to ensure that it was "personal" and had nothing to do with day-to-day NASA hardware. In eight years, I killed only one story. It was a proposed *Life* piece, with astronaut

Jim Lovell reviewing some lunar film he'd seen at the Jet Propulsion Laboratory in Pasadena. But this was before any other run-of-the-mill reporters had been given a chance to look at these pictures. In fairness, they might want to consider writing about the images themselves, and *Life* was not allowed to scoop the writing market on privileged stuff. Lovell wasn't happy with the decision. Boss astronaut Al Shepard huffed and puffed at me a bit, but the people running *Life* in New York agreed with me.

On another story, involving astronaut Ed White, I tried but failed to persuade him to tone down his patriotic personal approach to the story of his first spacewalk in *Gemini 4*. "I felt red, white and blue all over" when I stepped out the spacecraft, read the lead paragraph in the piece *Life* was proposing to run. "Come on, Ed," I chided. "Nobody feels *that* patriotic." But in the end, White *did* feel that patriotic. I read it in *Life*.

And then, of course, on the other side of the world, we had the cosmonauts. What did we know of them? Almost nothing, except for occasional leaked stories out of Moscow. The Russians, meanwhile, were maintaining their space lead over the United States. NASA's astronauts were flying the Mercury spacecraft, which offered half the volume of a typical phone booth. The Gemini two-man spacecraft was bigger, but it was still a squeeze.

After their success with the one-man Vostok craft, the Russians began flying multi-man crews in something called Voskhod. I seem to recall a Russian spacecraft moving over Florida that actually called in to Mission Control at the Cape when we were doing a flight simulation. There was the usual chatter to establish the line between ground and spacecraft; then the conversation was put on speakers in Mission Control. One of the Russians with excellent English (Alexei Leonov, I believe) commented about the weather over Florida. Then he said: "We're just getting ready for a game of handball. Got to keep fit, you know. Over." That may have been the high point of Russian–U.S. space competition.

Over time, politics changes everything. With Nixon in the White House, and just before the first manned landing on the moon in July 1969, the senior Public Affairs person in Washington was Julian Scheer. In the midst of preparing for the upcoming *Apollo 10*, he thought he had found a way to transfer me back to Washington and decided I should choose between heading the mission commentary team or doing the day-to-day job of running the office in Houston. I had been doing both for about ten years. It was a matter of identification. Few people knew who the head of CBS was, but everyone knew Walter Cronkite. Still, I thought over the choice and told Scheer I wanted to stick with the mission business, particularly as we approached the first of several manned lunar missions.

A few days later, I got a Scheer letter directing me to report to a new job in NASA's Washington headquarters. I asked him if I had made the wrong choice in Houston, if indeed I ever had a choice. No answer. So I had an option: I quit. The following day, I went to London at the invitation of the BBC. After a week of negotiation, I went to work for ITN, the commercial network in England, as commentator for manned spaceflights.

As I commuted between London and Houston, *Apollo 10* flowed into *Apollo 11*. The whole world stood still listening to Neil Armstrong and Buzz Aldrin land on the lunar surface on 20 July 1969. In time for *Apollo 12*, ITN opened a new broadcast center in downtown London. Queen Elizabeth herself showed up to dedicate the new facilities. During a break, the queen walked up to me and asked: "Mr. Haney, how is Mr. Neil Armstrong's health?" I didn't presume to ask why she was asking. I simply said, "Just fine, as far as I know, ma'am." That was the entire conversation.

Several weeks later, back in Houston, I ran into Armstrong at lunch. "A lady in London was asking about your health during *Apollo 12*," I told him. His face went whiter than usual as he said, "That must have been the queen!" I felt like I was being caught, but I couldn't tell by whom. After *Apollo 11*, Armstrong explained, he and his crew started a one hundred–day tour of all the world's capitals. By the time they got to London, about the fifth stop, Armstrong had an old-fashioned cold, so thick he couldn't talk. He thought about canceling a visit to Buckingham Palace to load up on cold medicine and sleep, but his wife Jan had other ideas. "She told me that if I had to be embalmed, we were going to the palace. She wanted to see the place," Neil recalled. After meeting the little princes and princesses and sipping white wine for maybe an hour, Armstrong said he was close to the queen's ear as they were leaving. She knew he had a cold. He said he tried to say something like "Thanks—we had a great time." Instead, he coughed in her face. Mortified, he again tried to mouth an apology and hit her in the face with another cough. She held up both hands in surrender.

Now, months later, she asks me innocently about Neil's health, knowing full well that I would go back to Houston and do what I did. It will forever be known as the "Queen Gotcha."

It is difficult to encapsulate or explain what this heroic era meant to me, particularly within the constraints of a few pages. With confidence, I leave it up to the authors of this worthwhile volume to flesh out the full story, to tell something more substantial about the men and women behind the missions. Knowing of their dedicated research and enthusiasm, I believe they will do eminent justice to this truly remarkable story.

Paul Haney
High Rolls, New Mexico

Acknowledgments

The authors would like to express their sincere gratitude to those who helped in the research and compilation of this book, which relates a number of complex and very personal stories. Such an undertaking would have been extremely difficult—if not impossible—without their support and assistance.

Our foremost thanks must go to the spacefarers, both Russian and American, who willingly gave their time and help. All have been interviewed hundreds of times over the decades, and yet, despite often demanding schedules, many took the time to invite us into their homes and confidences and gave fresh, insightful perspectives into their lives, characters, and experiences. They also provided us with introductions to others we had hoped to interview.

For answering all of our questions so helpfully and thoroughly, our thanks go to: Buzz Aldrin, Valery Bykovksy, Mike Collins, Konstantin Feoktistov, Valeri Kubasov, Alexei Leonov, Jim Lovell, Pam Melroy, Ed Mitchell, Pavel Popovich, Jack Schmitt, Valentina Tereshkova, Al Worden, and Alexei Yeliseyev. Dick Gordon also graciously allowed us to use fully an earlier interview we had conducted with him.

Going above and beyond the call of duty, Bill and Valerie Anders, Scott Carpenter, Gene Cernan, Walt and Dot Cunningham, Charlie Duke, Wally and Jo Schirra, Rusty Schweickart, and Tom Stafford not only provided extensive interviews but also proofread drafts and offered expert suggestions and amendments. In many instances, this process uncovered stories they had long forgotten, which appear here in print for the first time. Our special thanks also to Suzi Cooper and her husband, the late Gordon Cooper. Gordo was most generous with interview time and in responding to questions. Sadly, Gordo passed away soon after being sent some final manuscript drafts of this book, which he'd agreed to proofread.

Those who stay behind on the ground are often just as vital as those who get to fly into space. Key interviews and valuable editorial suggestions were made by spaceflight notables and family members Max Ary, Cece Bibby, Wally Funk, Lowell Grissom, Paul Haney, Gene Kranz, Jim Lewis, Dee O'Hara, and Guenter Wendt. In some cases the events covered were controversial, and we are most grateful that the participants found this project important enough to have the story told correctly, with history in mind. Special and abundant thanks are due to Kris Stoever, who devoted a great deal of time and effort to ensuring that we attained the utmost accuracy.

The aerospace history community is a surprisingly small one, and we were especially grateful that some of its principal researchers were so willing to help with this project. Our thanks go to Martha Ackmann, Jacque Boyd, Jim Busby, Igor Lissov, Tom Neal, Curt Newport, Tony Quine, Jake Schultz, Maciej Stolowski, and Larry Turoski. Photo researcher Jody Russell at the Johnson Space Center Media Resource Center assisted with photographs for this book, while University of Houston archivist Shelly Kelly was also most helpful in tracking down long-forgotten documentation. Others who provided much-needed assistance were Kerrie Dougherty, Elena Esina, Joseph Frasketi Jr., Ed Hengeveld, Lawrence McGlynn, Bob Northcutt, David J. Shayler, and Liz Warren. Erin French provided meticulous proofreading, and Sonia López assisted in reproducing old photographs.

Particular thanks must also go to noted space historians Mike Cassutt, John B. Charles, and British Interplanetary Society president Rex Hall, MBE. They devotedly read over each book chapter and provided insights that considerably sharpened the final drafts. We are grateful to them for their ongoing help and their enduring friendships. Another major contributor was Bert Vis, an extraordinarily dedicated researcher of the Soviet Union/Commonwealth of Independent States whose interviews and advice provided us with numerous firsthand quotations, revealing many previously unknown facts about the cosmonauts and their missions from the early days of the Soviet space program.

The authors would also like to acknowledge the skills of copyeditor Paul Bodine, whose subtle changes have lent even greater power to this work. We also appreciate the support and encouragement of the entire team at the University of Nebraska Press, particularly that of Director Gary Dunham and Assistant Managing Editor Linnea Fredrickson.

This book is a tribute to the welcome assistance of all those named above as well as other contributors, each of whom lent a willing hand in helping us research and complete this extensive and enormously worthwhile project.

Introduction

One momentous event begins this book, and another closes it. In April 1961 a human being rocketed into space for the first time, and in March 1965 another human floated out of a spacecraft on the first-ever spacewalk. The technological, political, and cultural momentum behind these two historic events, and the steps taken in between them, have already been well chronicled. This book certainly makes reference to these broader stories; it does not attempt to explain them.

Many books about this period of rapid spaceflight advances have described the space programs of the world in grandiose, lofty terms. Without doubt, enabling a human to leave the confines of Earth's atmosphere for the first time was a defining moment in human evolution, a bold step into a wider universe and understanding of our place within it. Space exploration has added to the general body of human knowledge, allowing us a new and remarkable view of who we are as a species and where we live. Today, it is almost impossible to imagine a time when we saw the Earth as the limit of our reach or when we had not seen photographs of our planet from above the atmosphere.

Into That Silent Sea describes in vivid detail the key moments when such visions of humankind's future and place were first realized, but it does not seek to give an overall view of what we have learned from flying into space. That is left to the large body of authoritative works on the subject. There are also histories that examine our push into space in the 1960s from a far more down-to-earth perspective. Undeniably, the pursuit of technological advances and achievement in that decade was due primarily to Cold War politics and national prestige. Both played an integral role in explaining the rapid advance of the so-called space race and why it stagnated when the decade was out.

Other authors have chronicled the timing that allowed innovative and even daring technology, ever-shifting global politics, and key political decision makers to come together in perfect symmetry, setting in motion a race to the moon between two resolute and unsubmissive superpowers. These events form a fascinating backdrop to the story, but the reader must seek the full details elsewhere.

What then is this book about? The answer is simple—it is about people. In particular some extraordinary people who, during that amazing era, found themselves propelled into space on top of unreliable rockets, ready to take a decisive step into a largely unknown realm. The massive and hugely expensive effort to hurl them into space on missions of rapidly escalating complexity was sustained by some extraördinary engineering technology, most of which was on the very leading edge of what was technically possible.

These pioneering space explorers ranged from test pilots to medical doctors, from spacecraft engineers to a textile mill worker, from backgrounds both military and civilian and from ages ranging from the early twenties to the first clutches of middle age. Most were engrossed in pursuing other careers when a fortunate combination of luck and good timing put them in the right place at the right time, and with the right qualifications. This book does not seek to mythologize them as Cold War warriors or as representatives of a new era in human evolution. It does not attempt to denigrate their achievements by questioning the costs associated with the space program, either in lives or in national budgets. Instead, it looks at their lives as people: who they were and what happened to them as they stepped voluntarily into mankind's greatest ever technological and scientific undertaking.

As Yuri Gagarin, the first person to travel into space, once said: "There is more than a bit of romance in being a space traveler, but by now we all know that the path into space is not strewn with roses. And those who have taken this path are not fanatics, robots or cogs in the cosmic mechanism, but brave and determined people. In each of them there is something unique and peculiar to them alone." From Gagarin onward, the astronauts and cosmonauts were always the first to tell interested admirers that their participation was just one facet of the space program. Supporting them were rocket designers, engineers, agency chiefs, and thousands of others who made the biggest decisions, and without whom there would have been no space program. Some of these individuals' stories are also told in this book, many for the first time, but the focus is always on those who flew the rockets. Whether the astronauts and cosmonauts were the most important or influential people in the space race is not the point; they were the people who willingly climbed into a spacecraft, accepted the odds, and risked their lives to pursue a dream. They underwent an experience that could not be fully predicted or imagined until it actually took place.

Where possible, both authors have talked directly with the space explorers from this era, their families, and others who knew them best. Memories of decades-old events can sometimes be hazy, and it is also a common human trait to want to tell a story in the way we'd like it to be remembered. The authors, however, wanted to step aside and allow these spacefarers to recount their stories through their own, often vivid, memories.

Every effort has been made to check each transcript against verifiable facts, and readers accustomed to official space agency accounts of these spaceflights may question a few surprising details. During interviews, several spaceflight veterans pointed out where official accounts of their missions were in error. The authors have chosen to transcribe and relate their stories in the same good faith in which they were given. This is by no means the first book to cover these flights, yet the subjects candidly told us many stories that have never before been recounted in print. Some will even come as a surprise to those who flew with them.

The reader will share the experiences of an unusual cross-section of highly motivated individuals. For the most part, they had the same strengths and failings as the person next door, but these men and women became part of an extraordinary and unprecedented undertaking. The space program changed the lives and philosophies of some; others used their newly acquired influence to change the space program more to their liking. A few fell victim to internal office politics or misplaced perceptions and prejudices. Others found that the respect and affection afforded to spacefarers were a way to reach even greater prominence and importance. Some left their space careers behind them and moved on completely. Others could not.

Whatever became of them, these pioneers of spaceflight history were the ones who did these extraordinary things, who went farther than humans had ever ventured before.

They were the first to fly—ever higher, ever longer, ever faster—into that silent sea of space, the new frontier.

Into That Silent Sea

1. First to Fly

I saw Eternity the other night,
Like a great ring of pure and endless light,
All calm, as it was bright

Henry Vaughan

When venturing into the unknown, the first step taken is often the biggest and the boldest. A young Russian pilot named Yuri Gagarin took humankind's first step into space. He died in his mid-thirties, so his image is fixed: a youthful icon symbolizing the first human journey above our planet. As President Lyndon B. Johnson wrote, "Yuri Gagarin's courageous and pioneering flight into space opened new horizons and set a brilliant example for the spacemen of the two countries."

Like many others who are remembered more for what they did than for who they were, Gagarin's life was far more complicated than the smiling photos in the history books would have you believe. "Gagarin is often spoken of as if he were an absolutely straightforward and simple person," his cosmonaut colleague Konstantin Feoktistov once said. "In fact, he was not at all as simple as it might seem at first glance."

Gagarin's life was tragically brief, yet he experienced more than most people ever will. His sudden and almost unprecedented fame brought its share of negative consequences, putting his basically honest and positive character through some severe tests. Yet he died while still trying to push his own personal development; despite numerous setbacks, he never gave up. It is not surprising that he had this strength of character considering he was lucky to survive his own childhood.

Anyone seeing the eleven-year-old Yuri Gagarin, just sixteen years before he made his historic spaceflight, would never imagine that he would be the one. The impish and mischievous young boy was a compulsive prankster, and his rudimentary schooling consisted of lessons culled from scavenged military maps and manuals left behind from a war that had devastated his homeland. It would scarcely have been possible to meet the ragged farm boy in 1945, living in a rural house made from the salvaged ruins of an earlier war-damaged home, and believe that in a few short years he would be flying a technological marvel. It would also have been difficult to imagine that his home country, shattered to the core by a ruinous war, could rise to such technological heights so quickly. Yet Gagarin and Russia pulled themselves out of the devastation and carried out one of mankind's greatest achievements.

Yuri Alexeyevich Gagarin was born on 9 March 1934 in Klushino village, near the eastern edge of the USSR's Smolensk region. Though the Gagarin family worked on the local collective farm, they had not always been farmers. Yuri's mother, Anna, had grown up in the big city of St. Petersburg and brought some of the city culture to her family's otherwise rural life. Yuri was the third of four children, and his mother ensured that her children either read or were read to every bedtime. "To this very day," Anna later wrote, "if I have an interesting book, I prefer to read it all night rather than put it down halfway through." The cosmonaut described his mother as a "well-read and well-informed person. She was always ready with an answer to her children's endless questions. I owe her everything I have achieved in life." Yuri learned to read by watching and listening to his older siblings doing their homework. In a village with no electricity, no running water, and only one radio, where much of the time was spent tending animals, the Gagarin children grew up with a slight educational edge over many of their peers.

Yuri's father, Alexei, was a carpenter. He maintained the collective farm's buildings and had built the family's house by hand from boulders and pine logs. His life had never been easy; his own father had been an alcoholic who often beat his mother before finally leaving for good when Alexei was four years old. This meant Alexei had never had much of a school education; he'd worked since childhood as a shepherd, providing a vital cash flow for his mother. "His entire education consisted of two years' schooling," according to Yuri, who described his father as "strict, giving us our first lessons in discipline and respect." Alexei did not value school education for his children, but did encourage his sons to develop practical skills. Yuri grew up around his father's tools and absorbed the concept of precise technical expertise. He was so used to being around carpentry that the young boy learned to distinguish different types of wood by smell alone. His father encouraged his interest and hoped that all three of his sons would one day work with him, as a team of carpenters.

Despite having parents who cared for him in their own very different ways, Yuri's childhood was neither happy nor easy. When the boy was seven years old, Hitler declared war on the Soviet Union. Anna cried when she heard that the country was at war; she remembered the suffering of World War I all too well. Yet even she could not have imagined what was to befall the family.

In 1941 the Nazis advanced rapidly into the Soviet Union, pushing deep into the country and capturing the areas near Klushino. "The war came nearer and nearer," Yuri remembered, "like the rising water at flood-time, until it reached our region." The advance was temporarily halted when, caught unprepared by the Russian winter, the invasion force was forced to gradually retreat. Those in the village able to fight left to join the army, taking the most valuable farm equipment with them and leaving the women, elderly, and children to run the farms. Soon, however, the precious animals were also taken deeper into Russia in the hopes of saving them. The front line of combat pushed forward and back across the Smolensk region, and was soon sweeping around Gagarin's village. There were battles in the surrounding woodlands, and Klushino's prominent landmarks were shelled.

Gagarin's older brother Valentin, too young to be a soldier, was put to work digging trenches around the village, but such efforts would not be enough to slow the invaders. Refugees soon appeared, pouring through Klushino to make their way to safety, drinking from the Gagarin family well until it ran dry. Retreating civilians were soon replaced by retreating Russian soldiers, who ate what was left of the farm supplies. Reports arrived in the village with increasing frequency of sons, fathers, and neighbors killed on ever-closer battlefields. It was an abrupt end to the innocence of childhood for young Yuri, who saw the cruelties of war played out in front of him on a daily basis. It was also, Valentin believed, when Yuri learned perseverance, courage, and the strength to risk his life for others.

Alexei decided that his family should flee the area too, but by the time they had packed up their belongings it was too late; the village was surrounded by German forces. By the end of 1942 Klushino was occupied by the Nazi invaders, who showed no hesitation in killing any civilians who offered resistance or otherwise questioned their authority. One particularly brutal German officer hung Yuri's younger brother, Boris, from a tree to die; Anna managed to save her child just in time. "We took him back," Yuri recalled grimly, "and with great difficulty brought him back to consciousness." Boris could not walk for a month, and his sleep was filled with nightmares. It may be that he never fully recovered; years later, he took his own life by hanging.

Boris was not the only family member to suffer under the foreign occupation. Anna's legs were badly scarred by a German soldier with a scythe, and when Alexei tried to sabotage the mill he had been put to work in he was beaten so badly he was permanently disabled. The entire family was forced out of their home by the soldiers and had to dig themselves a primitive shelter to live in. The shelter was never a safe place, with bombs shaking it until the dirt roof was ready to cave in. Valentin later said that he did not remember seeing his father smile during the entire duration of the war. The boy had little reason to smile himself; the Nazis put him to work as a manual laborer with the promise that he would be shot if he did not work hard. By 1943, Valentin and Yuri's sister Zoya had been taken by the SS to a slave labor camp in Poland. Luckily, both managed to survive the horrific living conditions long enough to escape. Finding their way to the Russian army lines, they were pressed into military service. Both survived the war, but their parents would not know this until the conflict was over.

With the Nazis using the village as a base, Russian forces repeatedly shelled the area, keeping the family constantly on edge. "We lived in fire and smoke," Yuri said of that time. "Day and night, something nearby was on fire." The Gagarins scavenged in the muddy fields for whatever food might have been overlooked, digging for the buried remains of rotting crops, surviving mostly on thistle root and sorrel soup, constantly hungry.

Despite the ever-present possibility of capture and death, Yuri and the other village children did what they could to disrupt Nazi war efficiency. They placed nails and broken glass on the road to burst army vehicle tires, stuffed potatoes and rags in exhaust pipes, and poured soil into tank batteries. "During the war, we boys felt powerless," Yuri remembered. "We did everything we could to annoy the Germans." It was risky, but it made Yuri feel like he was doing something to resist the invaders.

One dramatic event during the war had a particularly profound effect on Gagarin. An aerial dogfight over the village resulted in the downing of a Soviet fighter, which attracted a second pilot in a rescue aircraft. Yuri and Valentin were soon on the scene, along with other children. Valentin helped the pilots salvage the usable parts of the aircraft, while Yuri brought them some food. Young Yuri was fascinated by the two fighter pilots and spent as much time as he could with them before they left in the rescue airplane. He admired their medals, took pride in looking after their map case, and was thrilled when one of the pilots put him in the downed fighter's cockpit and showed him the controls. Meeting the pilots, only one of whom would survive the war, was a memory that stayed with Gagarin and changed the course of his life.

At long last, in spring 1944, the front line of the war came through the region again, and the Nazis were driven out for good. What they left behind, however, was a shattered village, its fields still peppered with mines, in a devastated nation. Almost everything of worth had been destroyed, and livelihoods had been utterly disrupted. The family hunted for edible mushrooms in the forests, the need for food outweighing the dangers of unexploded mines. With no animals left in the village, the women had to pull the ploughs themselves and plant what few seeds they had, hoping for a successful harvest. "Every one of us had seen the horrors perpetrated by the occupants," Yuri later recalled, "and had suffered the torments of hunger and oppression; we knew what war meant." The Gagarins would have to live with the memories of the horrors they had experienced and begin anew.

The family moved to nearby Gzhatsk in the spring of 1945 and built their own house by hand from the ruined remains of their former home. Gzhatsk was as pulverized as Klushino. "The charred remains of brick ovens loomed on both sides of the ruined road," Valentin later vividly described, "criss-crossed with ugly, water-logged ruts." As part of the rebuilding, the townspeople decided that it was important to construct a school. Yuri's schooling had ended during the long years of war, and now there were no materials left to learn from. He was, however, enthusiastic and willing to learn, and with the help of a volunteer teacher regained some of the years of learning he had lost. Pieces of old wallpaper and paper bags found in empty German bunkers were used as crude writing paper, while discarded flares were opened and their contents used to make ink. Without desks or chairs, the children improvised using long planks.

Yuri was still a mischievous boy who delighted in teasing his younger brother and soaking his sister with cold water. Yet he was also very bright; his tricks usually stopped short of annoying adults too much. Sometimes, his behavior did go too far; on one occasion, a paper airplane he made and threw from a classroom window broke a stranger's glasses. This incident did however demonstrate a new interest of his—flying. Before the war he had enjoyed making kites out of newspaper, but now Gagarin had become fascinated with making model aircraft, and he dreamed of flying a real airplane, just like the Soviet fighter pilots he had met in the war.

As Yuri grew older, what remained of his mischievous side seemed to anger Alexei all the more. His father resented Yuri's efforts to escape the limitations of rural life. He walked out of his son's school recital because he felt that Yuri was showing off; leaving would teach him a lesson about being too boastful. Alexei also hid his son's boots to stop him from playing hockey at school, exasperated that his son was wearing out boots faster than other family members. As Yuri later wryly commented, his father was "never effusive in his tenderness." Valentin agreed that their father was "not one to be crossed," the type who would quickly fly into a rage when upset.

Alexei had already made Yuri his own set of carpentry tools and grew annoyed when one of Yuri's teachers suggested the boy could do well in other careers. The stubborn father would grumble to his wife that the schoolteachers were filling his son's head with crazy dreams and ambitions. His sons should settle for the steady, safe job of carpentry, he thought; there was still much postwar rebuilding to be done in the area. No one could make him change the views that had guided him through his difficult life. "It was difficult to argue with him and useless to disagree with him, for he would rarely concede a point," Valentin later reflected. It is not surprising therefore that Yuri grew more and more eager to stay away from the house. His excuse was that he was studying nature, and he spent his free time collecting stones and grasses. He was also collecting picture postcards of Moscow and imagined traveling to the big city. His father had told him that, as soon as he had completed his basic schooling, he was to come along on carpentry jobs. That was the last thing Yuri wanted to do.

Yuri thought deeply about which career he wished to follow before it was too late and the decision was made for him. One night, he announced to his parents that he was not going to stay in the village and settle for being a carpenter; he was going to enroll at a college and try a different career. He wasn't sure what he wanted to do, but he knew he didn't want to stay at home any more. His mother began to cry. Having lost two of her children for so many years during the war, she did not want to let another one go. His father walked out of the room without saying a word. "They felt that I was a small child," Yuri later said of that turbulent time, "even though when they were my age they were already working like adults."

Eventually, his father tried to compromise, asking Yuri to complete one more year of village school before leaving. He hoped that, in that year, the boy would change his mind; he was, after all, only fifteen years old. But Yuri was not to be dissuaded, and made plans to leave town as soon as possible and stay with an uncle in Moscow who'd be able to help him obtain a college placing. As Yuri left home, his father's last instruction was a stern warning not to disgrace the family name.

Yuri decided to become a gymnast, but when he applied to the appropriate college all of the places had been taken. He did find an opening at the apprentice school of a steel mill in Moscow, where he could learn to be a foundryman, a trade that he imagined he might do for the rest of his life. It was difficult work, especially as he was small in stature, but he worked hard to learn and improve. Living in a big city was also a culture shock for the country boy, who had never seen so many people bustling around him before. But escaping from village life seems to have been exactly what Gagarin needed. He did so well in his studies that by the end of his first year he was selected to train at the Saratov Industrial Technical School. Here, he would learn all about agricultural machinery. By this point, he had won the opportunity to attend a physical training technical school and pursue his gymnastic ambitions, but he chose Saratov as offering a more realistic career. More important for Gagarin's future than the academics, however, was another attraction of the Saratov area: on the outskirts of the large industrial town was an airfield with a flying club.

As soon as Gagarin was permitted to enroll he joined the club. It would be a chance to see whether his childhood dreams of flying like the wartime fighter pilots matched the reality. His first flight, in a Yak-18, was all he had hoped it would be, and more. All the agonizing about becoming a carpenter, gymnast, or foundryman was now irrelevant. Even before he had landed, Yuri had decided that flying was what he really wanted to do with his life. It was going to be hard, however, to find the time to fly when he was still studying at school. Both the flying club and the technical school required long hours of academic study. After many months trying to do all of the studying simultaneously, Gagarin realized that it was time to make the giant leap. He chose to become a pilot, and in October 1955 he enrolled as an aviation cadet.

Yuri's father, once again, was displeased. After spending years in the village school, then studying as an apprentice, Yuri was now changing careers once more, and Alexei accused his son of wasting his country's money on training that he was now prepared to cast aside. He was, however, quietly proud of his son's stubbornness in pursuing his own goals and grudgingly impressed that his son had flown an airplane. The idea of flying a machine off the ground seemed unreal to the old man, something he could never imagine doing himself. For Alexei, technological marvels such as propeller-driven aircraft were part of a different world than the familiar Smolensk farmland.

Yuri's flying instructor grew so impressed with Gagarin's natural ability as a pilot that he recommended him for the military piloting school at Orenberg. This meant that Gagarin had to enlist in the military, but in return he would be able to dedicate himself to flying, with tough instructors to push him to even greater levels of skill. At Orenberg, however, he would just be one of many promising pilots—nothing special. He practiced hard and did well, though at first he found making smooth landings difficult. Eventually he determined that this was due to his height; he was too short to see the runway properly. He began sitting on a cushion while flying, and his landings improved dramatically. Years later, when Yuri was a household name, his flying instructors struggled to recall their former pupil. He'd been a proficient trainee pilot, but certainly not outstanding enough for them to remember.

One night at a school dance, he met a young girl called Valentina, who also worked on the base. "She was timid and shy," he said of that first meeting. Valentina, reflecting on that evening, remembered Yuri being one of many cadets, "shaven-headed, fidgety and excited." She was in fact more interested in the boys of the senior class, as were most of the girls. Not only did they look more serious, they also had hair, which made them far more attractive. Still, she enjoyed Yuri's simplicity and plain manner. He seemed genuinely interested in her, telling her all about himself and asking her question after question. He was not sophisticated, but he was honest and open—and a confident dancer.

Gagarin was not her idea of a perfect boyfriend, but Valentina found herself intrigued by the self-assured young man. For his part, he found himself falling quickly for Valya, as everyone called her, and also found her family just as inviting. As he was in a strange town, he greatly enjoyed spending time at her family home, where he was treated like a son. Valya's father was a cook, "by nature a very hospitable man" according to Gagarin biographer Nikolai Tsymbal, and Yuri reveled both in his company and his food. Yuri later said, "They were very warm towards me . . . I always felt completely at home with Valya's family"—perhaps more than he ever had with his own.

Yuri and Valya married in Orenberg a year after they met, in October 1957. Alexei refused to allow the Gagarin family to attend. According to tradition, marriages took place at the groom's family home, and Alexei was annoyed at his son for flouting this custom. He'd also hoped Yuri would finish his studies before marrying, having recently told him to "learn to make a living first, and only then have children." Yuri, it seems, paid little heed to these words. "My father mentioned his disapproval of our having celebrated the wedding in Orenberg and not in Gzhatsk," he later recalled. "I knew my father's character, knew that he brooked no contradiction, and so I kept quiet." Yuri also paid little attention to the launch of the world's first satellite that month; he was too caught up in his own wedding preparations and his final piloting examinations. "I was probably more occupied by how to seat all the friends we had invited to our wedding feast than by anything to do with the moon," he recounted. He did, however, wonder briefly about how long it would be until people would fly into space. He guessed that it would take another ten years.

Gagarin's first posting as a pilot was with a fighter-interceptor squadron based in the village of Luostari, in the North Murmansk district within the Arctic Circle, where he flew reconnaissance missions. Initially, to his frustration, he was not allowed to fly at night—and the polar night could last for six months. Valya had to make do with the gloomy room they were borrowing from a friend; there were no married quarters. "At first Valya could not get used to northern nature, the frowning, drizzling sky, and the dampness," Yuri remembered. "She would wake up in the middle of the night seeing that it was light outside, as though it was daytime." The couple endured heavy snowstorms, blizzards, icy winds, and, when the weather warmed a little, relentless rain. "Yuri's duties took up a great deal of his time," Valya recalled later, "but he tried to find an hour or two for me, the house, the family." It was an uninviting place for the newlyweds, but it was where their first daughter, Lena, was born in April 1959.

A few months later, all of the major air stations in the western regions of the Soviet Union received a visit from teams of officials. They asked many questions of the pilots, and it seemed that they were recruiting for some special project, but they did not say what it might be—not yet. Gagarin was one of those questioned, and as the number of people being asked to attend these closed-door interviews shrank from over three thousand down to around two hundred he was still among them. Together with the remaining candidates from around the country, Gagarin was ordered to report to Moscow for medical tests, and only at this point was he told the purpose behind them. The tests would determine who would become the nation's first cosmonauts. In Moscow he underwent many trials and examinations, which included being whirled around on strange-looking devices, talking with psychologists, and taking aptitude tests. His eyes were tested for night vision, squinting, color blindness, and any other slight imperfection. He was vibrated, subjected to changes in temperature and pressure, and spun on a centrifuge. Psychiatrists asked probing questions and tested his concentration. "The medical board was stricter than anything I had known before," Yuri remembered. "It was nothing like the usual air force medical examination that we went through every year. There were many doctors, and each of them was as strict as a prosecuting counsel."

Of the 154 pilots examined, 20 were finally selected, while the others were sent back to their units to continue their piloting careers. Yuri Gagarin was one of those who had passed the grueling trials. Without ever really choosing to be, he was now a cosmonaut. He was under strict instructions not to tell anyone else about his new assignment for now, including his wife. He instead told Valya that he had been chosen as a test pilot for a strange new flying vehicle. In a sense, it was not a lie.

The Soviets had thought about selecting submariners and mountain-climbers as well as pilots to be cosmonauts, considering them all to have the skills necessary to fly in space. Jet pilots under thirty years of age were eventually chosen, however. Not only was their training the most directly applicable; they were also trained parachutists and cleared in security screenings and fitness tests. It was believed that their occupation would require emotional and nervous stability, as well as strong willpower. Yevgeny Karpov, the medical doctor who became the first head of the Cosmonaut Training Center, later explained: "it was decided that paramount importance was to be placed on the intellectual scope and physical staying power of those chosen." None of the cosmonauts chosen were test pilots, but they were quite adequate for the spaceflights to come.

In March 1960, Gagarin moved with his family to the Moscow area. A brand-new and secret training facility, known as Star City, was being built for the cosmonauts in the middle of a forest outside of town. In the meantime, they would train at various scientific institutions around Moscow, and even at a Moscow sports club. As well as aviation medicine and rigorous parachute training, they also began an intensive physical fitness regime. The trainers soon realized that this was not going to be enough. It was important to motivate the group in other areas and not just let them feel they were medical test subjects. Classes were added in spaceflight theory. "It was an extensive program," Yuri stated. "We had to study the fundamentals of rocket and spacecraft technology, spacecraft design, astronomy, geophysics, and space medicine." As the cosmonauts and the program were both starting from similar levels of inexperience, some training elements were added as they were conceived. "On the whole," Gagarin later said, "it was a case of the blind leading the blind."

Just like their American counterparts, the team of pilots bonded closely as a cooperative group, despite their competitiveness to be the best. Unlike the *Mercury 7* astronauts, however, they were not known to the public. Without the pressures of instant fame, some aspects of life were easier for them than for the Americans. The secrecy, however, did lead to some strains. Many wished they could reveal to their families the exciting work they were embarking upon, and where it might lead them.

As the testing continued, the trainers constantly looked for strengths and weaknesses among the cosmonaut team. This was not only because they needed to choose the best for the initial flights; there were simply not enough simulators to train a group of twenty to the same level. Some would have to wait. Unlike American training, which used large aircraft to simulate weightlessness by flying parabolas, the cosmonaut team members were dropped in a specially adapted elevator in the Moscow State University building or flown in the back seat of a MiG. To test endurance, the young pilots were placed one by one in a sealed room, from which the oxygen was gradually removed, until they passed out. This was dangerous testing, but it was thought necessary to subject the pilots to such extremes in order to determine which of them were the elite candidates.

Some tests, like the isolation chamber, posed no difficulty for Gagarin. The chamber was partially pressurized with a 50 percent oxygen mix, and the lighting was altered so the test subject soon lost track of time. All sounds and vibrations were removed, except the ones the doctors wanted to introduce. The test could last up to ten days. Despite being locked in a room with nothing to help him occupy his mind or any idea of how long he'd be in there, Gagarin sustained his smiling composure, often singing to pass the time. The training on the centrifuge was less easy for him, and he found it hard to breathe during the higher g-force loads. Evidently he was good enough to pass. He wasn't the best candidate in many of the tests, but he was ranked highly enough in each to be a good all-round contender.

The cosmonaut team was kept in the dark at first about another secret figure, someone who would prove to be hugely important in their lives. He had helped select the twenty by closely monitoring their test results but had chosen not to meet the candidates until he felt the time was right. His name was Sergei Pavlovich Korolev, and he was the chief designer of the Soviet space program. He had enormous influence over all aspects of the space program in a way no American administrator could have. He not only designed the rockets; he also controlled the design bureaus and had an incredible grasp of politics, which allowed him to influence the Kremlin decision-making in his favor. Like the cosmonauts, his identity was kept secret from the public. His power and strong will created some nervousness in his underlings, but Korolev was also a good listener, and there were few who did not think he was the perfect man for the job.

When the chief designer did decide to meet the men he had helped choose as cosmonauts, he seemed to take to Gagarin immediately. Korolev admired people who did not let themselves be bullied by him, and Gagarin still had enough of the schoolboy rascal in him to stand up to authority figures. Yuri had seen his family abused by the Nazis; he wasn't afraid of anyone, and it showed. The duo impressed each other from the start. The moment that secured Korolev's affection for Gagarin, however, is said to have come later that first day. Korolev invited the group to look inside a Vostok spacecraft test vehicle, and asked who wished to go first. None of the cosmonaut team had seen the spacecraft before, and all were doubtless eager to see the interior. Yet Gagarin was the first to climb inside and, perhaps as a mark of respect for the spacecraft, he removed his shoes. The subtle gesture seemed to seal the chief designer's opinion; he later remarked to a colleague, "I like that brat!"

The spherical Vostok spacecraft was almost eight feet in diameter, and leaning in Gagarin saw a circular, padded, pressurized interior. Lined up directly with the round hatch was an ejection seat. Sliding in and trying out the seat for size, Gagarin looked over the instrumentation, which appeared simple and took up only a small area of the cabin walls. Directly in front of him was a square panel with dials showing spacecraft temperature, humidity, pressure, oxygen, and radiation levels as well as instruments showing the number of orbits and the mission duration. Other instruments and gauges indicated how well the spacecraft systems were working. "The things that were in that cabin," Gagarin remembered, "and none of them like the instruments in an aircraft!"

Most noticeable was a globe that showed where the spacecraft was in relation to the Earth. To the left of the seat was a control panel with three rows of switches, including those for the retrorocket sequence. On the right was a control handle that allowed the cosmonaut to adjust the spacecraft's attitude. Next to that were radio and telegraph controls and a food container. There were also controls for the window blinds and for regulating temperature. Three recessed portholes would allow a cosmonaut to look outside.

One was located just below eye level when looking forward in the seat, and was specially designed using internal mirrors to help the pilot orient the spacecraft in orbit. The lens of a TV camera stared right at the young cosmonaut's eager and quizzical face.

In all, it was a strange-looking piece of engineering to a pilot, nothing like the aircraft shape that some of the cosmonaut trainees had imagined. As Gagarin looked it over, Korolev had already decided that this young man might be ideal to test it for the first time. "I liked him the very first time I saw him," Korolev later said. "He is a born cosmonaut."

Pavel Popovich, also in the first group of cosmonauts, believed that Yuri's winning confidence was due to his difficult background. "You know his life," he told space historian Yaroslav Golovanov. "From the age of fifteen he was already fending for himself. Do you think it would have been possible for his character to take shape, for him to reach his civil maturity by the age of twenty-five if he had not been out and about in the world but had stayed under the protective wings of his mother and father?"

Learning to fly the spacecraft would be a challenge for them all. Although this was a brand-new kind of vehicle, very different from an airplane, Gagarin firmly believed that the decision to choose military pilots was a good one. He later reflected that while the cosmonaut trainees made numerous mistakes learning the new controls and instruments, they eventually mastered them. This was because they could transfer many of their piloting skills to these strange new flying craft.

By the early summer of 1960, six of the twenty candidates had been chosen as potential pilots for the first flight. Among them were Gagarin and another cosmonaut, Gherman Titov. The two had very different personalities—Titov was a more complex person than Gagarin, more educated and more opinionated—yet they managed to work well together. By the beginning of 1961, two of the six finalists had been replaced for medical reasons, but Gagarin and Titov were still in the group—in fact, they ranked in the top three. The duo trained together as a team, knowing that doing so would work to their mutual advantage and increase their chances of being chosen first. Both looked promising as candidates to fly the first mission, and both dearly wanted it. Yet they also knew that their characters were being scrutinized as much as their technical skills, and being cooperative was seen as an asset. It worked—the two were soon the only serious candidates for the first flight. There had been a third promising candidate for the first mission, Grigori Nelyubov, who had ranked second place in the group at the beginning of the year, but when Gagarin and Titov bettered him in the tests the confident candidate felt slighted. He evidently expressed this once too often, and was demoted to third place. While he would stand on the sidelines as a backup for Gagarin's flight, he would never get to fly in space himself. By 1966 he was dead, with some evidence pointing to suicide.

As the planned date for the first flight grew closer, the competition between Gagarin and Titov remained extremely close. Yuri and Valya had a second daughter, Galya, in March 1961, but the training was so intense at this point that there was little time for Yuri to get to know his new child. "I used to come home dead tired," he recalled. "I would play with my daughter for a while, then sit down and start dozing off." Although Valya was not supposed to know what her husband was training for, she had already guessed, and this added to the strains on their relationship. She later described how Yuri would frequently be away on training trips, and when he was able to stay the night he would return very late. He was not forthcoming with information about his work, and when she asked about it he would brush aside her questions with a joke. "It sometimes seemed to me," she recalled mournfully, "that his work was taking Yuri away from me more and more."

Titov and Gagarin were undergoing a barrage of physical and mental tests, designed to see which of them would make the better cosmonaut. Gagarin impressed many with his appearance of inner calm, but it was by no means certain that he would get the first flight. Some of the trainers preferred Titov over Gagarin, as he seemed more focused on the training and physically fitter. These strengths may have worked against him, however; the second Soviet spaceflight was planned to be much longer than the first, and Titov was seen by some as better suited to what could be a more difficult flight. There was also some political factors to consider. Like Soviet premier Nikita Khrushchev, Gagarin was from a farming family, and his background seemed to embody the ideals of Soviet equality. Titov, however, was the son of a teacher, and this middle-class family tie may have cost him the flight.

Before the first manned flight could be made, however, the cosmonaut team suffered the tragic death of one of its members. On 23 March, only a few short weeks before the first mission, cosmonaut trainee Valentin Bondarenko died under horrifying circumstances in the isolation chamber, where he had been locked for ten days in an endurance training exercise. Pavel Popovich describes the tragedy:

We had been participating in a test—I spent ten days in that pressure chamber myself. We were sealed in from the outside. There was a higher than normal oxygen level in the chamber; it was almost the same situation as the Apollo 1 *accident. We worked exactly according to our daily time schedule. The readiness to start registration of medical parameters was given by lights. I would apply the sensors; they would give the signal from outside. When we applied the sensors, we usually used a special paste to increase the contact. I would turn on a switch and a green light would light up to indicate I was ready. In case you experienced some chilliness—and sometimes it would get very chilly—they had installed a very simple, open heater with an open spiral coil. In my ten days, I didn't switch on the heater.*

Well, it was chilly for Bondarenko, so he switched it on: it was standing to his left on the floor. After all the data was recorded, you would remove the sensors, take a piece of cotton with alcohol and simply remove the paste from your skin. He did this and, without looking, threw it directly onto the heated coil.

In the oxygen-filled room the swab instantly burst into flames, and as Bondarenko tried to put the fire out his cotton training suit also caught ablaze. The flash fire burned him from head to foot.

"Seventy percent of his body was burned," Popovich remembers. "The doors of the chamber were sealed in the same way as submarine hatches. There were two of them, and it took time to open the first, then the second door. He lived for only four hours after that."

The critically injured cosmonaut was rushed to the hospital, but very little could be done. He soon succumbed to his horrific and painful injuries. A launch had not even taken place, and the cosmonaut team had already lost a member. Bondarenko had only been twenty-four years old, and for the whole team it was a difficult and sobering reminder of the dangers associated with their new profession. "The path of a cosmonaut is not an easy, triumphant march to glory," Gagarin later reflected. "You have to get to know the meaning not just of joy but also of grief, before being allowed in the spacecraft cabin."

Yuri Gagarin, the world's first spacefarer. Courtesy Colin Burgess Collection.

There was little time to reflect or mourn, however. That same week, Gagarin was taken to the cosmodrome from where he would launch. It was the first time he had been allowed to see it.

Although his name had been listed on the official authorization papers for the mission since the beginning of April 1961, it was only in the week before the flight that the State Commission made its final determination. Gagarin would be the prime pilot; Titov would be his backup. The flight was to be called *Vostok*, not *Vostok 1*; the Soviets did not wish to imply that there would be a series of continuing flights—not yet. The two pilots were brought together and informed of the decision later the same day. Gagarin was not sure how he felt about it. "Everything was clear and yet unclear—maybe very complex," he said of his emotions on that day. "I felt rather awkward. Why me? Why not him?" Titov had been utterly sure that he was going to be picked and was visibly disappointed that he was now the second choice. He did his best to keep his emotions in check, but he also did nothing to congratulate Gagarin.

The night before the flight, Gagarin and Titov were housed in a cottage near the launch pad. The two men played pool together before retiring. It would be the very last night in human history in which humans had not been into space. When Gagarin was about a year old, space theorist Konstantin Tsiolkovsky had written: "I have no difficulty imagining the first man overcoming the Earth's gravity and rushing into space. He is a Russian, a citizen of the Soviet Union; his trade, most probably, is a pilot. He is courageous, yet void of any recklessness. I see his frank Russian face." Gagarin certainly fit Tsiolkovsky's description, and he was about to fulfill his prediction.

At 5:30 a.m. on the morning of 12 April, the two men were awakened, examined by doctors, and declared fit to fly. "The time had come for me to dress in my space gear," Yuri recalled. "I first put on a warm, soft sky-blue set of lightweight overalls." Both of the cosmonauts were then fitted into their orange spacesuits. If Titov were needed as the backup, he would immediately be ready to go. "The helmet of the spacesuit," Gagarin recounted, "was very much like the headpiece of a knight of the Middle Ages, with its movable visor." While being suited up, Gagarin was asked for his autograph by the suit technicians, something that surprised him. It was the first small hint of the huge wave of fame that would engulf him for the rest of his life.

After final checks had been performed the two men boarded a modified city transit bus, filled with other cosmonauts and some medical personnel, for the long ride out to the launch pad. Gagarin took the right front seat, plugging his suit's air hose into a wall-mounted attachment. Titov, following the same procedure, sat behind him. On arrival at the pad Titov had to remain on the bus. He thought that this was silly—nothing was going to happen to Gagarin in these last few moments. Still, he waited as ordered in his spacesuit for an opportunity that did not come, only removing his spacesuit once Gagarin was secured inside Vostok.

Both the spacecraft and the modified R-7 rocket that would launch it had been taken from the assembly buildings to the pad by railroad and rotated into a vertical position the day before. The stout body was gripped on four sides by large arm assemblies that held the rocket suspended over the flame trench and accessible by retractable gantries and walkways. The rocket gleamed in the early sunlight, its white body covered in frost from the superchilled liquid oxygen inside. Toward the base, four strap-on boosters tapered elegantly. Together, the rocket and spacecraft were over 125 feet tall, and the pointed shape looked as if it would spring off the launch pad at any moment. It was a beautiful sight.

Korolev greeted Gagarin at the base of the rocket and remarked that he hoped the young pilot would be walking on the moon one day soon. Yuri then climbed the flight of stairs leading to the elevator that would take him to the spacecraft. At the top of the stairs, Yuri turned to look at those watching from below. "He waved to us and said, 'See you soon!'" Titov remembered, "and it was then that I suddenly comprehended that this was no training session, that the fateful and long awaited hour had arrived."

After the elevator had taken him up to the waiting spacecraft, Gagarin calmly grasped the top of the hatch and slid himself feetfirst into *Vostok*'s ejector seat, ready to be strapped in. Once in place, he went through the required communications and suit checks. *Vostok*'s guidance was automatic, and only in an emergency would Gagarin have to take over, by typing in a special combination to activate the manual controls. This number was written on a piece of paper inside an envelope in the cabin, to be opened only in a crisis. This rather cumbersome method was a compromise between the pilots who wanted to be able to control the spacecraft and the engineers and doctors who wanted as much control of the flight from the ground as possible. It was a difference of philosophy that would also affect the early American space program. One engineer who fully trusted Gagarin, however, was Korolev, who whispered the code numbers to him while the cosmonaut was being strapped into the spacecraft. What Korolev did not know is that he was just one of many people who had told Gagarin the "secret" code—the man who was about to make the first-ever spaceflight had the confidence of a great number of people, enough for many of them to break a rule that could have cost them their jobs.

After the spacecraft hatch would not initially seal correctly, it was finally bolted into place. "I was left alone with the controls," Yuri recalled, "now no longer lit by the sunlight outside but by artificial light. Of course, I was nervous—only a robot would not have been nervous at such a time and in such a situation." If anything had gone wrong on the launch pad, the ejector seat would have fired Gagarin out, but there was not enough time or altitude for a parachute to open. Instead, he would have fallen into a large net. Korolev talked to him over the radio as the final checks were made, ensuring that he was neither worried nor bored and going over the final details one last time.

Titov had been escorted from the bus and was now in the observation bunker. He had only been halfway out of his spacesuit when the technicians stopped helping him and walked away; they wanted to go and see the launch. Having come so close to making the flight himself, Titov now felt abandoned and forgotten.

Unlike the dramatic countdowns used in the American program, the Soviets simply launched at the appointed time. Observers heard a high, piercing squeal as fuel began pumping at tremendous speed through the rocket body. Next, a deep rumble was heard as the engines fired into life and growled to their full intensity. Finally, the clamps holding the rocket were released, the supporting arms swung away, and the rocket rose, sending an immense, rattling shock wave of sound through the surrounding bunkers and awed onlookers.

Vladimir Suvorov, in charge of filming the launch that day and dangerously close to the launch pad at the moment of ignition, recorded his impressions of the liftoff in his secret diary:

"Now the powerful engines come to life, the rocket is trembling and the white frost wraps it in a thin haze. There are reflections of the flame of the engines on the rocket body. At last after the final jerk it starts moving as if awakened from a long sleep." Inside the spacecraft, Gagarin found it hard to identify the exact moment when the rocket left the ground. He had felt the rocket sway slightly a minute before launch as the valves began to operate, but the difference in the rocket's shaking and noise at launch had been too subtle to pinpoint. In fact, the rocket seemed relatively slow and quiet as he began his mission, to the point where he did not feel like he was moving at all. Over the faint roar he heard Korolev wish him a good flight as he ascended, and he responded with a cheery "Poyekhali!" (Off we go!).

Prepared to eject if he needed to, Gagarin felt the shaking increase just over a minute after liftoff, then slowly decrease as the amplitude and frequency of the vibrations changed. As the rocket accelerated, the g-forces rose steadily, interrupted only by the first-stage booster engines falling away. It soon became hard for him to speak as his face muscles tightened, but following his extensive training handling the g's was no problem. When the second stage shut down, Gagarin was pushed forward against his seat straps with a jolt and a bang, making him fear for a moment that a piece had broken off the spacecraft. He became aware that the noise of the engines had suddenly ceased, and the sudden reduction in acceleration gave him the false sensation that he was weightless. Ignition of the third stage pushed him back into the seat again, but this booster stage was a quieter one. The bullet-shaped aerodynamic shield covering the spacecraft soon separated with another bang and sharp jolt. When the third stage also shut down, Vostok separated from it and began a slow spin. Nine minutes after launch, Gagarin was in orbit.

As his spacecraft slowly rotated, Gagarin looked at the Earth below him. His first view was of a mountain area with rivers, forests, and ravines, which he later believed to be the Ob or Irtush River area. As he continued to look at the revolving scene, he was struck by the beauty of it—a curved horizon above an Earth of surprisingly intense blue, the deep black of space, and a sun too bright to look at. It was not lost on him that he was seeing the Earth as spherical, with his own eyes—the first time any human had been able to do so. The horizon looked particularly beautiful to him, and he was entranced by the smooth change from the brilliance of Earth to the darkness of space, a myriad of colors filling the thin, filmlike layer that separated the two. He had never seen a blackness as deep as the infinity of space he stared into.

Looking down at the Earth again, Gagarin saw seas, cities, the coasts of continents, islands, and other sights never before seen by a human being from this vantage point. As the features slipped by, the speed of his journey around the world was vividly evident. As he observed a coastline change to ocean, he noticed how surprisingly gray the water looked, appearing "darkish, with faintly gleaming spots." From orbital height, the ocean's ripples reminded him of sand dunes. He was also fascinated by the sharp shadows of the clouds over the Earth's surface.

Over land, he observed the squares of fields, and could easily distinguish meadows from ploughed areas. Through gaps in the cloud, he saw the forested islands of Japan. For the first time in his life, whether on the ground or in space, he was looking at a foreign country. At times the view reminded him of the vista from a high-altitude jet, but there was so much more to it.

Back in Gzhatsk, his mother wept as she listened to the breaking news over the radio. "What has he gone and done?" she kept repeating, as if still talking about the mischievous childhood Yuri.

Gagarin kept his thoughts concentrated on the flight program; he wanted to carry out his assignments to the best of his abilities. He checked the spacecraft systems and saw that everything was functioning perfectly. He also reflected on how he was feeling; he was having no problems with the prolonged weightlessness. "It was an unusual sensation," he later reported. "Weightlessness is a strange phenomenon, and at first I felt uncomfortable, but I soon got used to it." The lack of up or down made him feel like he was hanging from the spacecraft ceiling by his straps. "It seemed as if my hands and legs and my whole body did not belong to me," he reported. Yet he was feeling no discomfort, and in fact found it easier to work in the cabin because of the lack of weight. He noted that he did not feel hungry or thirsty, but he ate and drank as part of the mission plan, with no problems. "On Vostok the water supply was kept in a polyethylene-lined container fitted with a tube with a special mouthpiece," he would later describe. "To get a drink I had to take the mouthpiece, press the button of a special looking mechanism, and then suck the water out." Some water leaked from the end of the tube and slowly floated around the cabin as he watched, fascinated by the tiny, gleaming sphere of fluid.

Gagarin enjoyed watching his pad and pencil float as he noted down observations on the flight's progress. He was very interested to discover that his handwriting did not change even when writing in zero gravity. In between note-taking periods, however, his pencil floated away and was lost to him. Luckily, he could continue his observations using his tape recorder. Television pictures beamed to the controllers also showed he was doing well; the same was also noted by U.S. intelligence operatives who were intercepting the signals. Gagarin came to the conclusion during his flight that weightlessness would not pose any difficulties for humans working in space. Later flights would prove that the matter was not so simple.

Half an orbit after launch, over the vast Pacific Ocean, *Vostok* glided serenely into the shadowed side of the planet following a beautiful sunset. Gagarin was surprised by how quickly darkness came on, a huge difference from the blindingly bright sunlight. He looked down at the Earth's surface but could see no lights, so surmised he must be traveling over water. As he looked to the horizon, more stars appeared than he had ever seen before, brighter and sharper than when observed from Earth. "Of all the nights I had seen in my lifetime," Gagarin recounted, "none was remotely comparable to night in space. I have never forgotten it. The sky was blacker than it ever appears from the Earth, with the real, slate-blackness of space." It would be the shortest night of his life, however. All too soon, Gagarin saw a vivid orange curve on the horizon, "a very beautiful sort of halo," which gradually changed to a multitude of colors, then blue as the blinding sun rose again, faster than he had imagined it would.

The spacecraft's automatic systems were functioning well, so much so that Gagarin never needed to touch the controls, only monitoring the systems to ensure that *Vostok* was doing everything correctly. Far from feeling isolated and alone in space, he felt like there was more attention focused on him than at any other time in his life.

For this first mission, only one orbit was planned. As scheduled, seventy-nine minutes into the flight, the retrorockets made a forty-second automatic burn, slowing the spacecraft. Gagarin closed his helmet, tightened his straps, and felt the bang and deceleration force as the rockets fired to bring him home. He was feeling great and would have been happy to continue the flight, but knew that wasn't within the scope of this mission. "After the retrorockets were fired I waited for the separation," Gagarin recalled later. "Just then, the spacecraft rotated." As soon as the retrorockets finished firing Gagarin felt a hard bump and the spacecraft began to spin, making a complete revolution about once every fourteen seconds in his estimation. He felt himself pushed back into the cabin seat. "The craft began to revolve on its axis at a very high speed. It was a 'corps de ballet' turned inside out, head-feet, head-feet."

To reenter Earth's atmosphere, the spherical cabin was supposed to separate cleanly from the equipment module that had provided the essential power and telemetry. No longer needed, the module was designed to be cast off and burn up in the atmosphere, leaving the ball-shaped cabin to descend safely. However, all was not going to plan. A cable between the two spacecraft parts had not detached, and Yuri found himself rapidly tumbling as the crew cabin and its unwanted encumbrance made a spinning, dangerous plunge into the atmosphere. "Through the windows I saw Earth and sky, from time to time the blinding rays of the sun," he recalled. "I waited for separation, but separation did not occur. The wait was terrible. It was as if time had stopped. Seconds felt like long minutes." Inside the spacecraft the lights flickered on and off disturbingly, and Gagarin grew concerned that he might overshoot his landing site and land in a foreign country. His concerns heightened as he realized the dangers ahead were potentially much worse. He heard an ominous sound as parts of the spacecraft never designed to face the fires of reentry head on began to heat and char. "The external skin rapidly became red hot. Something crackled, as if burning, a piece of the coating or some hardware, but very audible. It was clear to me that the temperature outside was very high."

Vostok was designed to align its heat shield by using the spacecraft's center of gravity, but with the extra module still attached the unplanned swinging was exposing the less protected areas. Gagarin could feel the heat begin to rise inside the spacecraft and watched helplessly as the glow outside increased. "The spacecraft was enveloped in flames," he later related. "Through the blinds covering the windows I saw the sinister, terrible crimson flame raging around the spacecraft . . . I was in a cloud of fire rushing toward Earth." As the spacecraft spun faster, blinding sunlight hit his face on each rotation. He put his feet up onto the porthole to block the sun, but did not close the blinds, as he wanted to keep watching Earth and blackness alternating in the window to get some clue as to what was happening to him.

After ten dangerous and uncomfortable minutes, the cable holding the modules together finally sheared and burned through under the intense forces with an audible bang, and the two spacecraft parts were whipped away from each other like spinning tops. In the dizzying spin, enduring a punishing force ten times that of normal gravity, Gagarin's vision began to blur and he came close to losing consciousness. As the sphere fell into ever-thicker atmosphere, however, the rotation slowed, and he regained his full senses. "The separation took place, and everything took its normal course," he recalled with relief. The rotation had not fully dampened, as *Vostok* was still oscillating back and forth, swaying almost ninety degrees. Gagarin could only hope it would not affect the final part of the dramatic reentry. Through the borders of the now-closed blinds, he could still see the bright red fire of reentry and could hear and feel the crackling thermal coating. He assumed the position for ejection, and waited.

The Soviets planned to win the world aviation altitude record with Gagarin's flight. The regulations, however, were written with regard to aircraft and not spacecraft, and stated that pilots had to land in their vehicles. The *Vostok* spacecraft had been designed to land hard; retrorockets to cushion a landing would have taken up more room and more weight. For this reason, the cosmonaut was to eject and descend by parachute. Though Gagarin could probably have survived a landing inside the spacecraft, there was a strong likelihood of injury. This fact had actually helped Yuri become one of the front runners for an early flight; the shorter the cosmonaut, the better the fit for the ejector seat. For now, however, the fact that he would eject had to remain a closely guarded secret.

At twenty-three thousand feet the spacecraft hatch automatically blew off with a loud bang, and the cosmonaut tensed, waiting for the ejection process to begin. "I sat there, thinking, what about me?" Gagarin recalled. "I slowly turned my head upwards, and at that moment the charge fired and I was ejected. I flew out with the seat." His parachute opened, and he steadily drifted down toward the ground. His emergency parachute suddenly and dangerously opened in addition to his main chute. Luckily for Gagarin, it hung limply below him and did not tangle with the other shroud. After the perils of reentry, he was safe at last, and as he descended he began singing to himself happily.

Gagarin landed near a village called Smelovka, in the Saratov region of the Soviet Union, not far from the air club where he had first taken to the skies. "It was like a good novel," he recounted. "As I returned from outer space, I landed in the area where I had started flying. How much time had passed since then? Not more than six years. But how the yardstick had changed!" As his feet hit the ground, he found that he could stand upright without any difficulties. The spacecraft landed under a separate parachute about two miles away. Some local children had been surprised to suddenly spot the charred, unearthly looking ball falling from the sky, and they managed to climb inside through the landed craft's open hatch, helping themselves to some uneaten tubes of food. The villagers who greeted the cosmonaut were similarly baffled, thinking at first he might be some kind of foreign spy. Gagarin, however, was elated. He introduced himself to them as the world's first spaceman.

Titov had spent the duration of the mission traveling from the launch site to an airfield near the landing zone. When Gagarin was brought to the airfield, he greeted his colleague warmly, hugging him so hard, a news reporter on hand noted, that it looked as though they were wrestling. They spent the evening in a woodland cottage in the Zhiguli Hills, talking about the future. The next day was taken up walking along the nearby Volga River, discussing the flight while every detail was still fresh in Gagarin's mind. Titov would be next, he hoped, and he wanted to absorb it all. He also knew that he would not have the opportunity to have Gagarin to himself for long. Yuri would have to face what Gherman referred to as "terrestrial overloads." Two days after the flight, Gagarin was flown to Moscow, and a different reality.

As the first, and for a short while the only, person to have journeyed into space, Gagarin was plunged into a new world the moment the official reception party came to pick him up from his landing site. He launched from a planet where he was an unknown air force pilot, and landed on a very different world, one in which he had suddenly become one of the most famous people alive. This was partly due to the intense secrecy surrounding the Soviet Union's space program. Even for many years after, the public did not know what the rocket or spacecraft looked like, nor did they know the identity of Korolev or the other geniuses who had pulled off such an incredible feat. Unlike the American spacefarers, who were known from the moment of their selection, Gagarin was for now the only cosmonaut known to the public. Without rocket launch photos or other images to satisfy the curiosity of the fascinated public, the only image of the Soviet achievement was Yuri Gagarin's smiling, happy face. Although the fanfare faded a little over the years, Gagarin's life would never be the same.

The scale of this change to his life became evident to Gagarin as his aircraft circled over Moscow. He could see a vast crowd of people gathering below, which he likened to a human river, and for the first time since his landing he felt apprehensive. The cosmonaut was about to discover what it was like to be a world hero, and for him this was more nerve-wracking than his launch two days earlier. "It's true, I was afraid to address thousands of people," he recalled. "I was ready for the trials of outer space, but I was not prepared to meet that sea of faces." Unknown to him, Khrushchev was taking second place in line behind Gagarin's family as they walked up to the receiving stand, something that would have been unimaginable the week before. Gagarin's father, who had worried that his son would shame the family name, was now witnessing the country's most powerful officials waiting eagerly to meet his Yuri. To Alexei's bewilderment, these officials were also greeting him with warmth and affection, and he bemusedly raised his cap to each.

All the way from the airport to Red Square the streets were lined with exultant people waving signs and banners, shouting and cheering. Yuri was being feted by his country in the most lavish way possible. Of all the tens of thousands of happy faces the cosmonaut saw that day, perhaps the most significant image for him was his father's broad grin. Yuri had finally made his father proud.

Fortunately, the world's first spacefarer had a surprisingly good personality and temperament for carrying the unexpected burden of such overwhelming fame. Despite being only twenty-seven, a very young age to have such recognition thrust upon him, he handled it with the skill of a practiced politician. From the moment he returned to Earth crowds of people, eager journalists, and seemingly every official in the country wanted to meet him, and the reaction abroad was identical. Unlike later space explorers, he had no one else's experiences and descriptions to draw comparisons to. He had to find the words to describe experiences no one else had ever had before. Yet Gagarin managed to answer all questions pleasantly with cleverly worded insights that shared the wonder of the view from space without giving away closely guarded technological secrets. British media correspondent Peter Fairley described Gagarin as "a master of the art of evading the awkward question." It didn't hurt that Yuri looked the part. The British and Australian writing team of Wilfred Burchett and Anthony Purdy met Gagarin shortly after the flight, and both were impressed by his confidence, as they later disclosed in their biography *Gagarin*: "The first impression was of his good-natured personality . . . an air of sunny friendliness. But the key to his character, perhaps, lay in two other points: his handshake, and his eyes. His hands are incredibly hard; his eyes an almost luminous blue."

When giving speeches, which Yuri had to do repeatedly, this pleasant personality came through, instilling a genuine affection in those who heard him. During one press conference three days after his flight, he was careful to make the distinction that he had not been "sent" on the mission—he had been "entrusted" with it. This was not just an exercise in semantics; it was an important distinction. The fact that Gagarin emphasized it not only demonstrated that he had taken his flight seriously, but that he also believed strongly in fulfilling his postflight duties to the best of his ability. He realized that the whole world wanted to know what his flight had been like, and he felt an obligation to tell them as best he could.

Gagarin always felt a little embarrassed by his fame, and like his American counterparts continually stressed that he was just one person out of tens of thousands who had made his flight possible. Unlike the American astronauts, he had no one he could publicly identify and acknowledge to back this claim up. Instead, he had to shoulder the burden alone. It was a huge responsibility for one person to bear, but he did so and incredibly well. He did not let the adulation go to his head, deferring political questions to the politicians and engineering questions to the engineers, while constantly reminding reporters that he was simply an ordinary pilot. Soviet premier Nikita Khrushchev would doubtless have used Gagarin as an icon of Soviet achievement in any event, but it seems the politician also genuinely warmed to the young man. The Soviets were not slow to trumpet the flight as an outstanding achievement for Communism, but they also ensured that the mission was celebrated as a triumph for all humanity. In a country where individual fame was not an obvious fit with the philosophies of collectivism, Gagarin had become a hero, and one the whole world could share.

The universal demand to meet the first spacefarer, and his usefulness as an instrument of propaganda, meant that Gagarin would have to begin a long and demanding world tour even before the month of his flight was over. Journalists referred to it as his second orbit around the Earth. His horizons were about to be considerably widened. "Although I had flown around the Earth," he commented, "I had never been abroad." In 1961 alone he toured ten different foreign countries, and traveled extensively within the USSR. For Valya, who had to accompany him, this was torture. She was still very much the timid girl he had met all those years ago, and the relentless crush of people exhausted and agitated her. Before meeting Yuri she had never left Orenberg, let alone the USSR, and the change of pace came as an unpleasant shock to her. Ironically, considering she was the wife of the world's most famous pilot, she suffered from air sickness. This proved to be an unwelcome hindrance during the months of global travel. Her husband "faced many years of an unprecedented, incomparable test by glory," Valya remembered. "It awaited him everywhere—it never hid or crept but just came roaring down on him. Usually on the third or fourth day of a trip he would say: 'I'm homesick.'"

Despite the pressures, Yuri seemed born to fulfill this demanding role. He skillfully downplayed the problems of the flight in public, never letting on that his reentry had been so dangerous. He also had to maintain the lie that he had landed while still seated inside *Vostok*. It was also not known to the public at the time, because of the secrecy surrounding the location of the launch site, that Gagarin had not completed a full orbit of the Earth. He had in fact landed over 900 miles short. This shortfall has always been regarded as so insignificant that his flight's status as truly orbital has never really been questioned. In 1961 it really did not seem to matter; the Soviets had pulled off an accomplishment long awaited in human history.

Ultimately, however, the unceasing demands for personal appearances and speeches did begin to affect Gagarin. When asked what he found most difficult in life, his answer was one word—fame. Being seen in a saintly light was, he stated, enough to make anyone feel sick after a while. "I never imagined that I might be met in this way," he admitted. "I thought I'd do the flight, come back, and that would be that. I never guessed that it would be like this. Sometimes I close my eyes, and still keep seeing endless lines of people. It's embarrassing to be made to seem like such a good, sweet little boy—you get to feel uncomfortable. I am made to feel like some kind of super, ideal man. Everything I have done has always turned out okay, but just like everyone else I make lots of mistakes."

When he finally did get a respite, he let himself go a little too much. In September 1961, along with other cosmonauts, the Gagarins holidayed at Foros in the Crimea, a favorite summer escape for Soviet leaders. Valya was unhappy almost from the beginning, as she observed her husband and Gherman Titov enjoying the local entertainment a little too much for her liking. Gagarin was having fun, but pushing it too far. In addition to alienating his wife with his relentless drinking and socializing, he hurt his hands one day while racing a speedboat, despite having been warned it would be unwise to go out in the choppy waters. After the most demanding year of his life,

Gagarin was doubtless letting off steam with the same intensity in which he had earned his fame. Eventually, however, the partying went too far.

On the last night of the vacation, when the partying reached its peak, Yuri is reported to have stumbled drunkenly into a nurse's room on the second floor. His wife came looking for him soon afterward. Trapped in an embarrassing predicament, Yuri felt he had no choice but to jump from the balcony. He landed badly, hitting his head and chipping the bone in his forehead. Barely conscious, the injured cosmonaut was rushed to hospital.

Gagarin's first worry when he regained his senses was whether he would ever fly again. For those who could make such decisions, it was a very serious question as well—the iconic hero had thoroughly disgraced himself. The impact had left him with a permanent scar on his face as a reminder of his drunken folly. To save his reputation, a story was concocted for the press to explain the very visible facial injuries. It was said that Gagarin had tripped while holding his baby and had hurt himself to save the child. Despite the successful deception, the incident still had an effect on what Gagarin could do in public. Khrushchev was most annoyed that Gagarin's scarring meant he could not be used to full propaganda effect at that year's Communist Party Congress. As Yuri later recounted, "my eye and eyebrow were all puffed up, and I could not go to Moscow looking like that. I was very upset, and stayed in the Crimea for a few more days until the wound was skillfully patched up." Up until this moment, Gagarin had been much liked by the premier. With this embarrassing incident, the relationship looked to be cooling fast.

After a few weeks, the incident officially blew over. Yet some party officials, jealous of Gagarin's closeness to Khrushchev, were not going to forget the cosmonaut's misconduct. They were looking for a way to remove Gagarin from favor, and as the momentous year of 1961 came to an end it looked like they were going to get it. It did not help that Gagarin's drinking and behavior toward his wife were not improving. Though Yuri seemed to accept the endless public appearances as part of his duties, he began to find their sheer number to be a great strain, and excessive drinking soon became his relief. He had begun the year as an unknown air force pilot with the hope of making the world's first spaceflight. He ended it as a public hero, but one in serious danger of losing his public and private future to the temptations and pressures of fame.

Luckily, Gagarin was a strong enough person to survive the whirlwind of change that had overtaken his life. Cosmonaut Alexei Leonov remembers that, despite everything, Gagarin continued to display the same generous nature he always had. Most importantly, he remained at heart an honest person. He did not capitalize on his fame, as he could have and as many expected him to do, turning down several luxuries offered to him by officials. He accepted that he had a role to play as a public person, yet he tried to stay in touch as much as possible with ordinary people. To Yuri, where he came from was important to the person he was now, and that knowledge helped him through a year that had altered his life forever. As the world touring finally began to wind down and his work as an active cosmonaut could resume, Gagarin had managed to keep his marriage, his health, and his good nature in one piece. He hadn't resisted all of the new temptations that had been thrown at him, but he had survived them.

Gagarin returned to Star City to supervise the training of the new female cosmonaut team. He had been promoted to head of the cosmonauts, and in 1963 was further promoted to deputy director of the entire Cosmonaut Training Center. Yet, for all his newfound status, he was not allowed to fly jet aircraft. Gagarin was seen as too important for the nation to risk, and he was promoted away from active flying. The bureaucratic and public relations demands on him had pushed him far behind in his cosmonaut training, and it was going to be very hard to catch up. As new cosmonauts were selected and rose in the ranks, many of them with engineering backgrounds that had far more relevance to spaceflight than Gagarin's, he felt more isolated than ever from the program and the chance to fly again. A few hours after his spaceflight, in a telephone conversation with Khrushchev, he had told the premier that he was ready to carry out more assignments for his country. The day after his flight, he told reporters that he wanted to visit Mars and Venus, something he later described as doing "some real flying." Now stuck behind a desk in Star City, it did not look like he was going to get the chance to even orbit the Earth again. "He lived through some very difficult moments when the question of whether he was allowed to fly or not was being decided," Valya remembers.

By 1964, Yuri was able to begin his studies in earnest once again in pursuit of an opportunity to pilot another mission. He had begun studying at the

Zhukovsky Air Force Engineering Academy in Moscow in 1961, but now he could devote his full efforts to it. He worked exceptionally hard, studying a variety of subjects related to spaceflight. As the first cosmonaut he had not been expected to know too much about how spacecraft worked, but now he was determined to learn the engineering theories and practices that would put him at the forefront of spacecraft knowledge.

It was an auspicious time for Gagarin to find himself a role away from the public spotlight, as his time as a political showpiece was about to slow dramatically. In October 1964 Nikita Khrushchev, who had welcomed Gagarin into the highest ranks of party favor, was deposed and replaced by Leonid Brezhnev. The new leader was determined to place his own stamp on Soviet society, and swept away many of Khrushchev's reforms. Gagarin's flight had been a crowning achievement of those years, and Brezhnev didn't want to be reminded of that. Though the Kremlin still used Gagarin for occasional diplomatic needs, the warm relationship was gone. Gagarin had been demoted from a world ambassador to just another cosmonaut.

Yuri threw himself more than ever into training, determined to regain his role as an active cosmonaut. His determination won him an assignment as backup for the first flight of a new generation of Soviet spacecraft—Soyuz. This meant that he now stood a strong chance of flying a future, longer mission, which would give him the opportunity to prove himself once again. Some cosmonauts resented Gagarin's jump back into the flight lineup, but he had the seniority to pull it off. He was determined to fly.

The trouble was the Soyuz spacecraft was not yet ready to fly, and Gagarin knew it. The ambitious first flight called for cosmonaut Vladimir Komarov to dock with another Soyuz to be launched the next day. The new spacecraft were being rushed through production and testing to meet a political, not an engineering deadline. Gagarin helped to prepare a document outlining the spacecraft's many remaining faults, which he and his colleagues attempted to deliver to the Kremlin decision-makers. However, no one in the political offices wished to be the bearer of such bad news, and the document was conveniently lost.

Komarov knew the spacecraft's problems just as well as Gagarin, if not more, but nevertheless accepted his assignment. In fact, both he and Gagarin jostled until almost the moment of launch to be the one to fly the mission. It seems that this competition was not for the honor of making the first flight of a new spacecraft but because the duo cared greatly for each other. Each wanted to take the risk of flying a deeply flawed spacecraft as it would allow the other to survive a potential catastrophe.

In the end, Komarov kept his assignment and flew the mission. The spacecraft began malfunctioning almost immediately, and Gagarin's colleague never made it back to Earth alive. Yuri was plagued with guilt; he felt he should have died in his friend's place. Not only was Komarov dead, but Gagarin's chance of flying in space a second time also died that day. No one wanted to lose the first cosmonaut—a national treasure—in a similar tragedy, and he was officially barred from ever flying another mission. Gagarin tried to have the decision overturned, but it was too late. The dispiriting news pushed him into another bout of wild socializing and heavy drinking.

Gagarin was, however, still deputy director of the center, and there was still a lot he could do with his career. Rather than relax into a desk job supervising the training of Soyuz crews, he felt he would keep the respect of the cosmonauts he was responsible for only by remaining as current as possible in his training. As journalist Boris Konovalov wrote, "he tried all the time to be an active cosmonaut and not a museum piece."

In February 1968 Gagarin passed the demanding tests of the Zhukovsky Academy, a great achievement that gave him further confidence and respect. He felt that the next step was to prove himself once again as a jet fighter pilot, an avocation he had had little time to pursue in the busy years of public appearances. Even the day after his spaceflight, Gagarin had told a reporter that he "used to fly planes," an indication of how much his flying time had been reduced when he had become a cosmonaut. Now, he felt, it was time to rectify that. "Space research is my profession," he said, "and I did not choose it just to make the first-ever spaceflight and then set it aside. What is most important in the training of flyers? Anyone who knows anything about aviation will say 'flying.' This is not to belittle the importance of theory, but as musicians say, in order to learn how to listen to music properly, you have to listen to music a lot. This is only the beginning; I am still young, and I do not intend to fold my wings."

On 27 March 1968, as part of a training program to qualify him for the newer MiG jets, Gagarin took off in a MiG-15 jet with Vladimir Seryogin, his instructor. Seryogin was an extremely accomplished pilot in his own right—a Hero of the Soviet Union and commander of the special air wing at Chkalov that supported Star City. The day's poor weather conditions rapidly deteriorated, and Seryogin cut the training session down to five minutes. He radioed air traffic control to say that the training was completed, and they were returning to the airfield.

Several miles away the weather was also affecting Alexei Leonov, who was attempting to instruct cosmonaut trainees in parachute jumping. It was a day he would never forget. Decades later, he would recall that it was a windy, gloomy day with a blanket of low clouds—certainly not ideal weather conditions for parachute training. Once his students had completed their first jump he decided to end training for that day. Then he heard the sound of a distant explosion. Immediately, he felt a deep sense of unease. Fearing the worst, Leonov hurried to the airfield where Gagarin and Seryogin had taken off. They had not returned and never would. Their remains were recovered from deep in the crater the jet had made in the nearby woodlands.

A later analysis of the shattered aircraft pieces revealed that the pilots had been in the process of pulling the jet out of a steep dive, and had almost succeeded. Leonov believes that the MiG was caught in the turbulent backwash of another aircraft that passed too close, unseen in the dense clouds. Witnesses that day did report seeing what appeared to be an SU-11 jet climbing on afterburner up into the clouds, and Leonov firmly believes that this aircraft was responsible for the turbulence that doomed Gagarin and Seryogin. Other accounts and investigations attribute the backwash to other MiGs that were also said to have passed too close.

Given the variety of accident reports performed over three decades, some with decidedly political agendas, and the confused air traffic control data from the event, what happened that day may never be fully known. The firm facts that could be pieced together from the bodies and the mangled remains of the controls ruled out, however, many of the more common causes of jet accidents where the pilots are clearly at fault. "He had just started down the path of his professional life," Gherman Titov reflected. "Yuri left us early, too early."

Alexei Gagarin lived five years longer than his son, passing away in 1973. There was not a day that passed after Yuri's death when he was not reminded of his famous son, as a constant stream of well-wishers visited the house to pay their respects to the family of the first spacefarer. Sadly, their gifts of alcohol forced Alexei out of politeness to join them in toasting his son, and the sheer amount of alcohol he consumed hastened his death.

Decades later, Yuri Gagarin's cosmonaut colleague Valentina Tereshkova expressed to her biographer Lady Lothian how much more Gagarin might have been able to share had he lived and been free of the constraints of propaganda: "Just after his flight in 1961 Gagarin wrote a book entitled *The Road to Space*. I believe that if he were alive he would have written another book, a more detailed one, with the title *The Way It Really Was*. He had been collecting information for it. Knowing him well, I think he would have written a very serious book, not only on the preparation and the spaceflight but also on the history of the Soviet cosmonauts."

In the moments before he entered the *Vostok* spacecraft on 12 April 1961, Gagarin had made a spontaneous speech, according to journalists at the time. In it, he declared that his entire life had been in preparation for that day. He felt joy, he felt pride, but he also felt something more: he felt the responsibility of accomplishing what many generations before him could only dream of, and becoming the first person in space. He could not imagine a task more difficult. It turned out to be a prescient statement.

Following Gagarin's return to Moscow after his spaceflight, Nikita Khrushchev had declared that Gagarin's name would live through the ages even more than Columbus's, becoming "immortal in human history." The premier described him in almost religious terms, saying "now we can touch the man who has come to us straight from the sky." Yuri Gagarin, however, was always careful to distance himself from such lavish praise, and instead took pride in his role in a wider quest. The year before he died, in a newspaper interview, he showed that he fully understood his flight was just the beginning of a bold new venture: "Spaceflights cannot be stopped. This is not the work of any one man or even a group of men. It is a historical process which mankind is carrying out in accordance with the natural laws of human development."

Some time after Yuri's death, Valentina reflected on her late husband. More than anyone, she had seen his good and bad sides. She insisted that he was just a normal person who had shared with her his moments of happiness and sadness. "Like everyone else," she explained, "he made many mistakes, and was always willing to correct them. Through the years, I came to understand Yuri much better, but I will be frank and admit there was something in him that still remains a mystery to me. The human heart contains a lot of things, more than anyone can imagine."

Perhaps Yuri Gagarin did not always live up to the iconic image so rapidly thrust on him, but no one should ever have expected him to. Three days after his flight, following a speech by the president of the Academy of Sciences that called Gagarin "the Columbus of Outer Space," Yuri began his speech by countering: "I am an ordinary Soviet man." He never asked to be a symbol of heroism, yet he lived up to the image more than most who achieve sudden fame ever do.

The first person in space was, after all, only human.

2. Lighting the Candle

A candle loses nothing by lighting another candle.

Erin Majors

In May 1959, an attractive twenty-three-year-old air force nurse based at Patrick Air Force Base in Florida began work at her newly assigned job. At the end of a curious string of circumstances, Dee O'Hara unexpectedly found herself at the very heart of perhaps the greatest engineering endeavor ever undertaken, and doing what would come to be regarded as one of the most enviable jobs in the world. She had just become the nurse for America's seven Mercury astronauts.

Dolores O'Hara was born on 9 August 1935 in the small town of Nampa, Idaho, some twelve miles west of the state capital of Boise. Although her parents, Edward and Genevieve, were born in the United States, in true Irish fashion they had simplified their names to Mick and Jennie. Likewise, their first son, William, would become Bill, while Dolores, born a year later, would assume the less cumbersome name of Dee.

Mick O'Hara could not find work in Nampa, so before Dee had even begun school the family relocated to San Francisco where her father had found a job with the Bethlehem Steel Company. Moving to a big city proved to be a mistake, and it was not long before O'Hara took his family north to the more familiar small-town environment of Crabtree, Oregon, where he had found employment with the local logging company. Dee attended grade school in Crabtree and high school in Albany, Oregon, for two years, and then transferred to nearby Lebanon High School for her junior and senior year.

Two weeks after her high school graduation, tragedy struck the close-knit family. Dee was at a friend's house taking piano lessons when a distraught neighbor came in. A news bulletin reporting that a worker had been crushed to death in a logging accident prompted her to make some frantic inquiries, which confirmed her fears. As Dee later recalled, it was one of the worst days of her life:

She broke the news to us that the victim had been my father. My mother wasn't home at that time, but I went home straight away, and the minute my mom drove up she knew what I had been told just by the look on my face. My mother had worked very hard to put me through school, but now I wanted to make it on my own and do it myself. I had a lot of self-pride, and would never have dreamed of having my mother support me.

I had always planned on being a teacher or social worker, but I gained my inspiration to go into nursing during my senior year at Lebanon High. A career day was held with people from different professions, and one of the speakers was a very elegant lady from the Providence Hospital School of Nursing in Portland, Oregon. Nursing had never occurred to me, as I would feel faint if I had to go into a hospital, and become ill at the site of an accident. But she looked so pristine in her white uniform and nurse's cap, and she really made an impression on me. She handed out brochures of the nursing school and I thought the photo of the entrance to the school was impressive. Then, as a clincher, I noticed a classmate at the back and I thought, "Well, if Luanne can do this, so can I!" I was selected for entrance, and the minute I walked into the school I knew I had made the right decision.

Eventually, Dee graduated from nursing school. One rainy day soon after, her roommate Jackie McMahan suggested that they join the U.S. Air Force Nurse Corps and see the world. Initially hesitant, Dee accompanied her friend to the recruiting office a few days later and almost before she knew what was happening had signed up. As it turned out, Jackie was sent off to Mobile, Alabama, while 2nd Lt. Dee O'Hara found herself assigned to Patrick Air Force Base at Cape Canaveral in Florida.

Soon after Dee's arrival, and before she'd really had a chance to settle in, the hospital commander, Col. George Knauf, called her in and asked if she would consider working as a nurse for NASA's astronauts. She would still be stationed at Patrick AFB, twelve miles south of the Cape, but on full-time loan to NASA from the Air Force Nurse Corps. She would work with Lt. Col. William K. Douglas (USAF), a surgeon who was also the astronauts' personal physician, and his assistant, Dr. Carmault B. Jackson, also a USAF colonel. (It must have amused them a little to know that they presided over seven of the healthiest men in the country.) Despite their job descriptions, none of the medical staff were actually NASA employees, and Dee would later learn that the space agency was always a little uncomfortable with the arrangement:

Although I wasn't really sure what the job entailed, I accepted. It turned out that I was to set up the Aeromed Lab, an examination area for the astronauts, and be with them as their nurse. The premise behind this was that they wanted someone who would get to know the astronauts so well that she would know whether they were really sick or not. The astronauts certainly were not going to tell the flight surgeon, Bill Douglas, because they knew the doctor had the right or capability of grounding them—the last thing these guys wanted. They were really fearful of doctors for this reason.

It was the very early, pioneering days of the space program, and everyone was in the same boat—not knowing exactly what to do or expect. "If I had been told back then," O'Hara recalled, "that I would be the one seeing off every American space crew right through to the first space shuttle launch in 1981, I would *never* have believed it."

Initially there were no permanent training facilities at the Cape for the seven astronauts, who were based at Langley Air Force Base in Hampton, Virginia. At Langley, the fledgling Project Mercury was managed by the Space Task Group, or STG, under the experienced leadership of Dr. Robert Gilruth. The STG was principally comprised of several scientists and engineers from Langley's National Advisory Committee for Aeronautics (NACA) who had stayed on when the NACA ceased to exist in 1958, its assets and personnel forming the nucleus of President Eisenhower's newly formed civilian space agency, NASA.

Early in their training, the astronauts would often fly down to the Cape during the week for their familiarization exercises and to watch launches. Later, they would also take part in spacecraft simulations and suit fittings, chamber tests, and other procedures in Hangar S, an immense converted airplane hangar with large sliding doors at either end that was situated in the heart of Cape Canaveral's industrial area. Once the astronauts had completed their tasks they would return to Langley for the weekend.

Just about every activity related to Project Mercury at the Cape was crammed into Hangar S. Even the Mercury spacecraft were kept there for final inspections and tests in a big, spotless room before being mated to their rockets at the pad. Dee O'Hara diligently took to the task of organizing the Aeromed Lab in the only space available, on the mezzanine level overlooking the hangar floor. Once the lab was in place and fully kitted out, she also helped set up and furnish other areas such as the pressure suit checkout room, a conference room, and the astronaut quarters in room S205.

Within the crew quarters was a lounge area that O'Hara had assembled. Though not plush by any means, it was furnished with a sofa, a television set, a pair of beige La-Z-Boy chairs, and two double-decker bunks in case the astronauts trained late or worked long hours in the capsules. Later, as an alternative to the bunks, they would often leap into their shiny new Corvettes and race each other along State Road A1A, down to the welcoming lights of Cocoa Beach. Of their duties at the Cape, Deke Slayton said, "Working down there meant long, tedious days—sitting around waiting for some hold in a countdown to be worked through, that kind of thing. We were always looking for ways to let off steam."

Back then, Cocoa Beach was a small resort community of squat motels, trailer courts, and pink or aquamarine bungalows dotting a narrow sand spit located between the Cape and Patrick AFB. Nevertheless, it provided an enticing diversion for the astronauts. They would often relax there in the relative comforts of the Starlight Motel and later the Holiday Inn, both managed in turn by the discreet and astronaut-friendly Henri Landwirth.

Gordon Cooper, for all his outward placidity, quickly became renowned as the wildest of the astronauts when behind the wheel of a Corvette, and his high-speed duels with Alan Shepard along the beach road became the stuff of local legend. Once they had reached Cocoa Beach the astronauts could enjoy a steak and Caesar salad at Ramon's or unwind with a bourbon or two in the Holiday Inn's bar. If, however, they felt like cutting loose a little, they could be found propped on dark-wood barstools at the Polaris Lounge or the Mouse Trap, or at any one of a number of entertainingly tacky, pseudo-Polynesian watering holes to be found along the main strip.

Setting up facilities in Hangar S was hard work, but the laboratory, exam room, and a long string of rooms on the second floor were nearing completion when O'Hara unexpectedly met the astronauts for the first time:

I happened to go down the hall into the conference room, and when I opened the door, there they all were! John Glenn was sitting on the table, and I was absolutely terrified! I said, "Oops, I'm sorry!" and I slammed the door, of course much too loud, as I backed out, and I went tearing up to my area. So anyway John, bless his heart, came up and asked me to meet the guys. So I went back down with him, and the only thing I could think to ask was if they'd like something to drink. And of course they did. Fortunately, I had some Cokes or other soft drinks, but that was my first introduction to them. For some reason I was very, very shy with them. But then I was twenty-three years old and not very sophisticated.

O'Hara eventually came to an understanding with the seven astronauts—a pact that she never shared with the flight surgeons. The seven could come to her with any medical concerns whatsoever, and she would only report it further if she felt it jeopardized their lives or a flight. Otherwise, it would be their secret. She has kept her vow to this day, and always will. O'Hara also extended her medical services to the astronauts' wives and their children, and this extra kindness was greatly appreciated by them.

When in 2003 the authors asked Dee O'Hara for her impressions of the seven astronauts, collectively and individually, she said:

They were a hell of a group. They were the cream of the crop, and of course in my eyes they still are. They were the best America had to offer. I'm not sure they would have passed muster today; they didn't have the scientific credentials and such that is part of being an astronaut these days. But you've got to hand it to these guys; they put their backsides on the line because it was such a new program—new everything—and they stepped up to the plate and did it.

All seven of them should be so terribly proud of themselves for what they did. I think they brought a lot of prestige and honor to their country; I really do. Those first seven led the way. They really are the true pioneers, if you will, of the space age. They're very *special, each in their own way. Not only to me, but I think to a lot of people. They are national treasures.*

Gordon Cooper: *I always liked Gordon very much. He was friendly, but he was also outspoken. He would always come up to my office when he was in the area and we would just talk; I got very comfortable with him. He was an excellent pilot, and he knew that. If* you *didn't know that it was* your *problem, and it was not in a bragging sense at all. He was from Oklahoma, and he had this wonderful kind of quiet way about him. He just went about his business. Gordon was very likable, and I think everybody liked him.*

Scott Carpenter: *Scott really is very sensitive, a poet. He could always see the beauty or the poetry in something, or feel the emotion. I'll never forget, he came into my office one day and said, "Dee, my lighter doesn't have a flint—do you have any?" Before I could find them he had to dash off somewhere, so when I found a couple I taped them under a piece of tape on his desk lamp. When I came in the next morning there was a piece of tape on my lamp, and underneath it was a note, saying, "A kiss is in here for you." Now see, that's Scott. It's hard to explain him to people, but that's a classic example of what he would say—the gentleness and the sensitivity. I've always had a special spot for him.*

John Glenn: *John, bless his heart; is a supernice guy. He is warm, friendly, and very approachable. In all the time I've known him he's just been such a wonderful man. John made me feel instantly welcome that time I innocently stumbled across the "First Seven" in the conference room in Hangar S. He came out the door after me and insisted on introducing me to the others. I do remember seeing him as a contestant on the* TV *show "Name That Tune" before he was selected, and he was absolutely wonderful and charming, but obviously I didn't know him at that time. What did stand out for me was that smile of his. John has two very distinctive and endearing features—his infectious smile and his freckles. He, like the others, was an excellent aviator and a strong competitor.*

Deke Slayton: *Deke was the first astronaut I met on my first visit to Hangar S with Colonel George Knauf of Patrick Air Force Base. Deke was well respected, very approachable, and very warm, but he wasn't over-demonstrative like Gordon would be, or Scott. Deke was just very reserved, but you knew if he liked you, and he put everyone at ease. He was a straight talker, a straight shooter, but tough; he didn't take stuff off anyone. But later, when he was in charge of the astronauts because of his heart problem, he was really very good with them, even though he sometimes had to be brutally honest with them. There were a*

Mercury astronaut Deke Slayton with Air Force nurse Dee O'Hara. Courtesy NASA.

lot of issues I'm sure that were very difficult for him to discuss with them, but he was extremely good about not playing favorites. The other guys knew it; everyone knew it. He had a tough job to do, but he was always fair and very friendly.

Wally Schirra: *There's no question that he was the personality of the group. Always ready with a quip or pun. He really has the most wonderful sense of humor, wit, and charm. I was prey for his pranks; he was* always *doing things to me! It did not matter what you said to Wally; you simply could not get ahead of him. Ever. Very quick on his feet, very sharp, very witty, but also deadly serious when it came to his work. Boy, was he serious! He did not tolerate any mistakes—no sloppiness, no nothing. You had to march to a certain beat . . . you'd better know he was in charge. But Wally, he could take any sort of a sit-*

uation and make it so much fun. You could always hear him laughing, coming down the hall. He's just a delight, a delightful guy. I just love him to death. Always has been a favorite.

Gus Grissom: *Gus was very quiet; it took time with him. He was very difficult to get close to, so I never really pushed it with him. He treated many people with a polite indifference. It may well have been because back then there were no women in the business. There was no one except the occasional one or two secretaries, and I was not wanted at all by the management of* NASA . . . *they simply didn't want me there. It was a total, total male world; it was all engineers, and they flat out did not want any nurse out there—particularly a female. I didn't know a lot of this . . . that I was not wanted at that point. I had no idea.*

But with Gus, he was just comfortable with other men and other pilots, and maybe it was the medical thing. It took a long time—so many months—for him to simply look me in the eye or ask me something. With the others it came very naturally, but not with Gus, and I don't know why. An interesting personality, and I'm not sure in which way.

Alan Shepard: *Alan was very difficult to get to know because he was intimidating, and I was very uncomfortable around him initially. Once you got to know him you didn't see that, but he was a very bright man, and his aim always seemed to be to find out how much you knew about what you were supposed to be doing. And he was brilliant at it. One time, as an example, he had to go for a checkup with a local dentist. He turned that poor dentist every which way but loose. He wanted to know all about the drill, its mechanics, what it did, the r.p.m. The dentist couldn't answer the questions, and of course Al's staring at him with those cold, ice-blue eyes, and the guy was just intimidated beyond words. But Al could do that to you. He wasn't threatening, but boy, was he intimidating!*

We got along fine eventually, but you had to prove yourself with Al. I went to him one day—I forget why I was so fired up toward him—and I said, "No more of this, Al, I'm not going to take this from you!" It ended whatever that was, and then I think he trusted me. . . . I think he figured, well, maybe I was okay. What most people didn't see, because he would not allow you to see it, is he really wanted to be liked. He really *wanted to be liked.*

Alan Shepard has often been described as having two completely different personalities, with not much in between. Fluctuating between a cool, cocksure determination and a quick, agile wit, it was hard to know what to expect from the naval aviator at any time, and he would often catch people unawares. One day he might be the jocular, smiling Al; the next he would be the renowned "Icy Commander," distant, stern, and abrasive. "He was hard to get to know," astronaut Gene Cernan said of Shepard. "But once you cracked the surface, he was your friend for life."

Paul Haney, from NASA's Public Affairs Office, would often rib Shepard about his perceived aloofness. He recalled one time when Dwight Eisenhower's son and family turned up for a tour of the astronaut training facilities, and Shepard agreed to take them on a walking tour of the center. "Shepard's outgoing manner caught the eye of a *Houston Post* reporter who was with the family," Haney recalled. "The reporter in a story lauded Shepard for being so 'nice.' I clipped the item out of the morning newspaper and sent it to Al, chiding, 'I never knew you were so nice.' The news clip came back to me, saying, 'I'm only nice once a year!'"

Alan Bartlett Shepard Jr., who could lay claim to being a Mayflower descendant and an eighth-generation New Englander, was born and raised in East Derry, New Hampshire, on 18 November 1923. The son of Alan Bartlett Shepard Sr., a retired army colonel and insurance businessman, and Renza (Emerson) Shepard, he grew up on the family farm and attended elementary school with twenty-four other students in a one-room schoolhouse, completing six grades in five years. He had what he would later describe as a "pretty normal" childhood, with chores that had to be done around the farm but also plenty of free time to play with his friends. Most things he wanted had to be earned, so to buy his first bicycle he began working a newspaper route and entering into a partnership with his grandmother to raise and sell chickens. It was a lot of hard work, but eventually he bought his bike.

Even at an early age he excelled in math at school, and in a 1991 interview with an archivist from the Academy of Achievement, Shepard spoke about his first teacher, Bertha Wiggins: "She was about nine feet tall as I recall and a very tough disciplinarian. Always had the ruler ready to whack the knuckles if somebody got out of hand. She ran a well-disciplined group. I think most of the youngsters responded to that. There were one or two that couldn't handle it, and obviously they dropped by the wayside. But that still sticks out in my mind. That's the lady that taught me how to study, and really provided that kind of discipline, which is essentially still with me."

Shepard showed an early interest in airplanes and in the renowned exploits of his great boyhood hero, Charles Lindbergh. Over time, his bedroom became festooned with lovingly constructed balsa and tissue paper models, and his greatest problem was keeping his little sister Pauline (known as Polly) from playing with them. He and some friends formed Alan's Airplane Model Club, and they would meet in a shed connected to the barn to discuss their mutual interest.

One unforgettable Saturday morning, when Shepard was twelve years old, the budding aviator got his first, albeit very brief and hazardous, taste of flying. His good friend Alan Deale had been given a small glider designed to carry a boy or other small person, and Shepard had pluckily volunteered to be the first to try it out on a nearby hill. Crudely designed and constructed, the glider had two separate wings and a tail made of light wood, all covered with thin canvas. The would-be pilot had to be carefully strapped in between the two rigid wings, with the glider's tail sticking out behind. Having donned the crude assembly, to the cheers of his friends Shepard began running down the hill, going faster and faster. Unfortunately, he'd made no allowance for a severe prevailing crosswind, and as he took to the skies the ungainly craft suddenly tipped sideways and crashed into the ground. The glider was a splintered and irreparable mess, but as Shepard extricated himself from the wreckage there would likely have been a huge, toothy smile on his face. It may have only been momentary, and he had only achieved an altitude of about four feet, but Shepard had flown. That fond memory would always remain with him.

In high school Shepard wanted to learn more about real airplanes, so he used to ride his bicycle ten miles to the local airfield, where he did odd jobs such as pushing airplanes in and out of the hangars and general cleaning. As a partial reward for his enthusiastic help he was given his first flight as a passenger, and this led to two or three more with the same pilot. Eventually, on one momentous flight, the pilot allowed the keen youngster to get a feel for the controls. "And that's when it really all started," Shepard recalled.

In 1940 Shepard graduated from the Pinkerton Academy in nearby Derry and spent the following year at Admiral Farragut Academy in Toms River, New Jersey. His record there reveals both his early leadership potential and an extremely high IQ of 145. The academy's 1941 yearbook was glowing: "He speaks words of truth and soberness." Shepard then enrolled in an accelerated wartime course at the U.S. Naval Academy at Annapolis, Maryland, where he was a member of the varsity crew, among other sports. He earned his bachelor of science degree at the academy in 1944, though he graduated only 462nd out of a class of 913. Shepard was remembered by an academy classmate as "undistinguished, but a real likeable guy." During his college days, Shepard met his future wife, Louise Brewer, whom he married after his graduation.

Shepard was commissioned an ensign and saw wartime duty as a deck officer aboard the destroyer USS *Cogswell*. In 1946 he entered navy flight training and won his wings a year later. While undertaking his flight training at Corpus Christi, Texas, and Pensacola, Florida, Shepard also paid for additional civilian flight training in his spare time because he felt he wasn't advancing quickly enough in the navy. After completing several tours of aircraft carrier duty as a naval aviator—first with Fighter Squadron VF-4B and then with the navy's 42nd Fighter Squadron flying Corsairs from the USS *Franklin D. Roosevelt*—Shepard was delighted to be admitted as one of the youngest-ever candidates to the Patuxent River Test Pilot School in Maryland in 1950. After graduation, he remained at "Pax" River as a navy test pilot, where he alternated between sea duty, flying F2HC Banshee night fighters with Fighter Squadron VF-193 at Moffet Field, California, as the squadron's operations officer, and strapping himself into some of the navy's hottest airplanes for high-altitude and in-flight refueling systems tests.

Shepard once described night operations from an aircraft carrier as "the hardest kind of flying I've ever done." Of flying as a test pilot in Maryland, Shepard told Ted Wilbur of *Naval Aviation News*, "I think that's when I realized I was the sort of person that was objective enough and dedicated enough to do a good job. Then there was the challenge to keep doing better and better, to fly the best test flight that anybody had ever flown." Shepard also carried out some of the first-ever landings on the navy's new angled-deck aircraft carriers.

The future astronaut remained at Patuxent River until 1953, when he was given a three-year assignment as operations officer for the 193rd Fighter Squadron aboard the carrier USS *Oriskany* on two tours of the western Pacific. Shepard then returned to Pax River in 1956 for a second tour of duty, participating in the test programs of the F-3H Demon, F-8U Crusader, F-4D Skyray, F-11F Tigercat, and F-5D Skylancer. He then spent the next five months as an instructor at the navy's Test Pilot School before attending the Naval War College in Newport, Rhode Island. In 1958 he graduated from the naval college and was assigned to the staff of the commander-in-chief, Atlantic Fleet, as aircraft readiness officer. It was while serving in this post that Shepard became the father of daughters Laura and Julianne and was asked if he wanted to be considered for duties as a NASA astronaut.

At five feet eleven inches tall, Alan Shepard was right at the maximum height for selection as an astronaut. Had he been just a quarter-inch taller, he would probably have seen out his career in the navy while another officer took his place among the Mercury Seven, perhaps to become celebrated as the first American to fly into space. It was a close thing for him.

Dee O'Hara, who came to know Shepard as well as anyone, believed that despite Shepard's apparent aloofness and stoic silence, he really wanted people to like him. "Sometimes he was hard," she recalled, "but he always took a very defensive position or posture to begin with. He had a wonderful sense of humor, and could be very kind, and then he could turn within a second completely the opposite way. It was very difficult to read him. It was an attitude with him, and I think the coldness may have been a defense mechanism for some reason, and it was easier to keep you at arm's length than let you in close. I don't know who coined the phrase 'The Icy Commander,' but that described him."

If Alan Shepard was something of an enigma to Dee O'Hara, Gus Grissom also left O'Hara unsure of where she stood with him. Many referred to Grissom as "Gruff Gus," quite often with good reason. He largely kept to himself; was not given to displays of gentler emotions, either publicly or privately; and had a marked preference for conversations on the hallowed subjects of airplanes and engineering.

It's true that Gus Grissom did not quite fit the archetypal American hero mold. Stocky, some would say compact, and just five feet seven inches tall, he was the shortest of the Mercury astronauts, and looked more like your suburban plumber or television repairman than a talented engineer who rode the hottest jets in the skies. But he was pragmatic in his outlook. As he once told a *Life* reporter, "Most people who know me know I'm not the hero type. Personal prestige I couldn't care less about."

Born Virgil Ivan Grissom in Mitchell, southern Indiana, the future astronaut grew up as the eldest of four children. The lifelong nickname of Gus came about during a game of cards when one of his young friends tried to read the name "Gris" upside-down on a scorecard and pronounced it as "Gus." The players all burst into fits of laughter, and after that the name—which Grissom really liked—just stuck.

Grissom first developed a moderate interest in flying at Riley Grade School, but later said it wasn't anything that dominated his life. Moreover, he would always be the first to admit that he wasn't a brilliant scholar, although he was said to have an IQ of 145. He had no idea what he wanted to do when he left school. "I guess it was a case of drifting and not knowing what I wanted to make of myself," he once suggested. "I suppose I built my share of model airplanes, but I can't remember that I was a flying fanatic."

After completing his elementary and high school education in Mitchell, Grissom entered the U.S. Army Air Forces as an aviation cadet in 1944. The following year he married Betty Moore, with whom he would eventually have two sons, Scott and Mark. With the end of World War II, Grissom was shuffled from what he felt was one second-rate, nonflying assignment to another, so he resigned from the army air force in November 1945. After graduating from Indiana's Purdue University with a degree in mechanical engineering a full semester early in 1950, Grissom found that no civilian jobs appealed to his sense of adventure and returned to air force cadet training. In March 1951 he received his wings.

At the age of twenty-five, Grissom joined the 75th Fighter Squadron at Presque Isle, Maine, before being shipped off to South Korea in December 1951. There, he flew a hundred combat missions as an F-86 Sabre Jet fighter pilot with the 334th Fighter-Interceptor Squadron. In March 1952 he was involved in driving off several MiG-15 fighters that had attacked a photo reconnaissance airplane he was helping to escort back to base. His courage and piloting skills that day earned Grissom a Distinguished Flying Cross. The citation read: "The superlative airmanship demonstrated by [Second] Lt. Grissom on this mission exemplifies his tour of duty, reflecting great credit upon himself and his comrades-in-arms of the United Nations and the U.S. Air Force." Grissom requested permission to fly another twenty-five missions as he had not shot down any enemy aircraft, but his request was denied. Leaving Korea that year as a first lieutenant, his six months' service was further recognized when he was awarded an Air Medal with Cluster. He then became a jet pilot instructor at Bryan, Texas. Three years later he studied aeronautical engineering at the Air Force Institute of Technology at Wright-Patterson Air Force Base in Ohio before being selected to attend the Test Pilot School at Edwards Air Force Base in California in October 1956. Eventually he returned to Wright-Patterson as a test pilot with the rank of captain and was assigned to the Fighter Branch.

Then one momentous day in 1959, the unit adjutant at Wright-Patterson handed Grissom a teletype marked "Top Secret," which requested him to report to Washington dressed in civilian clothing. Mystified, he traveled to Washington, where he was asked if he wanted to undergo qualification and medical testing to become an astronaut with Project Mercury. It was a big decision, as he was enjoying his work test-flying a wide range of airplanes, but the more he thought about it the more the whole idea appealed to him. "I figured that I had enough flying experience to handle myself on any shoot-the-chute they wanted to put me on," he later said. "In fact, I knew darn well I could."

Once at NASA, Grissom would carry that same determination into any sporting challenge. During a hard-fought series of handball games between the astronauts, for example, the only person who beat Gus on the court was Alan Shepard. At the time, *Life* reporter John Dille contradicted some inventive but well-intentioned reports that Gus had deliberately lost to his friend, knowing how desperately Shepard wanted to win. "This was unlikely, however," Dille wrote. "The astronauts compete in nearly everything they do. Competition comes naturally to them, and it would not be like Gus Grissom at all to *let* another astronaut beat him at anything. He is a no-nonsense pilot, a steady, dedicated professional who is completely absorbed in his work."

When Dee O'Hara was asked if the Mercury astronauts were as playful as Americans were led to believe, she laughingly responded: "Oh yes. For instance, they used to hide their urine samples from me. I'd find them in the shower stalls, the air conditioner, and the heater—anywhere! One time that great practical joker Wally Schirra was asked to provide a urine sample and he brought it to me in this gigantic container the size of a water cooler!"

Wally Schirra remembers the incident well. "Early one morning we filled a five-gallon bottle with warm water, figuring it would cool to body temperature by the time she arrived. [Gordon Cooper and I] added a bit of iodine to give it color, laundry soap to make it foamy, and put the bottle on Dee's desk. I tagged the bottle, writing the time of delivery in Greenwich Mean Time, and I attached a bunch of lollipops. Gordo and I were like little kids, peeking with glee around the corner of the doorway, when we spotted Dee first stop in horror, and then burst out laughing."

Schirra describes O'Hara as "like a sister to us, and a mother to our children. Dee could take care of us better than any of the doctors could. We allowed no one but Dee to draw our blood, not even the doctors, for fear that they would collapse a vein." This always amused Dee. "The guys got on this thing about nobody doing it but me. What these guys didn't realize was that I was really out of practice, and I was probably the worst person in the world to do it. God was good to me—I never missed a vein!"

Like many people in America at that time, Alan Shepard loved the character José Jiménez, created by comedian Bill Dana for Steve Allen's TV show in 1959. Although José was supposed to be pseudo-Mexican, Dana himself is actually of Hungarian-Jewish ancestry and was born William Szathmary in Quincy, Massachusetts, in 1924. Dana was swept up by the excitement of the public adoration for the Mercury astronauts. To express his own admiration in humor, he invented a slow-witted character named José Jiménez, a Mexican immigrant who had somehow become an astronaut. Today it would be considered a highly offensive stereotype, but in the late 1950s such crude humor was still very much in vogue. A television show was built around Dana's bumbling José, who soon featured in a series of comedy albums, and Dana also appeared in skits on such programs as *The Ed Sullivan Show*, doing a popular sketch known as the Reluctant Astronaut.

Shepard thoroughly enjoyed the Jiménez character, even though it poked fun at him and his fellow astronauts. He had a special fondness for the routine in which the interviewer asks José about the object under his arm. "Is that your crash helmet?" he asks innocently. Jiménez looks forlorn. "Oh, I *hope* not!" Soon everyone at the Cape and at Langley got used to Shepard playfully mimicking the character in a thick mock-Hispanic drawl, opening with the catchphrase, "My name José Jiménez." As John Dille wrote in the introduction to the Mercury astronauts' book, *We Seven*, "It was Al

Mercury astronauts Gus Grissom and Alan Shepard strike a playful pose. Courtesy Scott Grissom.

who introduced José Jiménez into the astronaut clan. . . . Al found the Jiménez brand of humor such a handy device for relaxing the troops when tension was building up before a flight that he once arranged for a tape of some of José's dialogue to be played in the Mercury Control Center during a practice run."

One night, when Dana had just begun his Reluctant Astronaut routine at a nightclub in Cocoa Beach, Florida, he was amazed to see a member of the audience stroll up on stage and, to wild applause, join in as the Man on the Street interviewer. It was Alan Shepard, who had been egged on to take the stage by Deke Slayton and Wally Schirra. "The place went wild," Dana later recalled, adding that he was absolutely astounded by Shepard's perfect performance of the routine that night. "He knew the whole thing."

The Mercury astronauts quickly adopted a delighted Bill Dana and his alter ego, José, into their ranks, and Dana soon became known as the unofficial "eighth astronaut." They would often attend Dana's live shows and sometimes join him on stage for a little impromptu fun with "José."

The question of who would fly first—take that first rocket ride into space—was proving a difficult decision for Robert Gilruth. He decided to ask the astronauts to poll themselves, and in December 1960 asked each of them to vote for the astronaut, excluding themselves, whom they thought was the best candidate. As Deke Slayton later recalled, "I think he just wanted to know if we agreed with his judgment."

Shepard certainly thought he was the best candidate. He once joked with interviewer Roy Neal that when he met his six colleagues for the first time, his initial reaction was, "I wondered first where these six incompetent guys came from!" For Shepard, the teamwork was always secondary to his aim to be the first to fly. "I knew there was a lot of talent there," he told Neal,

and I knew it was going to be a tough fight to win the prize. It was an interesting situation, because I was friendly with several of them. And on the other hand, realizing that I was now competing with these guys, there was always a sense of caution I suppose—particularly talking about technical things. Now, in the bar, everything changed, but in talking about technical things there was always a sense of a little bit of reservation, not being totally frank with each other, because there was this very strong sense of competition . . . seven guys going for that one job.

On the afternoon of 19 January 1961, the day before the inauguration of President John F. Kennedy, Gilruth placed a call to each of the seven Mercury astronauts, asking them to be present at an important meeting in their Langley office. Each of the men knew what this was about; they were about to find out who had been assigned to fly the first missions.

A mounting sense of apprehension hung in the air as the astronauts assembled just after five o'clock in their sparse office, and though they cracked jokes with each other while they waited for their project boss to appear, the tension was palpable. Finally, Alan Shepard leaned over to Deke Slayton and asked, "Deke, what do you think?" The other five quickly tuned in to hear the response. "I think I wish to hell he'd hurry up," Slayton grouched. A nervous Gus Grissom then blurted out, "If we wait much longer, I may have to make a speech!" At this the other six men burst into relieving laughter. The idea of the gruff, normally taciturn Grissom making any sort of speech was enough to crack up anyone who knew him. After all, this was the same fellow who, a year earlier, had been placed before a microphone in San Diego, where General Electric's Convair Division was constructing the mighty Atlas rockets. A warm speech of welcome had been delivered by a senior Convair executive before a crowd of around eighteen thousand crammed into a massive auditorium, and Grissom had been selected to respond on behalf of the astronauts. He truly hated giving speeches, but he also realized it was now part of his job, so he stepped up to the podium and waited for the applause to die down. He hesitated, and then stammered, "Well . . . do good work!" And that was it; he waved, stepped back, and sat down. A stunned silence ensued, and then the vast gathering began applauding wildly, as if Grissom had just delivered the Gettysburg Address. Those three words, "Do good work!" would actually become the plant's motto and were later stitched onto a huge banner that hung over the factory floor to provide inspiration.

When Gilruth finally made an appearance, he wasted no time. According to Shepard's recollection of that day, Gilruth closed the door behind him and said, "Well, you know we've got to decide who's going to make the first flight, and I don't want to pinpoint publicly at this stage one individual. Within the organization I want everyone to know that we will designate the first flight and the second flight and the backup pilot, but beyond that we won't make any public decisions. So . . . Shepard gets the first flight, Grissom gets the second flight, and Glenn is the backup for both of these two suborbital missions. Any questions?"

There was absolute silence in the room as the seven men digested the news. Gilruth looked around once at the seven faces, and then made his exit, saying, "Thank you very much. Good luck!"

For Shepard, it was a moment of pure joy and sweet victory, but he knew he had to temper his jubilation. His six friends had just been dealt some devastating news and were dealing with the fact that they had not been considered the best person for the job. It was a tough few moments, as Shepard later revealed. "I did not say anything for about twenty seconds or so. I just looked at the floor. When I looked up, everyone in the room was staring at me. I was excited and happy, of course; but it was not a moment to crow. Each of the other fellows had very much wanted to be first himself. And now, after almost two years of hard work and training, that chance was gone." One by one the other six, smiling to hide their obvious disappointment, came over and shook Shepard's hand before drifting out. Soon he was the only person left in the room.

Gilruth's intention had been to take the pressure off the primary astronaut—the one chosen to fly the first mission—before he finally revealed the selection of the three astronauts to the waiting press on 20 February. However, he refused to tell reporters who of the three had actually been chosen to fly the first Redstone, and to avoid overt speculation he announced the names in purely alphabetical order—Glenn, Grissom, and Shepard.

While Shepard was obviously elated, John Glenn was furious at not being selected, despite his famously pleasant countenance when dealing with the press. And just like many in the press, he believed he should have been the Chosen One. His patience seemed to be wearing a little thin when at a San Diego press conference he replied to the question, who would be the first to fly, by saying: "We would like to get away from the 'first' aspect. This is a beginning of many flights. Actually the second and third and fourth flights may accomplish far, far more scientifically than the first flight does. That first mission is going to be sort of an 'Oh, gee whiz, look at me; here I am, Maw!' type of deal, and you are probably going to get a limited amount of data back from it."

Glenn was never one to accept defeat without a damn good fight, however, and decided to take the matter head on. He felt that the change of administration at NASA after Kennedy's election might be of considerable advantage for him, and he put his case to the space agency's new bureaucracy. He hoped to persuade them to ignore such vague indicators as the peer vote and install him as the prime pilot. Though the new administrators sympathized with Glenn, they refused to intervene and overturn Gilruth's ruling, and Glenn finally had to admit defeat. It did not come easily, however, and for a time he withdrew into himself and refused to go out on weekends or talk with his neighbors. "Those were pretty rough days for me," he later admitted. "I guess I am a fairly dogged competitor, and getting left behind . . . was like always being a bridesmaid but never a bride."

As Shepard told Roy Neal in a 1998 interview, he never did find out exactly why Bob Gilruth selected him ahead of the others. "I asked him several times over the years, and he always said, 'Well, you were just the right man at the right time.' But I'm sure that he was very personally involved in that selection process. There were some suggestions from some of the other folks in the program that maybe he had made a mistake in the decision, that there might have been someone else who qualified better, but he did not change his mind. So he's one of my heroes."

Shepard, however, had two positive assets that might have been important considerations in his selection. Despite his sometimes stern demeanor, he actually possessed a lively sense of humor, which gave him a safety valve to gain an added source of self-control. He was, to use the contemporary American lexicon, "fully integrated." His second advantage lay in the fact that during training he had specialized in methods of recovery and missile-range network control, and getting the man back was always the first priority of manned space flight.

During one press event, Gordon Cooper was asked what would happen if the Russians beat America to the moon, and he gave a neat analogical answer. "This is a track meet, not a race," he responded. "You have to win a lot of events to win the meet." It was an analogy that would also have fit John Glenn well. He had lost this first race, but soon enough he would overshadow Shepard, becoming by far the most famous of the Mercury astronauts among the American public.

That public was aware of the risks these Cold War heroes were facing. They knew from television broadcasts that a lot of the rockets blew up on the launch pad or in the skies over the Atlantic beyond the Cape. For their part, the engineers and technicians also knew they bore a heavy responsibility for getting things right for those who would ride the rockets. Guenter Wendt, a mechanical engineer for McDonnell Aircraft, builders of the Mercury and later spacecraft, was given prime responsibility for all prelaunch activity in and around the spacecraft and its attending ground support equipment at the Cape. The astronauts came to trust Wendt implicitly, referring to their German-born colleague with fond irreverence as the "Pad Führer." Fun aside, Wendt and his team understood the heavy responsibility on their shoulders. "In the days of Mercury, everything took place at Cape Canaveral," he recalled in 2001:

We didn't have the Kennedy Space Center or any of that. We launched between fifteen and twenty rockets a week, but three out of five would blow up . . . there goes another nose cone! These failure rates would really get to us. We invited the seven astronauts down to watch an Atlas launch—that blew up! Shepard looked at Glenn and said, "I hope they fix that thing before we sit on it." These were the primitive things that we started out with in those days. Now, all of a sudden, you have a nose cone that you can't replace; there is a fellow in there that you know. Somebody you talk to, someone you have lunch with. All the rules changed quite a bit!

There is little doubt that the flight of Yuri Gagarin both shattered and infuriated Alan Shepard. He was absolutely livid, and he didn't care who knew. "That was a big blow, a *big* blow," O'Hara recalls. "The buzzing and the talking and the huddling and meetings that were going on. It was a big blow to everybody, and a great disappointment, because everybody was caught so off guard, and of course it was a PR fiasco—Gagarin's flight made us look like fools. Alan was bitterly disappointed, and I could understand that."

Shepard had been ready to go well before Gagarin soared into orbit aboard his *Vostok* craft. His flight had originally been scheduled for 24 March, but in January the Kennedy administration received a damning report on spaceflight activities from an eclectic science advisory group known as the Wiesner Committee. Among other recommendations, the committee supported delaying Shepard's flight. In fact, one of the committee's heads, George Kistiakowski, stated that launching Shepard too early would provide the astronaut with "the most expensive funeral man has ever had." The report was scathing in its evaluation of many aspects of NASA's manned space flight program and caused considerable grief for NASA administrator James Webb and Robert Gilruth. They felt they had a better grip than the committee knew on NASA's preparedness to fly an astronaut, but some of the more conservative aspects of the report prevailed. With unconcealed reluctance, Webb and Gilruth discussed the flight with their key Mercury personnel, following which they advised Wernher von Braun and his rocket team that a further unmanned test flight would take place on the date originally scheduled for Shepard's launch.

Von Braun was not unhappy with the decision, as he had already been pressing for another unmanned test flight. A "booster development launch" was then allocated to Shepard's launch slot, and if it proved fully successful then Webb and Gilruth insisted the astronaut would fly his MR-3 mission on 25 April. As an impatient Shepard sat waiting in the wings, von Braun had his final proving launch. Nineteen days later, on 12 April, a triumphant

Soviet Union successfully lobbed the first man into space and seized spaceflight history by the throat.

Less than three weeks after that, still smarting over the galling Soviet achievement, anxious Americans awaited with growing anticipation the moment when their first astronaut was finally launched into space. Newspaper, radio, and TV reports bombarded them with news of the forthcoming event. So what if Gagarin had been first? Here was NASA about to set things right.

There was only one thing the media could not tell them, and that was the name of the prime astronaut. Instead they teased the public by asking if it would be Glenn, Grissom, or Shepard, without being privy to the answer themselves. Only a few people close to the program knew, and they weren't saying anything. Public opinion heavily favored Glenn, and it would have surprised no one if the high-profile marine had been given the job. Speculation was high but often overspun. Robert Silverberg's anticipatory book *The First American into Space* would be released shortly after Shepard's mission, but the paperback's cover, obviously prepared and printed in advance, boldly featured a watercolor head-and-shoulders representation of the three astronauts—with a beaming John Glenn hogging the foreground and his two astronaut colleagues standing partially concealed behind him in the background. The artist may have picked the wrong rocket jockey, but like so many others at that time he had gone with popular opinion.

The first launch attempt was scheduled for the morning of Tuesday, 2 May 1961, but the evening before a heavy storm swept through the Cape, accompanied by lightning and rolling thunder. The Redstone rocket was fully loaded with liquid oxygen fuel, but at 3:30 a.m. a hold was called due to the continuing rain and lightning. Meanwhile, Shepard had been awakened at 1:00 a.m. and enjoyed a hot breakfast with Glenn and Grissom before taking his final preflight physical. Some time later, fully suited up and ready to go, he waited for the word to leave Hangar S and climb aboard the large transfer bus. Just at that moment someone called out for him to wait, so he took off his gloves and sat down to await any further news. He knew that if the launch was scrubbed with the Redstone already fully fueled, it would be at least forty-eight hours before they could try again. Soon after, at 7:25 a.m., he received confirmation that the launch was off.

There was disappointment all around, but in all the reports emanating from the Cape there came some interesting news: NASA had decided to disclose the name of the disappointed astronaut, and Alan Shepard was officially named for the first time. "I was relieved when they made the announcement," he later wrote. "It was getting to be a real strain keeping the secret." Two days later, at an afternoon meeting, the weather reports looked more promising, and it was decided to attempt another launch the following morning. Shepard and Glenn spent a relaxing evening chasing sand crabs along the beach, after which they shared a roast beef dinner. The two men then retired early to the astronauts' quarters and turned in. Shepard quickly fell asleep. "I woke up once, about midnight," he later wrote, "and went to the window to check on the stars. I could see them, so I went back to sleep."

Just after 1:00 a.m. on the morning of 5 May, Bill Douglas entered the crew quarters and gave the two astronauts a gentle shake. Shepard was almost instantly alert and clambered out of his bunk, keen to get going. He seemed calm, and Douglas later said that Shepard "acted just as if he were going out duck hunting or starting on a fishing trip." He may have appeared calm, but Shepard would say later that he had some "butterflies" in his stomach. Once again he showered and shaved and ate a hearty mid-evening breakfast, underwent a medical examination, and stood patiently as small electrical medical sensors were glued to his body. It was an extensive examination, lasting over two hours, and Shepard was deemed to be in good shape—apart, that is, from a slight peeling sunburn on his shoulders and a loose black toenail on his left foot where Gus Grissom had accidentally stepped on him. Glenn, meanwhile, had gone over to the pad to check out the spacecraft.

The psychiatrist, who spent nearly an hour with Shepard, reported that he had "never seen a man so calm. I tried to get him to talk about other things than the flight, about his family, for example, to see whether this would make him anxious, but I didn't succeed. All his mind, every nerve, was concentrated on the flight: nothing else interested him. Even while on his way to the suit room he was already a part of the spacecraft."

Next, aided by suit technician Joe Schmitt, Shepard began donning his $10,000 silvery space suit. It was just after four o'clock when Bill Douglas,

Alan Shepard, and Gus Grissom began the walk down to the large air-conditioned transfer bus that would carry them to Launch Pad 5 on the southeast edge of Cape Canaveral, three miles away. Dee O'Hara, nervously armed with her rosary beads, had accompanied them out to the hallway, but that was as far as she would go. Shepard turned, smiled, and said, "Well, Dee, here I go!" When asked years later what emotions ran through her that launch morning, her response came quickly:

I was frightened. Very frightened. Particularly when he left and went downstairs to get in the van. I didn't know if I was going to see him again. He'd come in earlier for his preflight physical and I drew a blood sample while Bill Douglas did the usual check. We didn't do a heavy physical because he'd been seen every day for a number of days, but Bill checked his skin very carefully because he was wearing these electrodes and wanted to make sure Al's skin wasn't cracked or there wasn't a rash or something that could become irritable. Granted it was only a short flight, but we didn't want any distraction for him such as an itch can do.

I remember when he walked into the suit room to get suited up, everything became dead silent. Joe Schmitt did his checks, and Alan got into his boots, but it was quiet. Deke was there, and everybody just sort of milled around, but there was hardly any conversation.

On the bus the three men were silent and somber, wondering if this would finally be the day. The quiet tension was too much for Shepard, and he turned to Grissom with a wide, toothy smile. Suddenly he was José Jiménez, the Reluctant Astronaut. "Hey Gus," he said in the familiar Jiménez drawl, "Do you know what it takes to be an ass-tronaut?"

Grissom fell straight into the routine. "No, José, tell me."

The smile on Shepard's face grew even wider. "You should have courage and the right blood pressure and four legs."

"Why four legs, José?" Grissom asked, also smiling.

"Because they *really* wanted to send a dog, but they decided that would be too cruel!"

Even Bill Douglas had to grin at this tension-breaking routine. Then, all too soon, they had arrived at the foot of the pad—the same launch complex from which *Explorer 1*, America's first satellite, had been launched three years earlier. It was 5:14 a.m.

As the first faint steaks of dawn began to color the eastern sky Shepard stepped out of the van carrying his portable air-conditioning unit. He was greeted by the spontaneous applause of some nearby fuelers and technicians. Ahead of him he could see eerie white clouds of vented liquid oxygen and water vapor swirling around the base of the frost-rimmed Redstone rocket. Shepard took a few steps, then paused for a moment to gaze up at the white-and-black rocket gleaming eerily in the arc lights, and then at his Mercury capsule, perched sixty-five feet above the pad. It was a moment he would never forget, as he later related for the book *We Seven*:

I sort of wanted to kick the tires—the way you do with a new car or an airplane. I realized that I would probably never see that missile again. I really enjoy looking at a bird that is getting ready to go. It's a lovely sight. The Redstone with the Mercury capsule and escape tower on top of it is a particularly good-looking combination, long and slender. And this one had a decided air of expectancy about it. It stood there full of lox [liquid oxygen], venting white clouds and rolling frost down the side. In the glow of the searchlights it was really beautiful.

Having watched a fully suited Shepard walk out of Hangar S, Dee O'Hara straightened the lab, then gathered up her records and anything else she felt she and Bill Douglas would need on Grand Bahama Island after the flight. "I then went over to this old Snark missile hangar where I'd set up a fully equipped mobile trauma unit, our Forward Medical Station. If there was a problem with the launch they could bring him there for first aid and then put him in a helicopter to a hospital, depending on the injuries. That's where I and my med technician and some others watched the launch. Bill was just awful . . . I mean he was just impatient, like a caged lion, walking back and forth."

After taking a last, lingering look at his rocket, Shepard, Douglas, and Grissom climbed into the elevator that carried them to the top of the launch gantry. On the way up, Bill Douglas handed Shepard a small present from the well-liked NASA engineer Sam Beddingfield. Bill Dana had worked up a routine in which José Jiménez is about to fly on a long mission and brings along a coloring book. The only problem was that he'd forgotten his crayons, and so refused to go. Shepard's gift was a box of colored crayons, which made him laugh out loud before he gave them back to Douglas. "Thanks," he said. "This time José's going to be busy. Take care of them for me."

John Glenn and the close-out team were already waiting in the canvas-covered White Room as Shepard approached the open hatch of *Freedom 7*, and he thanked them sincerely for their hard work on the flight. At 5:20 a.m. he disconnected the hose to his portable air conditioner, slipped off his protective overshoes, and made his way feetfirst into the spacecraft. Then Joe Schmitt strapped Shepard in hard and hooked his suit up to the craft's oxygen system.

Shepard began running his checks and removed the safety pins that had been placed as a precaution over several of the switches. He found a small note Glenn had left for him, reading "No handball playing in this area." Alongside it was an eye-catching centerfold from an adult magazine. He smiled at this tension-breaking bit of nonsense from the normally straight-laced Glenn, who with this gesture had unknowingly started a launch-day tradition of pranks and cards for space-bound astronauts. Because the capsule's interior would be filmed during the flight, Glenn wisely retrieved the card and centerfold, and at 6:10 the hatch was closed. "I was alone," Shepard later reflected. "I watched as the latches turned to make sure they were tight."

At the Mission Control Center (MCC), Deke Slayton ran an early communications check.

"José, do you read me?"

Shepard smiled at the nod to the Reluctant Astronaut, and despite the obvious temptation was all business in his response.

"I read you loud and clear, Deke."

Slayton, on the other hand, played his humorous parley to the hilt.

"Don't cry too much, José."

At 6:27, four long warning sirens sounded across the Cape, and seven minutes later the huge, rust-red gantry began to roll on its steel tracks back from the gleaming Redstone rocket. Once it was clear, a long-armed yellow cherry-picker crane driven by remote control was carefully positioned beside the rocket, and a long boom with a 116-foot reach and small square escape capsule at its top was nestled into position beside the spacecraft's hatch in case an emergency egress was required. It would remain there until one minute before the launch.

Then, at 7:14 came the first of the eventual four holds in the countdown. The weather was the first problem, as a front appeared to be moving in. Meteorologists were hastily consulted, and after a while they concurred that the ominous clouds would soon clear and the launch could proceed. The countdown resumed. Fourteen minutes later, as the clock wound down to T-15 minutes, the countdown was halted yet again because of a problem with the inverter in the Redstone. Gordon Cooper, the prelaunch voice communicator in the blockhouse, advised Shepard that the launch gantry would have to be brought back to the rocket in order to change the inverter. It would take, he guessed, about an hour and a half. Shepard had been lying on his back with his feet up in the air, strapped into his couch, for so long that he was experiencing an increasingly urgent need to urinate.

"Gordo?"

"Go, Alan," Cooper responded.

"Man, I got to pee."

"You what?"

"You heard me," was Shepard's retort. "I've got to pee. I've been up here forever! Gordo, would you check and see if I can get out quickly and relieve myself?"

Three or four minutes then passed as Cooper consulted with Wernher von Braun. Finally, he came back online and reluctantly told Shepard that he was under instructions to stay put inside the capsule.

"Well alright—that's fine," Shepard responded, "but I'm going to the bathroom!" He was reminded that there were wires all over his body, and the leads might short-circuit if they got wet.

"Don't you guys have a switch that turns off those wires?" he inquired. There was some more consultation, and then Cooper confirmed there was such a device.

"Tell 'em to turn the power off," Shepard snapped. A bemused Cooper responded a few moments later.

"Okay Alan, power's off. Go to it!" At this, Shepard happily relieved himself into his thick cotton undergarment. "It was either that or completely scrub the launch, and take him out, de-suit him and let him go pee," O'Hara remembers. Liquid quickly pooled in the small of Shepard's back. His heavy undergarment soaked up the urine, and with 100 percent oxygen flowing through the suit he was soon dry. Power was reintroduced to the medical wires without incident, and the countdown resumed. But not for long—technicians were briefly concerned about a computer that would plot the trajectory of the spacecraft and predict its impact point in the ocean. It was soon resolved and the countdown resumed. Once again the red-and-white gantry rolled back from the pad.

When a bright ray of early morning sunlight began pouring through his spacecraft's optical periscope, annoying him, Shepard reached out and flicked a switch that would shield the periscope with a thick gray filter. Problem solved. Much to his consternation, at T-minus two minutes and forty seconds and counting, Shepard once again heard the word *Hold!* This time it was an abnormally high reading on a fuel pressure regulator. The technicians did not like this and called for a hold while they discussed the implications and possible solutions. Shepard's exasperation was evident. He'd played the game and endured the delays, but now it was time for the Icy Commander to take control of the situation. "I've been in here more than three hours," he said, with a noticeable bite to his voice. "I'm a hell of a lot cooler than you are. Why don't you just fix your little problem and light this candle?!!"

The problem was fixed, and the countdown proceeded toward liftoff at 9:34 EST. Just before the launch, communications would switch from Cooper in the blockhouse to Deke Slayton, seated at his MCC panel. Meanwhile, Wally Schirra and Scott Carpenter were circling the launch area in a pair of F-106 jets to record the launch and acceleration on movie film. They saw the yellow cherry-picker edge away from the rocket, leaving the slim Redstone poised for liftoff sixty seconds later. Thirty-five seconds before liftoff, a controller on the propulsion console firmly pressed a red button that threw into life an automatic sequencer that closed valves, turned on power, and ignited the fuel. The spacecraft's umbilical, which had continued to feed freon and power into *Freedom 7*, was detached in this process, and a door automatically closed over the periscope to protect it during the ascent.

Interestingly, however, there was a final and essential human element to the launch. In the moments between ignition and liftoff, the fate of the mission lay solely in the hands of a slender, scholarly propulsion engineer. Observing the flame belching from the foot of the rocket, Dr. Kurt Debus had to decide through long experience whether the color, shape, and intensity of the flame hinted at problems. If he foresaw disaster—any major aberration that placed the astronaut in imminent danger—he could press the first of two buttons beneath a raised plastic shield, arming the sixteen-foot escape tower mounted on top of the Mercury capsule. This tower was equipped with a powerful solid-fuel rocket motor atop three exhaust nozzles. It also had an automatic abort sensing system designed to register any impending booster catastrophe, ignite the motor, and drag the spacecraft up and away from the launch pad. The rocket would only fire for one second, but this was sufficient to propel *Freedom 7* a safe distance from the booster, after which the tower would be jettisoned, allowing the spacecraft to descend under its deployed parachute. The rocket motor could also be fired by the astronaut or by an electronic signal transmitted from the blockhouse, using the second button that Debus was now contemplating. Debus poised his fingers over the buttons as he intently watched the Redstone burst into life and, seeing no abnormality, slowly pulled them away and lowered the plastic cover over the buttons. He then breathed a deep sigh of relief. The agony was over; he had done his job, and America's first manned space flight was under way.

In all the excitement, Shepard later said he failed to hear much of the closing countdown, with the exception of the firing command at 9:34 a.m. During this period his pulse rate rose from 80 per minute to 126 at the liftoff signal. Then he heard the welcome, laconic voice of Deke Slayton, uttering the first words ever relayed to an American astronaut blasting his way into space: "You're on your way, José!"

"Roger, liftoff," Shepard responded, "and the clock has started." Over at the Forward Medical Station, Dee O'Hara had been busy with her rosary beads. "I don't think I breathed for quite a few minutes," she recalled. In his postflight report, Shepard stated that he had tightened his grip on the capsule's abort handle and probably braced himself "a little more than necessary" during the final moments of the countdown. He justified this by pointing out that no one really knew the amount of shock and vibration he would experience as the Redstone left the pad. To his relief, the ride was not as rough as he'd imagined it might be, and he relaxed somewhat as the rocket soared ever higher into the deep blue skies. "I knew I was going up," he later remarked, "but the sensation was very gentle."

A minute into the flight the Redstone reached transonic speed. It began to shake and vibrate with dynamic pressure as the forces of speed and air density came together. Shepard knew he just had to ride it out, although his vision blurred with the excessive buffeting. Soon after, the vibration began to diminish, and Shepard was able to report to Slayton that the ride was "a lot smoother now, a lot smoother." As the ascending Redstone continued to gather speed, Shepard reported to Slayton that he felt "okay" with "all systems go." No longer needed, the escape tower fired and detached itself from the top of the spacecraft. Shepard heard it ignite and tried, unsuccessfully, to observe its departure through a small porthole.

Thirty-seven miles up and 147 seconds into the flight, *Freedom 7* separated explosively from the spent Redstone booster and moved away under the gentle push of posigrade rockets. It continued to climb as the empty aluminum shell serenely tumbled away. "Cap sep [capsule separation]," Shepard reported. "Periscope is coming out and the turnaround has started." Now weightless, he felt, he later recalled, the capsule "begin its slow, lazy turnaround to get into position for the rest of the flight. It turned 180 degrees, with the blunt or bottom end swinging forward now to take up the heat." He then initiated a sequence that would shut down the capsule's automatic pilot and give him control over his craft—something that Gagarin had not been permitted to do. Suddenly, *Freedom 7* was no longer a capsule—it was a spacecraft, and Alan Shepard was no longer a passenger—he was a pilot.

Though he knew he was weightless, it was difficult for Shepard to perceive this as he was strapped into his couch so tightly, but he had visual confirmation in a loose washer and some particles of dust floating before his eyes. Now, as America's first person in space, Shepard allowed himself a few moments to take in the view of the planet below. Then he discovered that he had inadvertently left the gray filter on the periscope window. He reached out to rectify his oversight, but the switch was perilously close to the abort handle, so he gave up. Nevertheless, he reported his use of the periscope by saying, "Oh, what a beautiful view," and after the flight he would relate to the press how spectacular were sights and colors he had witnessed below him. But it was all an exaggeration. When the authors asked Wally Schirra about astronauts' reports regarding what they could see from space, he admitted it was never a big thing for him on any of his own missions but that all astronauts felt they had to say something nice about the view from space for public consumption. "It was just the game that people play," he commented, then adding, "I'll never forget Alan Shepard, on the very first manned American flight, saying something to the effect of 'What a beautiful view.' I asked him later, did you see anything at all? He said 'I couldn't see a damn thing through that periscope—but I had to say something nice!'"

The spacecraft was about to come back under automatic control, allowing Shepard to continue his instrument checks, monitor the spacecraft's fuel and electrical systems, and report back to Mission Control on his physical condition. Over the next five minutes, he continued to experience weightlessness, which he later described as "pleasant and relaxing." He had now reached the maximum altitude of 115 miles four-and-a-half minutes into his flight. There wasn't any time to waste on such a short mission. Shepard continued to report his instrument readings and his own reactions to spaceflight. "All okay, all okay, all okay," he repeated at intervals.

All too soon, Deke Slayton began reading out the countdown sequence for retrofire. Five minutes and fourteen seconds after he had lifted off, Shepard fired the first of the three retrorockets into life, followed at five-second intervals by the other two. These braking rockets were actually superfluous for this flight, but they needed to be tested for the upcoming orbital missions in which deceleration would be necessary to bring the spacecraft back down into the atmosphere. *Freedom 7* arced earthward and began its plunge back into the atmosphere. "Everything is going smoothly," Shepard reported.

Eight minutes into the mission, and once again controlling the capsule by manual or fly-by-wire means, Shepard was experiencing a crushing reentry force of more than 10 g's, but he tightened his muscles and later described the strain as "tolerable." He then added: "All the way down, as the altimeter spun through mile after mile of descent, I kept grunting out 'okay . . . okay . . . okay,' just to show them back in the Control center how I was doing." Soon he could feel the g-forces slowly dropping as the spacecraft slowed, and he switched *Freedom 7* back to automatic control. The beryllium heat sink shield behind him had endured temperatures of around 1,230 degrees Fahrenheit, but it had worked perfectly, protecting the capsule and its human occupant.

At 21,000 feet a small stabilizing parachute was deployed from the top of the capsule, and it soon dragged out the main ringsail parachute. Shepard later recalled his thoughts as he waited for the main parachute to open. "If it failed to show up on schedule I could switch to a reserve chute of the same size by pulling a ring near the instrument panel. I must admit that my finger was poised right on that ring as we passed through the 10,000-foot mark. But I did not have to pull it." At the 15,000-foot level the "snorkel," an ambient air valve, popped open right on the money, and cool, fresh air began flowing into the capsule's interior. Then, just as planned, the 63-foot, candy-striped main parachute deployed and blossomed out fully at 7,000 feet with a reassuring jolt. *Freedom 7*'s speed rapidly dropped to around twenty miles an hour. Next, Shepard dumped the remaining control system peroxide fuel, noted from the green light on his instrument panel that the heat shield landing bag had been correctly deployed, and waited for the moment of splashdown. Watching the spacecraft's oscillating descent, but keeping a good distance, helicopter pilots prepared to pluck the astronaut and his Mercury capsule from the ocean.

At 9:49 a.m. the spacecraft's heat shield hit the water, the impact cushioned by the extended porous landing bag, and the parachute's shroud lines were automatically cut loose. *Freedom 7* listed over for a few moments before very slowly assuming an upright position in the water. Fluorescent marker dye was automatically vented to mark the capsule's position. The first American spaceflight was virtually at an end. *Freedom 7* had landed 302 miles down range from the Cape, and in his fifteen-minute ballistic flight Alan Shepard had traveled at 5,100 miles an hour, soared to an altitude of 115 miles, and enjoyed five minutes of weightlessness.

Just seven hundred feet from the splashdown point, Lt. Wayne Koons was holding his olive-drab U.S. Marine helicopter stationary over the water. Once he and his copilot, Lt. George Cox, had established that the spacecraft had righted itself to the correct attitude in the water, Koons, following procedure, calmly asked Shepard if he'd like to be picked up. The reply was similarly calm and unhurried: "Roger. Come after me."

Shepard removed his lap belt and loosened his helmet so he could quickly remove it when he went out through the hatch. He then began making a final reading of all his instruments as the helicopter crew tried to engage the shepherd's hook under a sturdy loop at the top of *Freedom 7*. Shepard heard the hook engage, and then Koons said, "Okay, you've got two minutes to come out." Shepard could see water outside of the small window in his capsule, so he asked Koons if he could raise the capsule a little as a precaution. Once this was done, Shepard radioed that he would exit the spacecraft in thirty seconds.

Removing his helmet and setting it aside, Shepard disconnected the communications wiring linking him to the radio set and took a long, last look around the interior of *Freedom 7*. He then manually removed the hatch, which fell into the ocean, and squeezed out, assuming a sitting position on the open sill. Cox, at the open door of the helicopter, lowered a horse-collar sling, which Shepard grasped and slipped over his head.

Five minutes after his capsule had splashed down in the Atlantic, Alan Shepard was sitting in a comfortable bucket seat aboard helicopter number 44. "Boy, what a ride!" he remarked with a wide grin to Cox. There was no other conversation. All members of the recovery crew had been given strict instructions not to converse with the astronaut unless he spoke to them first, in case idle chatter somehow distorted his recollections of the flight.

The pilot's job had not ended with Shepard's safe arrival aboard the helicopter. Koons made sure his recovery line was secure, then slowly lifted *Freedom 7* from the water and with infinite care turned in the direction of the aircraft carrier *Lake Champlain*, flagship of the fleet of seven recovery ships, four miles distant.

"On the way to the carrier I felt relieved and happy," Shepard later wrote in *We Seven*. "I knew I had done a good job. The Mercury flight systems had worked out even better than we had thought they would. And we had put on a good demonstration of our capability right out in the open where the whole world could watch us taking our chances." It would take just seven minutes to transport the astronaut and his spacecraft to the deck of the waiting carrier, and as the helicopter lowered itself to the flight deck Shepard waved in delight at the crowd of cheering, hat-waving sailors below.

Koons slowly nestled the dripping spacecraft squarely onto a waiting pile of mattresses. Once it had been unhooked he drifted to the side a little and landed. Seconds later, as the engines shut down, the door of the helicopter was thrown open, and a smiling Alan Shepard leaped out onto the deck to a riotous roar and applause from the carrier's 2,600-strong crew. He waved at them, then reached out and shook hands with George Cox. "Thank you very much," he said as Drs. Robert Laning and Jerome Strong made their way over to greet him and escort him to the recovery task force commander's cabin for a quick medical once-over. Then Shepard remembered something. "Just a moment," he told the waiting doctors and jauntily walked over to the open hatch of his spacecraft. He reached in and extracted his helmet, which he shoved under his arm, and enjoyed one more look at *Freedom 7* before rejoining the doctors. Still smiling, he said to Laning, "I don't think you're going to have much to do."

He would later say that he was "deeply moved" by the warm reception he'd received from the ship's crew. "I'd made hundreds of carrier landings before, but to see all the sailors dressed in white up on the flight deck as a welcoming committee, that was a little emotional." Meanwhile, photographs of an exultant Shepard emerging from his helicopter onto the deck of the *Lake Champlain* were flashed around the world. In Virginia Beach, a beaming, relieved Louise Shepard ventured out onto the porch of their ranch-style home, and a waiting throng of reporters and cameramen moved in. "I don't have to tell you how I feel," she said. "It's just wonderful. It's beautiful . . . just wonderful." Shepard's hometown of Derry, New Hampshire, erupted into a spontaneous cacophony of noise—church bells rang, sirens wailed, and cheering people swarmed out onto the streets in relief.

Though they expressed similar elation, Western analysts soberly agreed that Shepard's epic flight fell short of Yuri Gagarin's impressive orbital mission in the larger *Vostok*. However, Shepard could comfort himself that he had actually taken control of his spacecraft and conducted maneuvers, however briefly, whereas Gagarin had been little more than a passenger aboard his ground-controlled spacecraft. Second, though Gagarin's mission had been veiled in secrecy, Shepard's entire flight from takeoff to recovery had been broadcast live, watched by millions of people around the world.

An hour after Shepard boarded the *Lake Champlain*, relieved doctors reported after a quick preliminary examination that Shepard was "disgustingly normal" and in excellent spirits. He had been perspiring noticeably, had lost a little weight, and his pulse was a little high, but there was nothing abnormal about this given his natural excitement. A short dose of solitude and relaxation were all he would need. A more detailed, two-day battery of tests and a debriefing by medical and technical personnel would take place when Shepard reached the special one-man hospital set up on Grand Bahama Island. After the examination and a short debriefing questionnaire, Shepard dictated his first impressions of the flight into a tape recorder.

The *Freedom 7* spacecraft was in excellent condition, and the engineers who went over it declared that it could easily be used again. One *Life* reporter, on hearing that the doctors had declared Shepard to be in top shape, both physically and psychologically, wryly observed that "the commander could also be used again."

President Kennedy was keen to congratulate the first American to fly into space, and soon the two men were hooked up by radio telephone. Shepard expressed his pleasure at the way the flight had gone, saying it worked "just about perfectly, and it was a very rewarding experience for me and the people who made it possible."Navy aviator Ted Wilbur had meanwhile watched with pride as Shepard arrived on the *Champlain* after his historic spaceflight, and two hours later was ready and eager to fly Shepard to his island debriefing on a Carrier Onboard Delivery (COD) Grumman aircraft: "America's first man in space hopped aboard my C-1 Trader, and we were off to Grand Bahama Island where he would undergo an extensive debriefing and physical examination. No sooner had I cleared the bow than he was out of his seat in the cabin and up into the cockpit, with that big wide grin spread across his face. Shouting above the noise of the COD's engines, he described his morning's monumental adventure, and it was easy to see he had been on top of the world, literally."

Gus Grissom had been in the control center at the Cape when Shepard's flight began, but he had been flown over to Grand Bahama Island where, as everyone waited for the man of the moment, his comments were eagerly sought by reporters. They wanted to know how Grissom felt, and if he would be the next to fly. "I'm very happy," Grissom stated. "You can underline that. I wanted to be the one chosen for this shot, and I certainly want to be chosen the next time. Everything went perfectly, just like we practiced it a thousand times."

A bevy of technical people and medical personnel were waiting for Shepard on Grand Bahama Island, including Bill Douglas and Dee O'Hara, the latter recalling later that her joy in hearing of Shepard's safe splashdown was "overwhelming . . . I was so relieved to see him." Deke Slayton had also flown in from the Cape and was on hand with Gus Grissom to greet Shepard after he disembarked from the C-11. As they walked over to a waiting air force station wagon belonging to Capt. Hugh May, the commander of the island's missile-tracking station, the three men were in high spirits. Photographs show them laughing and joking with each other, relieved of the stress of recent hours. When they finally pulled up at the medical facility Shepard was quickly whisked inside, clearly ignoring the shouted questions of reporters.

As well as undergoing a full medical examination by Bill Douglas and his team and a psychological evaluation by Dr. George Ruff, Shepard was subjected to an intense grilling on medical matters, his in-flight activities and performance, and the spacecraft system's performance. In all, some thirty-two specialists took part in the debriefing, including physicians, program managers, operations engineers, public relations personnel, and official photographers.

Another witness who still retains vivid memories of that day is Bob Northcutt Jr. His father, Col. Robert E. Northcutt (USAF), was the base commander at Patrick Air Force Base. Not only was he responsible for all base activities, but he was also third in the air force command structure for Cape Canaveral.

Bob Northcutt Jr. had dropped out of college but later decided to reenroll ("My dad said the first go-around at college was on him, but the second was on me. Smart man."). Saving up to pay his college fees, he worked as a laborer in the supply department at Patrick, but in November 1960 he transferred to the department's downrange tracking station as a supply assistant ("a grunt") on Grand Bahama Island, where the pay was a whole lot better, "plus room and board."

On Grand Bahama, Northcutt would often lend Dee O'Hara a willing hand after work as the medical team set up their small facility. "Mostly I moved things and unpacked boxes or whatever manual labor was needed," he recalled. "At twenty-one I was strong enough to move mountains, so helping set up the Mercury Hospital was easy, and being part of history made it exciting." It would become even more exciting on the evening of 5 May 1961:

That night base personnel were invited to the "Grand Bahama Yacht Club," our on-base watering hole. It wasn't much of a club, and we surely did not own any yachts, but it did have a dartboard and card games to complement the liquid refreshments.

As I remember, there were press personnel at the front gate of the base, but they were not allowed in to take photos or have interviews. In the early evening a buddy and I decided to go to the club and see what was happening. I clearly remember walking toward the front entrance and being followed by some other men to whom I paid little attention, because there was noisy excitement in the club. We walked through the front door, and the whole club started to applaud us. Naturally, my buddy and I were pleased to be greeted with applause until we looked behind us and recognized the men who followed us: Mercury astronauts. I can't remember which ones they were, but it made you feel proud but a little embarrassed. I turned right and continued into the bar and ordered a beer.

I was standing at the bar when Dee O'Hara, sitting at a booth, called me over to sit down. She introduced me to an air force colonel sitting next to her, who also knew my father at Patrick [O'Hara believes this was Dr. Bill Douglas]. We talked for maybe 10–20 minutes when some guy slid into the booth next to me. I didn't look at him until Dee said, "Well, Alan, how are you feeling?" I turned and there he was—Alan Shepard. It took me a minute to recover, and Dee said, "Oh Alan, this is Bob Northcutt; his father is commanding officer at Patrick." Alan shook my hand and also said he knew my father, but I was still in shock. When I recovered I did what any normal twenty-one-year-old would do—I asked him for his autograph. He said "Sure," but we didn't have any paper, so I pulled out my Grand Bahama Yacht Club card and he signed and dated it for me.

The rest of the evening is a blur, but that encounter has stuck with me—a memory I will never forget.

During that same day, Gus Grissom also had a chance to relax with Shepard over a meal and ask him about the flight. With the safe conclusion of MR-3, Grissom was next in the flight line, and his spacecraft was already undergoing final checks and tests for the July mission. Grissom's capsule was to be very similar to Shepard's *Freedom 7*, with two major exceptions. First, *Freedom 7* had only two small portholes, as well as the pesky periscope. Grissom's capsule, which he named *Liberty Bell 7*, would instead have an enlarged, trapezoidal window to enable better observations. Second, instead of the awkward latch-operated hatch in Shepard's craft, Grissom's would feature an explosive side hatch that could allow the astronaut a rapid egress if the spacecraft started sinking.

Finally, the day and the excitement began to wind down, but like Bob Northcutt, Dee O'Hara retained one lingering memory:

That evening, I remember, we were relaxing in an island bar with a little TV sitting on a plank in a corner. Alan, Bill Douglas, and I were able to sit back and finally watch all the news of his flight. Alan was in such a good mood—it had been quite a day!

It seemed so unreal, sitting there watching TV with Al, and the reporters were saying how the astronaut was now on Grand Bahama Island, and probably enjoying a glass of iced tea. And there we were, knocking down a quiet glass of Scotch and water!

I used to smoke back then, and when I lit one up Al leaned over and said, "Ah, could I have a puff of your cigarette, Dee?" He also smoked back in those days, and he hadn't had a cigarette since before his flight, so I said "Sure!" He took one puff, had a swig of my Scotch and water, and I remember thinking to myself, "Oh, public—if you only knew!"

If Alan Shepard had enjoyed fame before his flight, it was nothing compared to the public adulation that engulfed the astronaut after his flight. He was honored with ticker-tape parades in Washington, New York, and Los Angeles, and received NASA's Distinguished Service Medal from President Kennedy in a ceremony at the White House. During the presentation, the medal accidentally fell from its lined box as James Webb was passing it to the president. Jackie Kennedy, standing nearby, was mortified. "Pick it up, Jack!" she urged. Shepard always recalled with a laugh that both he and the president instinctively bent over to retrieve the medal and nearly bumped heads. It would have made for an interesting news item. As it was, Kennedy saw the humor in the situation, and when he finally handed Shepard the medal, he said with a chuckling voice, "This decoration has gone from the ground up!" It was a quip worthy of Wally Schirra. "We had a big laugh out of that," Shepard recalled.

Buoyed by the incredible public reaction to Shepard's flight, President Kennedy would stand before a joint session of Congress just three weeks later, on 25 May, and publicly declare the national goal of landing a man on the moon and returning him to Earth before the end of the decade. Shepard's flight had been brief, but its success was enough to give America the confidence to aim for a new, far more complex goal.

Alan Shepard recalled that when he was selected as an astronaut in 1959, his mother had been "delighted," which was quite true. In 1991, in an interview for the Academy of Achievement, he also recalled the moment he broke the exciting news to his father. The immediate reaction was one of stunned disbelief. "When your old man says, 'You're gonna do *what*, son?' there is a little pause of reflection," he admitted. Deeply concerned about this unheralded deviation from a highly successful career path in the navy, Shepard's father urged him to reconsider. He pointed out to his son that Alan was almost certainly sacrificing future promotional opportunities in order to become an anonymous human test subject on risky rocket flights for a newly established civilian organization.

"Fortunately, in my case, he lived long enough to see me go to the moon and back," Shepard recalled. "One evening, we'd had dinner, the ladies had retired, and we were having a drink in front of the fire, and he said, 'You remember when I said "What are you going to do, son?"'

I said 'Yes sir, I certainly do.'

And he said, 'Well, I was wrong.'"

In an interview for the Hall of Science and Exploration, Alan Shepard was asked what he thought was his proudest accomplishment. He nominated being chosen to make the first manned American flight into space. "That was competition at its best," he stated, with his usual unapologetic candor. "Not because of the fame or the recognition that went with it, but because of the fact that America's best test pilots went through this selection process down to seven guys, and of those seven, I was the first one to go. That will always be the most satisfying thing for me. During the actual process of flying aircraft, or flying the *Spirit of St. Louis*, one doesn't think of oneself as being a hero or historical figure. One does it because the challenge is there, and one feels reasonably qualified to accomplish it." After a pause, he added, "I must admit, maybe I am a piece of history after all."

3. The Pursuit of Liberty

Ignorance is preferable to error, and
he is less remote from the truth who believes nothing
than he who believes what is wrong.

Thomas Jefferson

Hot on the heels of Alan Shepard's triumphant flight, a second manned mission was launched from Cape Canaveral. This one, however, came perilously close to ending in tragedy. At mission's end NASA would lose a precious spacecraft and come within a hair's breadth of losing an astronaut. Worst of all, that astronaut would almost drown within sight of a circling fleet of recovery helicopters. A death so early in America's man-in-space program would have severely curtailed the nation's space plans.

Eight weeks before the flight of *Liberty Bell 7*, President John Kennedy had made a dynamic pledge to send an American to the moon before the end of the decade. Gus Grissom, who dreamed of being the first person to set foot on the moon, almost became the first astronaut casualty in America's mammoth undertaking.

Although those who came into contact with Virgil (Gus) Grissom freely admit he was not an easy person to know or confide in, this was not because the Mercury astronaut deliberately shunned people. Quite simply, Grissom was a perfectionist—particularly when his life and those of others depended upon his professionalism, perseverance, and cool-headed experience. His younger brother Lowell reflected on their shared upbringing and influences:

To know about Gus, it is important to know about our parents, Dennis and Cecile. Dad worked for the Baltimore and Ohio Railroad for forty-seven years, as a signal maintainer. He was one of few who had a job during the Depression. They were very giving and generous people; although they had modest means they were always very willing to share what they had with others in need. It seemed that when I was growing up there was always a relative that was living with us. We were blessed with parents who exhibited emotional stability and a sense of security.

Surprisingly Gus was not an outstanding student in high school; in fact he probably would have been classified as an underachiever. The high school principal did not endorse his application to enter Purdue University. I don't want to give you the wrong impression . . . he did excel in math and sciences; I guess he just didn't see the importance of those other classes.

When Gus entered high school he was five-foot-four and weighed about a hundred pounds—not quite what the high school coaches were interested in for the athletic teams, but he was well-coordinated and one of the most competitive people that I have ever known, and he tried even harder. As a mild extrovert, he could still surprise you with his wit and humor, and it appeared when you least expected it. He was also a man of few words.

Jim Lewis of Marine Air Group 26, 2nd Marine Wing, first met Grissom at Langley Air Force Base while visiting the Space Task Group in early 1961. He was there for a meeting on the recovery of Mercury spacecraft from the ocean. Lewis's involvement in Grissom's flight was expected to be a small one. Spacecraft recovery would never be routine, yet for an experienced crew it was seen as relatively easy. After the dangers of a ballistic rocket launch and reentry into Earth's atmosphere, plucking the capsule from the ocean was not considered by planners to be a matter of great concern.

However, the marine pilot would eventually play a far more important role in the events discussed that day than anyone could have anticipated. When meeting Grissom at Langley, Lewis could hardly have guessed that they would both be indelibly linked in the history of spaceflight because of an event that would threaten both their lives and challenge their professional skills to the fullest.

From the outset, Lewis was favorably impressed by the short, stocky Mercury astronaut, although he later recalled little personal contact: "I think my impressions were like most. Gus was a serious guy, and the more one had the opportunity to work with him, the more one appreciated how good he was. He worked technical problems well, penetrating to the core, and making sure he and all of us took care of any peripheral concerns. In other words, I really appreciated how comprehensive his work ethic was. I imagine that's one of the things that helped him survive his combat missions in Korea." The son of an army warrant officer, Jim Lewis was born in Shreveport, Louisiana, on 10 November 1936, a date that also marked the 161st anniversary of the founding of the United States Marine Corps. Gus Grissom had been born ten years earlier in the small midwestern town of Mitchell, Indiana. But while Grissom elected to train and fly in his nation's air force and later flew with distinction in the Korean War, Lewis joined the Marine Corps.

At the end of World War II, Lewis's family moved to Oklahoma City, settling in 1947 in Houston. Joining the Marine Corps' Platoon Leaders Program during his junior year at the University of Houston, Lewis was commissioned a second lieutenant on graduation. He applied for flight school at Quantico and was accepted, which meant he could begin taking flying lessons as soon as he received his bachelor of science degree (in psychology) in 1957. He elected to stay with the Marine Corps because they had "a great reputation, and I wanted to be one."

Following his graduation from flight school Lewis spent several months serving as a Sikorsky helicopter pilot with Marine Air Group 26 in Japan and Okinawa, then on carriers flying supplies and equipment to Vietnam in 1959 and the first few months of 1960. On Lewis's return home, a good friend from flight school offered him some advice that would change his life. Wayne Koons told him he should request an assignment to the 2nd Marine Air Wing in New River, North Carolina, where Koons was involved in NASA's Mercury capsule recovery program.

It sounded like a heaven-sent opportunity, so Lewis applied and was accepted. Eventually, he became the prime recovery pilot for America's second manned mission, MR-4. His friend Koons had meanwhile become the recovery pilot for MR-3, and successfully retrieved Commander Alan Shepard and his *Freedom 7* capsule. Lewis later observed that the Marine Corps was involved in the Project Mercury recovery program for several reasons: "One was that our helicopters had the payload capacity to lift the capsule. Similar navy models contained a lot of sonar search equipment that reduced their payload considerably. In addition, one of the Marine Corps' missions with the aircraft was to deposit heavy external loads in small, tight, jungle-type areas surrounded by trees—a task which required a fairly high degree of precision. While most pilots could accomplish this after training, marine pilots had been practicing it as part of their normal duties for quite a while."

On 21 July 1961, Gus Grissom became America's second man in space, riding a 4,022-pound Mercury capsule he'd named *Liberty Bell 7* to 118 miles' altitude on a sixteen-minute flight. Liftoff on flight MR-4 occurred at 7:20 a.m. EDT after a now-familiar routine of technical hitches and launch scrubs had delayed the flight by several days. Grissom's 83-foot Redstone rocket and Mercury capsule were not significantly different from Alan Shepard's, although, as noted, he had the added luxury of a nineteen-inch window, replacing the meager ten-inch porthole in Shepard's capsule. *Liberty Bell 7* was also equipped with an explosively actuated hatch, which had not been ready in time for Shepard's flight. Shepard's hatch had been latch-operated—a potentially time-consuming process when every second counted. NASA's engineers and astronauts had agreed that all future spacecraft should have a more rapid evacuation capability in the event of an emergency on or below the water. It should also provide a means for rapid escape if the main chute had problems during descent.

"A percussion-activated, explosive primer cord surrounded the new hatch, and the astronaut had to remove a cap and pin in order to easily depress and trigger the mechanism. When fired, the explosive would cause the bolts to fail and the hatch to rapidly shoot away from the spacecraft. The recovery crew could also blow the hatch from the outside, by removing a small screw and pulling on a wire lanyard. Designed to save an astronaut in danger, it seems that this new hatch design instead almost cost Grissom his life."

Grissom was given fewer duties to perform than Shepard, who had complained after the flight that he had been overloaded with tasks during his all-too-brief mission. Consequently, while still busy, Grissom would have time to peer through his window to test an astronaut's visual abilities in a weightless condition. After the Redstone booster separated minutes into *Liberty Bell 7*'s flight, Grissom looked through the window and was immediately captivated by what he could see. Moments before on ascent he had been impressed by the fast transition from blue to black skies. "It's such a fascinating view out the window you can't help but look that way," he told Mercury Control, where Alan Shepard was duty Capsule Communicator

(CapCom). Shepard smiled as he recalled his own brief impressions of Earth from space, just forty-seven days earlier. "I understand," he responded.

Grissom excitedly reported that he could see a star, which he believed to be Capella, thus clinching a bet he had made with John Glenn. When this was later determined to have been the planet Venus, Grissom lost his wager. Taking manual control of his capsule, Grissom experienced what he later described as "sluggish" responses when conducting yaw, pitch, and roll systems tests. As the capsule swept southeastward, clouds largely obliterated the astronaut's view of the Earth. "I can see the coast, but I can't identify anything," he reported. He would later say he was captivated by the sight of the horizon coming into view and the surrounding bands of color that changed outwardly from light to dark blue before merging into the total blackness of space.

On such a short flight, however, there was little time to look. His mission almost over, Grissom prepared his spacecraft for reentry and manually fired the bank of three retrorockets, right on schedule. "It was a strange sensation when the retros fired," he would later write. "Just before they went, I had the distinct feeling that I was moving backwards—which I was. But when they went off, and slowed me down, I definitely felt that I was going the other way. It was an illusion, of course. I had only changed speed, not direction."

Despite problems with manual attitude control, Grissom kept *Liberty Bell 7* steady during the twenty-two-second period of retrofire. The reentry, which he had thought might present some challenges, began smoothly, but as the spacecraft hurtled into the thickening atmosphere he observed what he later described as "some interesting results." At one point he noticed what appeared to be a contrail skipping off the heat shield as *Liberty Bell 7* buffeted its way through the atmosphere, convincing him he was witnessing shock waves. Although he was alert for any problems, there were few surprises, as he later recalled for the book *We Seven*: "We were really bouncing along at this point. I was pulling a few Gs—they built up to 11.2. But they were no sweat. I had taken as many as 16 on the centrifuge, and this seemed easy by comparison. I could also hear a curious roar inside the capsule during this period. This was probably the noise of the blunt nose pushing its way through the atmosphere."

Grissom was elated, and justifiably. He did not know it then, but his flight had marginally outperformed Shepard's. Because the Redstone booster had burned eight-tenths of a second longer, he had been propelled thirty miles an hour faster than Shepard. As a result, he penetrated one and a half miles deeper into space. He also outdid his colleague by experiencing weightlessness for five minutes and eighteen seconds—thirty-seven seconds longer than Shepard. Ever a keen competitor, Grissom doubtless enjoyed those statistics.

Hurtling back to Earth at nearly four hundred miles an hour, the spacecraft's six-foot nylon drogue parachute deployed at around twenty-thousand feet. Activated by barometric pressure, it stabilized the capsule's fall before the main sixty-three-foot ring-sail parachute deployed at ten thousand feet. This occurred as planned while the capsule was still traveling at two hundred miles an hour. "There was a slight bouncing around when the big one bit into the air," Grissom wrote of the deployment, "but this was no problem."

Floating down to the Atlantic, 303 miles downrange from Cape Canaveral, Grissom began preparations for his recovery after splashdown. He reported opening his faceplate and then had problems inserting one of the door pins that held the hatch to the side of the capsule, a procedure designed to prevent the hatch from traveling too far if fired once the spacecraft was brought on to a ship's deck. This safety measure had nothing to do with actuating the explosive feature of the hatch that was soon scheduled to come into play.

Apart from this minor annoyance, everything was going according to plan. While *Liberty Bell 7* continued its descent Grissom heard from the pilot of the radio relay airplane, designated *Card File 23*. "We are heading directly toward you," the pilot announced, as the spacecraft passed three thousand feet. Then the first of the rescue helicopters, flying under the code name of *Hunt Club-1*, was in contact. Pilot Jim Lewis told Grissom they were about two miles southwest of the projected splashdown site.

When the landing bag deployed below *Liberty Bell 7*, Grissom felt a slight jar as the heat shield dropped down about four feet. This revealed a circular fabric collar filled with air holes, designed to absorb some of the approaching shock of impact. After splashdown, by filling with water it would also help stabilize and right the bobbing craft.

At 7:36 a.m., just fifteen minutes and thirty-seven seconds after liftoff, *Liberty Bell 7*'s landing bag hit the water, and America's second manned spaceflight was at an end. Or so it seemed.

When asked how it had been communicated to him on the aircraft carrier USS *Randolph* that the MR-4 launch had taken place, Jim Lewis vividly recalled the facts:

My logbook shows that we flew two missions that day, the first being a checkout flight of just over a half hour. As I recall, our plan was to lift off the carrier at the same time the booster lifted off from Cape Canaveral. We were waiting in the cockpit with engines running and received word that the launch occurred via ship-to-aircraft radio. All that remained was to engage rotors and take off once clearance was granted from flight control.

Once MR-4 had lifted off, we had about fifteen minutes to get there and begin recovery operations, and I believe the carrier was standing off the impact area about five miles. We flew at about ninety knots, so getting into the primary recovery areas quickly was no problem. I was initially occupied with observing the sky above, searching for the Liberty Bell 7 *parachute. Beyond that, I was intent on executing the mission procedures and plan. I finally saw the spacecraft on its chutes. I couldn't say what altitude, but it wasn't very high.*

Gus Grissom was surprised that the only sensation he felt on splashdown was a slight jolt. The spacecraft heeled over as expected, and in moments his window was completely underwater. He did, however, hear a disconcerting gurgling noise as *Liberty Bell 7* slowly rolled upright and the upper part of his capsule drew clear of the water. A quick check reassured him that water was not entering the spacecraft. His next action was to eject the reserve parachute by clicking a recovery aids switch, allowing the craft to fully right itself. Grissom now completed his final checks, as he later reported in his postflight debriefing:

I felt that I was in good condition at this point and started to prepare myself for egress. I had previously opened the faceplate and had disconnected the visor seal while descending on the main parachute. The next moves in order were to disconnect the oxygen outlet hose at the helmet, unfasten the helmet from the suit, release the chest strap, release the lap belt and shoulder harness, release the knee straps, disconnect the biomedical sensors, and roll up the neck dam. The neck dam is a rubber diaphragm that is fastened on the exterior of the suit, below the helmet-attaching ring. After the helmet is disconnected, the neck dam is rolled around the ring and up around the neck, similar to a turtleneck sweater. This left me connected to the spacecraft at two points—the oxygen inlet hose which I needed for cooling, and the helmet communications lead.

At this time I turned my head to the door. First, I released the restraining wires at both ends and tossed them toward my feet. Then I removed the knife from the door and placed it in the survival pack. The next task was to remove the cover and safety pin from the hatch detonator. I felt at this time that everything had gone nearly perfect and that I would go ahead and mark the switch position chart as had been requested. After about three or four minutes I instructed the helicopter to come on in and hook onto the spacecraft, and confirmed the egress position with him.

I took the pins off both the top and bottom of the hatch to make sure the wires wouldn't be in the way. . . . I took the detonator cap off and put it down toward my feet.

On *Hunt Club-1*, Jim Lewis had responded to Grissom's transmission, letting him know that a rescue sling would be lowered and waiting for him outside the capsule once the hatch had been blown. Each of the rescue helicopters carried a two-man crew; on *Hunt Club-1* Lewis's copilot was Lt. John Reinhard. They worked well together as a team. Despite the noise within his helicopter, Lewis was later emphatic that communications with the spacecraft were "normal and excellent, as the system was designed for that acoustic environment." He clearly heard the astronaut ask for a couple of minutes to power down and record instrument and switch positions. "All our voices were calm," he remembers. "We had rehearsed these procedures and activities in the bay off Langley AFB with the astronauts, and there was no reason not to be calm. That's what good training does. In addition, Gus had a low resonant voice, which was pleasant to hear. It was all very calm and professional."

Recovery procedures assigned Lewis and his copilot the task of cutting off the spacecraft's 4.2 meter high-frequency (HF) whip antenna. This was accomplished using shears that were attached to the end of a long pole, to minimize the risk of the antenna interfering with the helicopter's main rotor. They would then hook a recovery cable onto a strong Dacron loop at the top of *Liberty Bell 7* and lift the spacecraft a little so the hatch was well clear of the water, allowing Grissom to emerge. They would also lower the rescue sling to a position just above the spacecraft and relay this information to Grissom, who would cut spacecraft power, blow the hatch, remove his helmet, look for the rescue collar, and then carefully exit through the hatch and insert himself into the rescue sling. Apart from the explosive hatch, this was the same procedure that had worked so well after Shepard had splashed down.

As he waited for instructions, Grissom unhooked the oxygen inlet hose and lay back on his couch. He later stated that he had momentarily turned his attention to the knife in his survival pack, idly wondering if he should take it with him as a souvenir. Suddenly, he heard a dull thud as, without warning, the spacecraft hatch blew. He looked up in disbelief, not only seeing blue sky, but the unnerving sight of saltwater spilling in over the doorsill. Grissom reacted instinctively, lifting the helmet from his head and discarding it, then reaching for the right side of his instrument panel and hauling himself out on his back through the open hatch. Moments later he was swimming in a tumult of noise and a roiling Atlantic swell, realizing that in his haste he had neglected to bring along his survival pack.

Still in shock, Grissom swam backward away from his spacecraft, watching with mounting horror as seawater poured in through the open hatch. He became snarled in a dye-marker line attached to the spacecraft but quickly disentangled himself in case it sank and dragged him down. Grissom was a good swimmer, in fact the best of the seven Mercury astronauts, so while he was treading water his concern was centered not on himself but on the pilots' desperate efforts to hook onto *Liberty Bell 7*, which was settling ever deeper in the water. "As I got out," he later wrote, "I saw the chopper was having trouble hooking onto the capsule. He was frantically fishing for the recovery loop. The recovery compartment was just out of the water at this time and I swam over to help him get his hook through the loop."

Grissom had anticipated that his air-filled spacesuit would keep him buoyant, but he suddenly realized that in his hurry to evacuate he had forgotten to lock the oxygen inlet port on the midsection of his spacesuit. Not only was air bleeding out, but water was seeping in, making him increasingly heavier. Another unwelcome factor in his struggle was a number of souvenirs he had carried with him in the left leg of his suit: two rolls of fifty dimes, some small models of his spacecraft, and a wad of dollar bills. "They were added weight I could have done without," he later admitted.

It is generally agreed that one innovation helped to save Grissom's life that day—a special watertight neck dam that had been designed and tested by Wally Schirra for the precise purpose of preventing water from getting into a floating astronaut's spacesuit. Schirra had recently persuaded Grissom to wear it.

Meanwhile, Jim Lewis was calling on all of his professional skills to try and save the spacecraft. Before getting on station over the capsule, he and his copilot had seen the hatch blow prematurely, and watched in alarm as it hit the water a couple of feet away, skipping over the surface before it sank. He vividly recalls what happened next:

I was not worried about Gus being in the water. Because we had trained on these procedures at Langley AFB *and the Space Task Group we knew the astronauts floated very well in their suits. They were sealed and had neck dams at the top to prevent any water ingress. At that point we no longer had communication, so there was no way for any of us to know there was an open port in his suit.*

My last call to Gus before the hatch blew was that I was turning base. That meant I was downwind and had to do a 180-degree turn into the wind and complete the approach over a distance of one hundred feet or more to get there. We saw the hatch blow, which means we had completed the turn, but still hadn't closed the distance. My plan at that point was to have my copilot cut the HF *antenna, then try to snag the capsule before it sank. There was probably a minute or less from the time the hatch blew until the capsule disappeared below the surface.*

During this time I could see Gus in the water trying to help in the recovery process. He later said he wanted to help my copilot make the connection between the aircraft and capsule if he could, so he was close by. I had to put the wheels in the water—the aircraft was not designed for this—after my copilot cut the antenna so he could reach the recovery bale on top of the capsule. By the time he had made the connection between the helicopter recovery line and capsule recovery loop, the top of Liberty Bell 7 *had actually disappeared below the surface. I began attempting to lift it out of the water at that point, although I knew that the combined weight of the capsule and water was more than the lifting capacity of the* UH34D *helicopter.*

Helicopter pilot Jim Lewis desperately tries to raise *Liberty Bell 7*. Courtesy NASA.

According to procedure, copilot John Reinhard then began lowering the hoist for Grissom so they could bring him aboard the aircraft. Lewis, meanwhile, was hoping that he might be able to drain sufficient water from the capsule and landing bag to allow it to be lifted so he could fly it and Grissom back to the carrier. However, a considerable volume of water lay below the hatch level and was never going to drain out. Even when they managed to drain some water from the landing bag a wave would catch it, pulling the helicopter down enough for water to flow back into the bag. As Lewis recalls, it was a tense and potentially disastrous situation:

I was using maximum power at this point—2,800 RPM and 56.5 inches of manifold pressure. Shortly after I began this process, I saw the chip detector warning light on the helicopter instrument panel illuminate. This light indicated there were metal filings in the oil system. Our standard operating procedure for this event said that the engine would probably last about five minutes with metal being distributed throughout the engine before it failed. Because of this, I ordered the copilot to cease lowering the hoist for Gus and to bring it back up be-

cause we had a sick bird, and I didn't want to lose the aircraft with Gus aboard. Water egress from a helicopter down in the water with rotors turning overhead is neither a risk free nor an easy task.

I called the backup helicopter, told him I had a chip detector light, and to come in and pick up Gus, and also said that I would drag the capsule clear of Gus so he could come in and make the pickup. Dragging it away was not that easy, but we managed to get it clear in a couple of minutes.

Capt. Phil Upshulte, the pilot of the backup helicopter, managed to maneuver in behind Lewis's craft as the capsule was being dragged through the water, then lowered a rescue sling for the astronaut.

Meanwhile, the entire drama was being captured on film by photographers aboard a third helicopter, whose pilot was under clear instructions to keep well clear of the recovery scene. Their dramatic photographs show the hapless astronaut flailing around in the heaving seas, his body sinking ever lower in the turbulent water. As Grissom swam around he noticed the photographic helicopter in the distance, "and I could see two guys standing in the door with what looked like chest packs strapped around them. A third guy was taking pictures of me through a window." These photographers could very easily have been capturing his last moments, for as he revealed in *We Seven*, he now found himself in a desperate struggle just to stay afloat: "At this point the waves were leaping over my head . . . I was floating lower and lower in the water. I had to swim hard just to keep my head up. I thought to myself, 'Well, you've gone right through the whole flight, and now you're going to sink right here in front of all these people.'"

Then, as Lewis's helicopter was attempting to haul his spacecraft away, Grissom saw to his relief that the backup unit was moving in, dragging a sling across the surface chop of the water. For a short time, however, the rotor wash between the two aircraft and the weight of Grissom's spacesuit prevented him from getting close enough to the sling. Upshulte was finally able to close in and position the collar near Grissom, who, despite his frightening predicament, spied Lt. George Cox in the doorway. The marine had earlier been involved in successful ocean recoveries of both Alan Shepard and the chimpanzee Ham, who that January had flown a similar, precursory Mercury flight. "As soon as I saw Cox, I thought, 'I've got it made,'" Grissom later stated. He was able to reach the lifesaving collar, but was now close to complete physical exhaustion. "I had a hard time getting it on," he stated at his postflight technical debriefing, "but I finally got into it. A few waves were breaking over my head and I was swallowing water. They pulled me up inside and then they told me they had lost the capsule."

Jim Lewis had made a heartbreaking decision:

I was pointed into the wind at this stage, and the backup helicopter was behind me, also pointing into the wind to give added lift, so I could no longer see Gus. After close to five minutes of pulling max power, the cylinder head temperature began to rise and the engine oil pressure began to drop, and I made the decision to release the capsule so that I could set the helicopter down "normally" in the water if the engine died. That was accomplished by pulling a trigger in the cockpit that caused the hook to open and release the recovery cable from the helicopter. I couldn't see the line released, but could feel the result of a reduced payload. I declared an emergency at that point and proceeded back toward the Randolph, *and was able to land aboard the carrier.*

The moment he was aboard the helicopter and had shrugged off the rescue sling, Grissom seized a nearby Mae West life preserver and donned it. This was normal over-water military procedure, but he was also unwilling to face another ordeal in the Atlantic. He spent the entire flight to the *Randolph* buckling up the lifejacket.

For his part, Jim Lewis was less concerned about the loss of the spacecraft than with getting his helicopter and crew back to the carrier. Once they had landed safely, his attention turned once again to the well-being of the astronaut, as he and his crew anxiously waited for the second helicopter to arrive. Shortly after, Grissom was clambering onto the deck of the carrier and asking for a glass of *fresh* water.

A ship's officer would later hand Grissom his helmet, which had been retrieved from the water. To the astronaut's astonishment, the man told him it had been plucked from the sea right next to a circling ten-foot shark. Although he was a good swimmer Grissom had never really liked the water, and this extraordinary news probably sent a small shudder through him. However, he was in a good mood, and a photo taken aboard the *Randolph* shows him looking quite elated after his recovery.

Helicopter pilot Lewis was understandably pleased that the recovery phase of the operation had gone well. "Gus was aboard that aircraft and on his way back to the *Randolph* in just over five minutes after the hatch blew on *Liberty Bell 7*, so our contingency procedures worked perfectly," he explained. "This is exactly why we had a backup helicopter close at hand." Once the excitement had died down, and while the astronaut was undergoing a preliminary physical examination, Lewis made an official report to the ship's captain, Harry E. Cook Jr., on his part in the recovery effort. Then, along with Reinhard and Grissom, he boarded a twin-engine S2F for the flight across to Grand Bahama Island, where the nation's newest hero and the helicopter crew would complete a full postflight debriefing. Following a request from Captain Cook, the navy copilot had happily relinquished his left-hand seat (and with it command of the aircraft) to Grissom for the catapult-launch flight, while Lewis and Reinhard buckled themselves into two seats in the rear of the airplane.

Once they were safely on the island the three men exchanged some final brief pleasantries. It was only then that Lewis discovered that Grissom had actually come close to drowning. Once he had completed his own debriefing Lewis was able to meet some of the other waiting Mercury astronauts before flying back to the *Randolph*. Despite the loss of the spacecraft, there was still cause for him to feel a certain amount of pride and satisfaction:

It's interesting to me how well this all worked. We barely hooked on to the capsule—the loop was disappearing beneath the surface as John hooked onto it, while I had our wheels in the water—and we were able to drag it away just *in time for Phil to come in with the backup bird and pick up Gus.*

It was a matter of seconds either way that would have prevented this scenario from working: that is, if the capsule had sunk before we hooked on, we would have picked up Gus in our helicopter in less than a minute, and we wouldn't have had the engine problem. He would have had very little water inside his suit, and the backup helicopter would have stayed off station. If it had taken me longer to drag the capsule away, Phil might not have been able to get there in time. If we had brought Gus on board our helicopter and not made it back to the carrier, who knows the result?

This was probably another way of saying our contingency plans and procedures worked in just an excellent manner. For all the possibilities, the correct decisions were made quickly and executed well by all concerned.

Jim Lewis's connection to the space program did not end that day. In fact, he became more deeply involved than ever. Following the MR-4 mission, he became a member of the civil service when he joined the Space Task Group at Langley. Fortunately, NASA granted him leave to return to the University of Houston, where he earned his master's degree in 1962. By this time, the Space Task Group had itself moved to Houston, where it became the Manned Spacecraft Center (MSC), a move that Lewis found quite provident. "I guess I was destined to be in Houston no matter what," he wryly reflected. While working at MSC, initially as a flight systems technician, he also undertook postgraduate studies in physics at the University of Houston from 1962 to 1965. During this time, he was assigned to NASA's Engineering Support Team, working on the first two manned Gemini-Titan (GT) missions. Gus Grissom would command the first of these, GT-3.

Though his work at MSC brought Lewis back in contact with Grissom, he never recalled discussing the *Liberty Bell 7* incident with him:

I think we'd moved on, and both of us knew we had followed nominal and contingency procedures properly. The friendship was largely professional, but there were times when other things were the topic of conversation. I had received two letters of commendation for my actions that day, and Gus was subsequently selected to command the first Gemini and Apollo missions. No greater vote of assurance could have been given to him.

We both knew he had done nothing to cause the hatch to detach itself that day, and we both knew we would not be able to find out specifically what had happened, so there was little to discuss. Our conversations revolved around the present and future, and we had plenty to keep us occupied. The work in those days was exhilarating, intense, long, hard—and great fun.

Lewis later served on an engineering support team for Gus Grissom and John Young on the first manned Gemini mission (GT-3). "Our job was to provide engineering support to assigned flight crews while their spacecraft were being readied for flight," he reflected. "So we would cover meetings and tests that they were unable to cover due to the sheer volume of activity, and coordinate the results with them." Lewis and Grissom both had offices in Building 4, so they saw each other a lot.

After the successful completion of the GT-3 mission, Lewis continued working as a support team member for GT-4, and then as support team leader for *Gemini 6* and *Gemini 9*. He would later be assigned to work with the *Apollo 2* (alternately known as *Apollo 258*) crew of Jim McDivitt, Dave Scott, and Rusty Schweickart. On 27 January 1967, Lewis was working as support team leader for this flight in a test of command module CM-101 when he heard some appalling news: Gus Grissom and his *Apollo 1* crew had just perished in a launch pad fire back at the Cape. "We were conducting a closed-hatch test on the *Apollo 9* command module at the North American facility in Downey, California," he recalled. "The test was stopped, the hatch opened, and mission commander Jim McDivitt was called away to the phone. He returned and told us what had happened. My reaction, and I think that of all of us, was stunned, misty-eyed disbelief. Gus was one of the best, and it just didn't seem possible at the time that any such thing could have happened."

After the fire, the Apollo flight order was eventually reshuffled. The first manned Apollo flight would be commanded by Mercury astronaut Wally Schirra. Jim McDivitt's *Apollo 2* crew eventually flew *Apollo 9*, and CM-101 was reallocated as the command module for *Apollo 7*. "I was subsequently assigned to the hatch redesign team led by Frank Borman," Lewis noted. "As I recall, that took about six months and most of it was spent at the Downey facility. All of us went about that task with a vengeance to ensure that kind of thing would be precluded in the future."

Jim Lewis was not, then, just "the guy in the helicopter" on the day of Grissom's first flight. He had played a role in both of that astronaut's flights and, sadly, his final connection with Grissom was working to fix the spacecraft design flaw that had killed him. There is a tragic irony in the fact that Lewis saw one spacecraft lost to a hatch that opened too quickly and a second to a hatch that did not open fast enough.

Lewis's role in the space program continued after his work on the Apollo hatch. He returned to graduate school for a second time and earned his doctorate in human factors engineering in 1979, then managed the development and integration of crew-to-spacecraft interface requirements as the head of NASA's Crew Station Design Section. In this capacity he generated the first crew station design specifications. In this period, he also founded three of the Flight Crew Support Division's major facilities and the majority of the division's research programs. Today these capabilities represent a unique world-class system for human-machine integration. Lewis followed these duties by becoming division manager for the proposed space station *Freedom*, managing and directing the design, development, test, evaluation, and verification of its crew-interface hardware. His final role before retiring from government service in December 1999 was serving as chief of NASA's Human Factors Branch. Though in the early part of his NASA tenure Lewis had continued to serve in the Marine Reserves, in 1983, after twenty years of service, he retired with the rank of major.

In the 1983 movie of Tom Wolfe's best-selling book *The Right Stuff*, Gus Grissom was characterized as a panicky man who may have blown the hatch prematurely. Far more accurate portraits of Grissom and his spacecraft are given by people who knew both very well. Gene Kranz joined the NASA Space Task Group in 1960; he was assistant flight director for Project Mercury, and later flight director on *Apollo 11*. In an interview with the authors he dismissed the unflattering speculation about Grissom with a few terse words: "I spent a lot of time with Gus. Everybody alleges that the guy panicked. Gus was *not* the kind of guy who would panic—he was a very controlled person. I also knew that we had an inherently different hatch design, from the standpoint of a release mechanism, to the other one. I knew the limitations in the testing, and if Gus says he didn't do it, then he *didn't* do it. It's that straightforward." This was an opinion shared by NASA mechanical engineer Sam Beddingfield, who was responsible for the pyrotechnics and recovery system on the Mercury capsule and who had known Grissom before he had become an astronaut, while both were test-flying at Wright-Patterson AFB. "I knew if he had done that he would tell me," he recalled. "But Gus did not fire it. In the air force we had tested enough airplanes together, that when anything went wrong we knew we had to tell each other about it. They put me on a team to go interview Gus as to what happened, and he told me he did not fire that hatch. We were fairly comfortable in that. He was a very, very good pilot, but a lot of people don't realize what a significant astronaut he really was." While helping to investigate the cause of the incident, Beddingfield actually worked out two ways in which the hatch on *Liberty Bell 7* could have blown by itself as Grissom described.

When Jim Lewis was asked, as someone who knew Grissom well and had witnessed the events misrepresented in *The Right Stuff* movie, what he thought of the allegations about Grissom he was adamant:

I am sad that the media apparently chose to ignore what, to me, is the obvious. To answer a question I've been asked many times, never did I have the feeling that any kind of failure had occurred on that day—only a feeling that we had been wildly successful. I was truly amazed when the media tried to build other possibilities into the happenings of that day. What I can *personally recollect, looking back, is happiness at being selected to be there as the prime recovery pilot, then frustration at not being able to recover the spacecraft. But then there was pleasure that we had executed the contingency plans properly and well and that they had worked as we had intended. And there was great contentment in the fact that Gus was okay, though I had never doubted that he was.*

In addition, think about this: MR-4 *was the first spacecraft to have such a large window adjacent to the hatch. When the capsule was floating, Gus looked right out that window and could see water above the hatch sill and above the lower edge of the window. Do you think anyone would have purposefully released a hatch under those conditions? I would add that since we had practiced such things, he also knew that I wasn't there yet, and obviously hadn't lifted his spacecraft clear of the water.*

So then, did he accidentally hit the release? NASA *records show that every astronaut who used that plunger to release a hatch got a bruise or skin abrasion from the rebound of the plunger. Gus's postflight physical documented that his body was totally unmarked. This is positive evidence that he did not "accidentally" hit that plunger. Had he done so, he would have been even less able to escape its rebound than any of those who activated it on purpose.*

Those kind of things do not happen to a "screwup." That kind of person would never have survived combat or being a test pilot, and would not have been selected to be the first in line to blaze the way for new space programs such as Gemini and Apollo. NASA *obviously had confidence in him.*

Recovering *Liberty Bell 7* from the ocean floor was something NASA had never attempted or even considered because of the prohibitive cost. In addition, the technology needed to recover a lost spacecraft resting on the ocean floor simply did not exist until the mid-1980s. But on the thirtieth anniversary of the first moonwalk and one day before the thirty-eighth anniversary of Gus Grissom's suborbital flight, his spacecraft was finally reeled in and hauled aboard a recovery ship by an underwater salvage team. Despite nearly four decades on the ocean floor the titanium capsule was in amazingly good shape, and would later be restored to almost pristine condition.

Curt Newport was the driving force behind the recovery of *Liberty Bell 7*. An experienced underwater salvager, he had been developing and operating remote operated vehicles (ROV) for nearly twenty-five years. He had worked on such intricate seaborne projects as the recovery of sections of the space shuttle *Challenger*, TWA flight 800, and RMS *Titanic*. He was the first to do the research and planning needed to define the possible location of the lost Mercury spacecraft and also conceived and designed the recovery methods and equipment needed to raise *Liberty Bell 7* once it had been located.

Advancing technology was yielding better detection methods and recovery tools. Newport began seeking sponsors to fund search and recovery expeditions for *Liberty Bell 7*. But the eventual recovery of the spacecraft would become possible only after the innovation of sonar processing software, which had a resolution of half a meter, and ROVs that could function despite the crushing pressure miles below the ocean surface.

The recovery concept Newport devised was certainly well ahead of its time. When he began putting his plans together, the massive liner *Titanic* had yet to be identified and explored using ROVs. *Liberty Bell 7* was not only smaller than one of *Titanic*'s boilers, it also sat three thousand feet deeper. Eventually, the Discovery Channel became the first to offer financial support. It also provided widespread television coverage of the successful recovery effort.

On 19 April 1999, the 180-foot salvage ship MV *Needham Tide* set off from Florida's Cape Canaveral, heading southeast on a three hundred–mile journey. Newport was prepared to spend two weeks to locate *Liberty Bell 7*, sweeping waters three miles deep with the latest sonar technology. "We have a pretty good idea where to look for it," he told reporters before setting sail. "To say I'm cautiously optimistic is probably the right term."

Six days later the team lowered the ROV in an area ninety miles northeast of Bahama's Great Abaco Island to check a potential target—one of several detected by sonar scanners. Incredibly, they hit the jackpot on their first attempt. As the ROV moved in on its target, with Newport watching via closed-circuit imagery, the shadowy mass began to assume the bell-shaped appearance they were seeking. Within seconds there could be no doubt—they had located the lost spacecraft.

Perched upright on a small sandy knoll with its periscope extended, the capsule was in astonishingly good condition. Some crystallized corrosion from the heat shield and landing bag were evident at the base, but the words *United States* and *Liberty Bell 7* and the famed bell's crack painted on the spacecraft's side were all plainly visible. The video imagery was so clear that Newport could even make out marks on the capsule's side made by the explosives that blew the hatch. The salvage team's triumph soon turned to deep frustration, however, when the cable to the ROV snapped during heavy swells, and one very expensive submersible was lost. The team had to return to port empty handed, but they were now determined to finish what they had begun.

Two months later, on 6 July, Newport's team set out again on the *Ocean Project*, a larger and heavier vessel. After a frustratingly protracted search, the spacecraft was relocated using a second ROV. This time cables were successfully hooked up, and the lengthy process of lifting *Liberty Bell 7* began. In the middle of the night, under the glare of floodlights and to the cheers of the crew, the spacecraft finally broke the surface and was hoisted aboard the *Ocean Project*. Fittingly, Curt Newport became the first person to lay a hand on the missing craft in thirty-eight years.

Liberty Bell 7, filled with a stinking accumulation of muck, was quickly placed into a specially designed container of seawater to prevent the onset of corrosion. The team did not have the time left in the expedition to also locate the missing hatch.

Jim Lewis had been asked if he wanted to be part of the recovery mission, and he readily accepted. Accompanying him on this second voyage was the legendary pad leader of Projects Mercury, Gemini, and Apollo, Guenter Wendt, together with Newport's friend and longtime supporter Max Ary, from the Kansas Cosmosphere and Space Center. For Lewis, it all proved to be a great adventure: "Once the spacecraft was located in 1999, Curt invited me, Guenter Wendt and Max Ary to join in the recovery effort. It was truly an honor to be there and to be able to observe *Liberty Bell 7* emerge from the depths at 2:00 a.m. on July 20, 1999, almost thirty-eight years to the day from when I released it. As Gus's capsule broke the surface of the Atlantic I had this inner presence of surrealness, and a feeling of completeness."

Another who experienced great joy in the recovery of *Liberty Bell 7* was Gus's brother Lowell. He recalled it as a time of mixed emotions:

Standing on the dock in that hot July sun, waiting for the spacecraft to be moved from the ship to the dock, I also wondered what Gus would be thinking and feeling as that tiny craft came swinging over onto the dock. I know I had many emotions that arose, from deep sadness that Gus wasn't there to see it to immense pride in knowing that the only craft he had flown and lost had now come home. Just like they had said that man could not fly in space, they said Liberty Bell 7 *was so deep that it could never be recovered. Gus was always up for a challenge, and I think he would have been very pleased that those who said, "It can't be done" had, again, been proven wrong.*

The next step in the process, according to Jim Lewis, was the difficult task of restoring the spacecraft to its former glory. *Liberty Bell 7* was removed from the recovery ship at Port Canaveral and transferred to Max Ary's Kansas Cosmosphere and Space Center in Hutchinson, Kansas—the only private museum authorized by the Smithsonian Institution to restore American manned spacecraft. "Max Ary and his group . . . undertook the restoration job," Lewis explained. "Max was kind enough to invite me to participate in the first week of a restoration activity that took over six months to complete. The Cosmosphere did a remarkable job in restoring the capsule from the condition in which we found it on the ocean floor, and in preparing it for a three-year tour of the world before it is returned to the Cosmosphere, where it will be on permanent display."

Max Ary had been trying to stir up interest in recovering *Liberty Bell 7* since 1978, but the technology to do so was just not there. As he told researcher Lawrence McGlynn, his celebrations as a member of the recovery team were soon tempered by the sobering realization that he and his restoration team had a long and intricate job ahead of them:

The video we saw of the interior of the spacecraft, when it was still on the ocean floor, looked promising. Everything still seemed to be in place. That turned out to be just a façade; everything was in place but was just hanging on by a molecule or two. As soon as it started to be raised, virtually any aluminum-based components crashed to the bottom of the spacecraft. They had deteriorated quite badly through what we later termed to be electrolytic action. I have to admit, when the spacecraft arrived on deck and we realized the commitment we had made to Discovery to restore the spacecraft, I was a little concerned about how we were going to be able to deliver on our promise.

Despite the extraordinary complexity of their task, Ary's team actually completed the restoration in just over six months:

The spacecraft was disassembled into an estimated twenty-five to thirty thousand pieces. Each piece was methodically cleaned, preserved, and sealed and then completely reassembled. Many of the pieces required extensive detective work . . . trying to figure out where they came from. There were hundreds of components lying in the muck at the bottom of the craft such as tiny nuts, bolts, gears, [and] electrical components that had corroded away and become detached from the original hardware. We had to figure out where everything went.

The primary challenge we had with LB7 was not so much the fact that it had been sitting in salt water for thirty-eight years, but that it had been subjected to over seven thousand pounds per square inch of water pressure. Many sealed areas of the spacecraft had filled with salt water because of the pressure.

There was also an unexpected surprise in store for the restoration team when they began to find souvenirs—coins and dollar bills that the spacecraft's technicians had secreted aboard to retrieve after the flight:

We, of course, were very mindful of the dimes that Gus carried with him in his boots as personal souvenirs. We had not expected the fact the McDonnell technicians had smuggled their own souvenirs on board. We found five or six in the muck at the bottom of the spacecraft on the evening of recovery.

We recovered fifty-two Mercury head dimes from the spacecraft. Based upon the location where we found most of them . . . the roll of dimes was placed in the pocket on the front of the periscope. As the paper dissolved, the dimes slowly dropped out of the pouch into the bottom of the spacecraft. All of the coins were Mercury head dimes, and some had personal inscriptions or initials placed on them.

Delighted beyond words that *Liberty Bell 7* has now been restored to near pristine condition and will find a permanent home at the Kansas Cosmosphere, Jim Lewis's one wish was that the reputation of the astronaut who flew her could also undergo a factual restoration, that the record be set straight for Gus Grissom:

NASA and a few of us had it figured out right from the start. Maybe one day we'll get history written correctly. Gus was a consummate pilot who flew a hundred combat missions in Korea and was a successful test pilot. He had to be in order to be selected as an astronaut. Many applied and few were chosen. He was a very bright person and a skilled engineer who had everyone's respect.

No one who knew him could or would argue with that statement, and that is how he should be remembered.

4. Flight of the Eagle

From the pale horizon,
The chorus of stars goes down,
The edge of the earth softly lightens,
And gradually the day arises.

Alexander Pushkin

Gherman Titov awoke gradually, after a deep sleep. If he had been dreaming, those dreams quickly vanished as he tried to make sense of a confusing rush of sensations. At first, he had no idea where he was, and a nagging headache wasn't helping as he tried to orient himself. He looked at his arms; to his astonishment, they were floating gently in front of him, and he could also see a couple of objects lazily floating by them. Then he remembered where he was. He was inside a craft named *Vostok 2*, with some of the most spectacular views he would ever witness passing by his porthole, an endless procession of breathtaking beauty. He was in space.

True to his independent character, Titov had overslept. "When I woke up and saw the time," he later recounted, "I thought: there will be trouble about this." Yet, to his relief, there wasn't; ground control did not mention anything to him about oversleeping. Throughout his military career, Titov had pursued a risky habit of questioning orders but usually came out on top. Once again, luck seemed to be on his side.

During the flight of *Vostok 2*, Titov spent just over a day in space. After Gagarin's single orbit and the two American suborbital missions, Titov's seventeen-orbit flight was an incredible leap forward in the history of space exploration. The three flights that had preceded his, while providing substantial steps forward for their respective programs, now seemed like mere tentative hops by comparison. Titov would travel a greater distance than the journey to the moon and back, and spend sufficient time in space to explore in detail for the first time the interaction between a human and the weightless environment. Titov didn't just go into space; for a whole day, he lived there.

Though the spacecraft, the training, and the selection process for *Vostok 2* were identical to those for the first Vostok flight, Titov was by no means a clone of Gagarin. Astronaut Frank Borman, who stayed at Titov's home in Russia on a visit in 1969, describes him in his autobiography as a "short, stocky man with an ingratiating grin and a dedicated sense of humor . . . the Soviet version of [astronaut] Jim Lovell." Titov's ease in social and official situations, however, masked an incredibly complex character. To many, he seemed outspoken, stubborn, haughty, and distant; some were surprised that he was on this historic flight at all.

Gherman Stepanovich Titov was born on 11 September 1935, in a log cabin in the rural Soviet village of Verkhneye Zhilino. Situated in a remote western Siberian region known as Altai, his birthplace is more than two thousand miles from Moscow; in fact, the border with China is closer. Titov later wistfully remembered that "you can see wolves and hares, and squirrels climbing the trunks of cedar trees . . . the woods there are rich in game, the rivers in fish and the meadows in flowers." It is indeed a beautiful region, renowned for its lush green forests, sparkling lakes, and spectacular waterfalls. There is, however, an annual price to pay for this beauty and serenity—the area's inhabitants have to endure the bitterly cold winter months. Titov remembered that they were the kind of winters only Siberia suffers. "They were very cold, with blizzards. In winters like that, we were always in the snow, with red faces and our clothes soaking wet."

His father, Stepan Pavlovich, was a schoolteacher who not only taught Russian language and literature, but also organized a drama group for the village. Titov recalled his father playing the violin, painting artworks, and leading a household in which a love of the arts was always encouraged. Stepan always hoped that Gherman would emulate his interests in music and painting.

Most space histories state that Stepan named his son after a character from the Pushkin short story "The Queen of Spades." Even Gherman himself once stated "My father gave me that name in honor of the character created by Pushkin, and he named my sister Zemfira, likewise in honor of Pushkin." Yet when Stepan was interviewed in 1961, he remembered it differently, stating that while it was true he loved Pushkin's work, the character of Gherman was not a particularly pleasant one. He actually selected the name, he said, to be original, to choose "something different to the Ivans and Stepans."

It is not surprising that Stepan did not want his son identified too closely with Pushkin's work. In "The Queen of Spades," which explores the nature of obsession, the character named Gherman is cold, greedy, and manipulating. The tale deals with the themes of social status and misdirected inferiority and ends with an open question—are a person's actions a result of his or her character or of fate? These are questions that could just as easily be posed to the life of Gherman Titov, a man whose flying career was almost ended on many occasions, and yet who very nearly became the first human to travel into space.

A very athletic child, Titov had learned to ski by the age of three and also found great enjoyment in soccer and skating. He had no thoughts of becoming a pilot at that time. Stepan described him as a clever, energetic, bold, and tenacious boy. Yet his chosen activities did not always lead to enjoyment. "Once I decided to show off," Gherman recalled of a skating incident on a local pond, "and several times crossed the clear water that had just been covered by a thin sheet of ice." The ice shattered, and Titov plunged into the water up to his neck. It could have been a humbling, even disastrous, lesson for the young boy, but the next day he was back out again, happily skating on the pond with the other children.

Stepan was not only Gherman's father; he was also his schoolteacher. In the classroom, he was careful not to show his son any favoritism. In fact, he did the opposite, giving him more homework and harsher punishments than the others. Being treated as if he were an adult taught young Gherman to grow up fast, and it also had a positive side—his father treated him as an equal.

The world war was still a long way from Siberia, but its effects were already being felt in the village. As the younger men were called away to fight, Stepan was given the responsibility of school superintendent. He took his new position and duties seriously, but for young Gherman having his father now overseeing the entire school did not make life any easier. Eventually, however, even Stepan was conscripted, first as a truck manufacturer, then as a truck driver at the front. Altogether, Gherman lost his father to the war for three years. As he later recalled, when his father left Gherman found he suddenly had to take on many of the family duties previously entrusted only to the adults. He remembered those years as ones in which he had to force himself to think as an adult at all times. He was less than ten years old.

One of Gherman's uncles was a pilot who flew during the war and came to visit his family on leave once the conflict was over. "He spoke a lot with the lad about flying and flyers, the life they led," Stepan recalled, adding that the visit made quite an impact on his son. "All that he could talk about after that was flying." According to Titov, it was this family event that first inspired him to become a pilot. More than anything, he found he admired his uncle's air of sophistication; while he was impressed with the daring flying stories, he was more impressed by his "shining boots and luxurious trousers."

Continuing his schooling after the war, Gherman found that moving into a new class also brought a new challenge—a bully. His immediate response was to fight back, even when, as he recalled, his nemesis slashed him across the face on one occasion. "The scar serves as testimony that I stood up to my first bloody combat in life without flinching. After that time, no one bothered me any more." Such incidents set a pattern for his life, and he rarely backed down from a challenge ever again.

There was another, less impulsive and more thoughtful side to Titov in his school years. This was demonstrated when he joined the school's literary club and was soon following his father's example by absorbing the classics of literature. Without reading, Gherman would later maintain, he did not believe that people could be fully educated, especially when it came to enjoying life and the beauty of the world. Later, when he became a cosmonaut, he delighted in reciting many classic poems he had memorized. His extensive knowledge of literary works written before the era of communism distinguished him from many of his colleagues—and would cause some to question whether he was suited for the propaganda aspects of being a cosmonaut. Titov was always going to be too clever to easily control.

A fierce confidence in his own abilities led to an unfortunate event that should have permanently ended any flying ambitions: Gherman broke his wrist in a cycling accident at the age of fourteen. "I always liked to ride a bike," he said of the incident. "I was very good at it. Once when I was going very fast—flying down our village street—I fell and hurt my hand rather badly. There was a sharp pain in my left arm. I could not get up." Rather than have the injury treated, Titov chose to conceal it, even from his family. It was a peculiar decision, but one that exemplified just how self-assured and self-contained he could be. There was another, truly practical reason for maintaining his silence—he'd already applied to aviation school and was worried that disclosing the injury might result in his rejection. His wrist continued to hurt long after the bones had mended, so he secretly undertook exercises that he felt might strengthen it. "He would come home white as a sheet," his mother Alexandra remembered, "but never say a word." It was agonizing at first, but the young man was determined. "The bone knit slowly, and I began to exercise my arm stubbornly," Gherman explained. "Thus, I was drafted into gymnastics. Regular training began, and soon I was as keen on gymnastics as I was on riding my bicycle. I conditioned my arm; I followed a regimen. I was determined to fly!"

When he was finally called up before the medical board for flight school, Titov was relieved to find that no one noticed the break. To his delight, he was also recommended for the air force. Luck had been on his side, and he eagerly awaited his first opportunity to fly. At the age of eighteen he headed to the Kazakh city of Kustenai and the military air school there, where he would spend two years undergoing training. This was followed by a further two years of training at the Stalingrad Higher Air Force School. As his flying career progressed, Titov remained anxious about the injury to his wrist. It would still disqualify him from piloting if discovered, so he spent an hour in the gym each morning, diligently exercising on the parallel bars.

Titov was never going to be an obedient military cadet. Given his bold and impetuous nature, he was bound to run into trouble before too long. In a letter written to a friend, he admitted that he was still learning to watch his own steps now that his father wasn't around to keep him on the right path. He quickly found himself in hot water when he complained about the basic training, which he remembered as being incredibly boring: "Our first job was to dig dugouts. For the time being, all flying was out of the question." He was instructed to learn the articles of war, "which I, the future flyer, immediately declared to be something useless . . . of course, I had to pay several times over for that kind of an attitude." Titov was given extra duties for his impertinence.

Soon the training changed to subjects more to Titov's liking, especially when he began making flights in a Yak-18. He was so busy training all day both in flying theory and practice that by evening he barely had the energy to stagger to his dugout and fall asleep in his makeshift cot. At first, Titov's flying lacked consistency, and he had to work particularly hard just to keep up. His greatest weakness was making night landings, but he slowly improved. Then he was briefly grounded by his superiors for not ensuring that the cadets in his group were maintaining their uniforms in top condition. He'd spent all of his energies concentrating on flying, which he considered the most important duty, and discipline and neatness took a distant second place. Titov grew increasingly frustrated with the less glamorous duties of military life and briefly considered giving up flying, finishing his military obligations in the infantry, and then going back home. While pondering such a drastic move he felt that he had nothing to lose, and this soon led to another clash with a superior officer. Titov shouted rude remarks at a colonel, an action he said was "incredibly foolish for any cadet." The outraged officer immediately called for Titov to suffer the severest punishment—an immediate and permanent end to his flight training. The incident, Titov later admitted, was caused by "my youth and hotheadedness. . . . I was too quick with outraged reaction, too young to appreciate the need to bend . . . too brash to recognize a requirement for tact . . . being discharged from the academy would have been a catastrophe for me." Fortunately, his flight instructor, who had been monitoring Titov's steadily growing flying skills, was able to persuade the colonel to give the rash young man another chance. The instructor later took Titov aside and berated him soundly and at length for giving in to his "stupid temper." For the rest of his flight training Titov was able to subdue his outspokenness, and on his twenty-second birthday he graduated with distinction.

In November 1957, in his first postgraduate assignment, Titov was sent to the Leningrad district as a combat pilot, based in the village of Siverskaya. Flying in two different regiments of the 41st Air Division, he managed to master several of the difficult maneuvers required for combat flying, including high-g turns that required all of his strength. "The air regiment I was in paid special attention to physical training," he recalled. "Modern planes need flyers that are physically fit and can endure great strain. The flyers worked out in the gym and many took up acrobatics. It called for skill and the ability to control every muscle, something that could only be achieved after intensive training and strenuous daily practice. Two of my friends and I set up our own act: we performed at the regimental club and at the local bread-making plant. Needless to say, we had a long way to go to compete with professional acrobats, but our audiences gave us a warm reception."

In the spring of 1958, the dashing young pilot began dating Tamara Vasilyevna Cherkas, a Ukrainian cook at the air base. "I had no idea of getting married," Titov said of the meeting, "and women hardly looked at me as a suitable fiancé." Yet, in another of his impulsive decisions, the two were married within a couple of months. Stepan was worried it had all happened too quickly, and Gherman admitted that he and Tamara had not really taken the time for a proper courtship. Knowing so little about each other, the couple soon had their disagreements. He was not happy about the "new furniture and knickknacks" she bought, and Tamara would leave the room whenever he read poetry out loud to himself. Slowly, however, they grew to know each other better, and their lives settled somewhat—just in time for a career change that would have a profound effect on both of them. "The life we had only just started had to be overturned," Titov later wrote.

It was while he was in the air division that Titov was interviewed as a potential cosmonaut candidate. At first, he had reservations, not wanting to disrupt his aeronautical ambitions. "I was taken completely by surprise, and was a bit dubious at first . . . I said, I have to think it over. . . . I didn't see why I should change my job . . . I was really still quite new at advanced jet flying and thought it would be better to master that first." He told the interviewers that he was interested in theory, but no more. He also knew not to ask them for more details. "You do not ask brash questions of colonels," he remembered. "I had learned that lesson well."

After further deliberation, however, Titov decided that he really *was* interested. He was therefore pleased when he was included in the ongoing selection process, but in the midst of it he caught a fever. Knowing they were looking for superfit candidates, he later confessed that it was an anxious time for him. "I awaited my sentencing," he recalled. "Could this really be the end?" Titov found it bitterly ironic that he had spent much of his childhood playing in deep snowdrifts, and now an exciting career might be suddenly terminated because of a bad cold. He was sent back to his regiment, expecting to hear that he had been left out of the selection process. Fortunately for Titov, his physical prowess had made quite an impression on the selection committee. He was called back and informed that he had been selected as a candidate for the first cosmonaut team. In the meantime, he had become intrigued with the possibilities of spaceflight. When he returned, he was more curious than ever about flying into space. Although the training seemed hurried to him and he had many questions that were not being answered, he felt a new sense of kinship with his colleagues. "We were all young people," he recalled, "all volunteers, and we were united by the idea of flying into space; that brought us together. There was nothing especially romantic about it, just a normal professional interest to fly higher, to fly more modern vehicles."

The long hours spent in the gym overcoming his wrist injury had turned Titov into an excellent athlete, much like America's Scott Carpenter. "My general physical health and my training in gymnastics and acrobatics were of great help in passing the medical committee prior to being enrolled in cosmonaut school," he admitted. Gagarin joked with him that he should consider joining a circus instead of flying into space. His physical stamina had impressed those making the selections, and Titov's rigorous regime had made his wrist injury difficult to detect. The doctors did not notice it; had they done so, he would have been rejected. "They'd have dropped me like a hot potato if I'd said anything," Titov admitted.

In fact, the twenty-four-year-old Titov was so physically impressive that he had been selected despite being technically ineligible. The lower age limit had been set at twenty-five, but Titov, along with Georgi Shonin and Valentin Bondarenko, was younger. Though he was in superb physical condition, Titov said his enthusiasm did not necessarily extend to all of the training exercises. He particularly hated running; when told by an instructor that he'd grow to like it, he curtly responded: "You can't force me to."

The testing did not end after his selection. If anything, it grew steadily more intensive, and the process still held the possibility of disqualification. Once again, Titov's pride almost got the better of him during the psychological tests. He argued that he was being asked "a lot of silly questions," and this naive haughtiness almost cost him his place in the cosmonaut team. Yet his belief that he could perform on tests better than others was frequently correct. Having outwitted the doctors regarding his wrist, he also impressed the psychologists during the isolation chamber test. Unlike more impulsive candidates, he carefully thought out how best to use his limited rations of water and food and came up with the most sensible solution. Over thirty years later, Titov was still talking with pride about how he'd been "clever about it."

Astonishingly, he also managed to talk the psychologists into allowing him to take a book into the chamber. "I wooed them," Titov explained with his characteristic self-confidence. The book he chose was *Onegin*, by his beloved Pushkin. "By the fourth day I knew a chapter by heart," he later recalled proudly. As he smilingly recited the poetry to the observing doctors, one can only speculate whether they noted the ironic relevance of some of the lines:

The enthusiasm of youth
Could never hide a single thing
Love, hatred, pain or gladness
It will blurt out quite readily

The young man was, indeed, so sure of himself that he felt these tests were a simple preliminary step to his becoming the first person in space. Decades later, he soberly reflected on that time. "What did I know," he admitted, "a young lieutenant with eyes full of courage and scarcely a single sensible thought in my head."

Titov soon found that the cosmonaut group was made up of equally diverse personalities. "Some of us were a little hurt by criticism and sulked," he explained. "Others were too brash, while others brought enthusiasm and salty humor. We were all very earthly people . . . men who had kept the fire of youth." Despite the diversity, Titov still seemed different from the others because of his somewhat more privileged and educated background. It was a noticeable characteristic that generally did not work in his favor. Some felt his pride in his background and upbringing was merely an excuse to be arrogant, hot-tempered, and opinionated. This impression was reinforced by Titov's looks. Gagarin described him as having "a handsome, thoughtful face, high forehead with soft chestnut curls above it," and journalist Evgeny Riabchikov also noted that he had "a sharp nose and strong lines of the mouth and eyebrows." These elegant features, combined with a meticulous devotion to keeping his uniform perfectly pressed and smart, gave the impression of a higher-class individual—and class distinction was not going to benefit a career in communist Russia.

Titov, however, was more worried about his wife than what others thought of him. "I embarked on the difficult task of preparing my wife for a new line of work," he later recounted. "Tamara listened to me and sometimes agreed, but more often I read distrust and alarm in her eyes. I admit it was not easy for her; she was about to become a mother." Their son, Igor, was born in July 1960, but in the first two months of his life the doctors discovered the baby had a congenital heart defect. Igor's health quickly deteriorated, and he passed away a few months later. Because of the couple's tragic loss Gherman was briefly removed from the cosmonaut team. "This was a great grief for us. When the baby died, she took it terribly hard," he said of the devastated young mother. "Fate had struck us a cruel blow." He relied on the support of his new colleague, Yuri Gagarin, and this drew the two cosmonauts' families closer to each other. "Yuri offered genuine support," he remembered. A grateful Titov began to realize he had found a close friend, whom he'd call "Yura" out of affection.

When Titov returned to the cosmonaut team, the workload steadily mounted. He was soon selected as the youngest member of the first six who would undertake accelerated training for the first flight, undergoing what he later described as an enormous, difficult schooling of both physical and theoretical preparation. By January 1961, he was ranked number two behind Gagarin for the first flight. "The six of us were equally well trained, and each could pilot the Vostok spacecraft," he recalled, but "in conversations, we were all inclined to say that Yuri would fly. We knew he was good, having gained great respect from his colleagues."

But the reasons for Gagarin's selection cannot entirely be answered by Gagarin's good points alone. Though he had performed superbly in the training tests, Titov's personality had simply not impressed the right people. His temper was, in fact, very similar to Sergei Korolev's, the Vostok chief designer. As Titov later admitted, he and Korolev frequently clashed like "two lions in the same cage . . . a difficult relationship." Titov was actually in a far more precarious position than he realized at the time; when Grigori Nelyubov, a fellow cosmonaut in the top six, displayed similar overconfidence he lost his opportunity to fly a Vostok mission altogether.

Titov nevertheless said he smarted at the fact that he was number two in the running. "I wanted to be the first into space. Of course, I wanted to be chosen. Why shouldn't I? Not just for the sake of being first—we were all interested to see what was out there." Yet he later recognized why it was not to be:

It was Gagarin's character that mattered most. Yura turned out to be the man that everyone loved. Me, they couldn't love . . . I'm not lovable. I have a very explosive character. I could easily say rude things, offend someone, and walk away. I wasn't a very convenient person for the leadership; I had my own opinion about things and knew how to insist on things. This did not always stir up warm feelings . . . but Yuri could talk freely to anyone—he could speak their language. The first man in space had to be a nice, attractive person . . . they were right to choose Yura.

Yet even at this late stage, early in April 1961, Titov still had a slim chance of becoming the first Soviet cosmonaut. The final, official decision was still to be made. On 7 April both men trained inside the Vostok spacecraft that would be the first manned production model to fly. Meanwhile, the influential general Nikolai Kamanin was still favoring Titov over Gagarin. Kamanin, one of the first designated "Heroes of the Soviet Union," headed the cosmonaut training program. "In the last few days I hear more and more people speak out in favor of Titov," Kamanin wrote in his diary, "and my personal confidence is growing in him too."

Being favored by Kamanin, however, would actually turn out to be a mixed blessing for Titov. In fact, the general's diary notes reveal that he had something else in mind for Titov. "The only thing that keeps me from picking [Titov] is the need to have the stronger person for a [subsequent] one-day flight." Gagarin agreed, writing before Titov's flight: "He was as well trained as I and was probably capable of a better performance. Maybe he was not being assigned for the first flight because he was being kept for another, more complex journey."

Titov therefore lost the opportunity to make humankind's first spaceflight, both because of his strengths and his weaknesses. When asked how he felt when he was officially informed of Gagarin's selection, Titov replied, "Why even ask! Painful or not, it was at least unpleasant. I was frustrated, of course, because up until the last minute I thought that my chances were high." On 12 April 1961, he accompanied Gagarin to the launch pad as his backup and watched his friend thunder skyward into the history books, leaving him behind. "I was just to be a spectator," he mused. "It was a tense moment."

Following Gagarin's successful flight, Soviet leaders decided to make another bold step—bypass a more modest three- or four-orbit mission and aim for a flight that lasted an entire day and at least sixteen orbits. Most of the cosmonauts supported this move because they saw it as a way to thoroughly test the spacecraft. The new mission's other objectives would be as bold as its length. "Dozens of theories had to be confirmed or disproved," Titov explained, "since this was the first long flight into space." Because Titov had come so close to flying the first Vostok mission, there was never any real debate about who would take this second flight. Titov was soon formally assigned. Unlike Gagarin, Titov would assume manual control of the spacecraft and move it in flight. He would also eat solid foods, and even go to sleep. Medical advisors were unsure what would happen to a sleeping cosmonaut in weightlessness. Sleep, by effectively taking the cosmonaut out of operation in the middle of the mission, might also pose risks from a piloting standpoint if quick manual control was needed. But if space exploration was to progress, these were the kinds of questions that needed to be answered. Titov, in top physical health, seemed the ideal candidate for such a mission.

His training began almost immediately after Gagarin's return from space. Though Titov had closely supported Gagarin for his flight, he now found that his colleague was gone—quickly hauled away for a grueling world tour. The difference in their status had become starkly apparent to Titov following Gagarin's flight, when he'd had to stand on tiptoe at the back of Red Square, anonymous in the crowd of Muscovites, simply to catch a glimpse of his friend. Titov was still just "an unknown understudy," as Kamanin put it. For Titov, the adulation for Gagarin that followed proved interesting, if not slightly alarming. "I believe even Yura didn't think he would become so world famous after a 108-minute flight," he commented. "We realized, of course, that we would be honored—but that it would be on such a tremendous scale? He was honored by presidents and kings . . . frankly, we'd never dreamed of that." Titov missed the support of his friend and found the arduous training exhausting. But with assistance from his backup, Andrian Nikolayev, Titov pushed ahead.

Titov did get the chance to query Gagarin on one vital point. He was intrigued by the concept of weightlessness and eager to know how it felt. "Weightlessness was a mystery," he explained. "That was the only thing I was interested in, I didn't want to know about anything else. He said it was okay." It turned out, however, that Titov would experience zero gravity very differently from Gagarin.

On 6 August 1961, Titov awoke in the same bed he had occupied when serving as backup pilot to Gagarin four months earlier. This time, however, he had not slept so well; the summer nights were much hotter than they had been back in April. He was still only twenty-five years old and to date the youngest person ever to be launched into space. In fact, some of the technicians at the pad had not believed the young man could possibly be a cosmonaut when they saw him before launch. Journalist Alexander Romanov, who was at Baikonur in the days before the launch attempt, had a similar reaction: "We would probably have never paid any attention to the boy, who looked like any other worker, had the man next to me not whispered: 'That is the cosmonaut . . . .'"

Initially, Titov's launch day began much as the one on which Gagarin had made history, but it would soon change. He had felt acute disappointment at being the one left behind that day in April, but this time he was the prime pilot and the one who climbed into the waiting spacecraft. "It was a beautiful morning," he remembered. Romanov was able to observe the launch from a special viewing area. "The rocket slowly, it seems very slowly, leaves the Earth," he later wrote of the spectacle. "Gathering force it streaks more and more quickly upwards like a roaring fiery globe." To Titov's relief, the moment of launch was so gentle that he barely felt the difference in vibration. He reported that he felt fine all through the ride into orbit, and even stole a quick glance out of the porthole. Looking back years later, Titov was not sure he could ever have anticipated what was to come. "I can't say I was ready for any of it. We couldn't train for malfunctions, because with so few flights behind us nobody knew what kind of things might go wrong. When we'd seen another launch that exploded, we knew that technology and equipment was something that could fail. I always hoped that things would go well, but knew that there could be complications."

At the moment the last rocket stage cut out, Titov noticed two profound differences. First, the noise of launch was replaced by a near silence. Second, weightlessness felt particularly *odd*. It seemed as if he was flying upside down, sometimes even as if he were somersaulting, and it was difficult to shake this disorienting sensation. It also made it hard for him to read his instruments at first. "I couldn't understand why I felt this way," he remembered. "For the life of me I could not determine where I was. I was completely confused, unable to determine where Earth or the stars were. It seemed as if the somersault had carried me completely around and that I was floating upside down, attached to nothing. I could not figure out what position I was in. Everything whirled around in a strange fog that defied all attempts to separate order from the sudden chaos. Something had gone drastically wrong with my sense of balance; my sense of orientation vanished abruptly and completely. Zero gravity is a serious business."

The sensations Titov experienced are ones that still affect spacefarers today. Space Adaptation Syndrome, or "space sickness," is still not fully understood, and it affects almost half of those who travel into space. On what was only humankind's second orbital flight, it was much too early to know this, but those monitoring Titov's condition via telemetry on the ground were very attentive to this strange new phenomenon. They were not alarmed—disorientation and sickness in space had been anticipated—but they did not know how much worse it might get and were concerned Titov might become incapacitated. The unpleasant feelings became more pronounced over the first two orbits, so much so that Titov even considered requesting that the mission be cut short. In addition to nausea, he was suffering from vertigo, headaches, and a pain behind his eyes. However, his stubborn determination kicked in, and he chose to press on, telling ground controllers that all was well. He'd save the frank descriptions for his debriefing and continue with the flight program.

It helped a little to look out of the window. He noticed almost immediately that he could observe his motion about the planet as the Earth passed below. Soon he crossed into the shadowed side of the globe, witnessing a sunset "to delight the eyes and souls of poets . . . night rushes at you like the darkness of a tunnel you enter in a car." He was fascinated to witness the planet seemingly glowing a dark gray, with a slightly brighter horizon, "like a glittering sickle." Soon he was treated to his first orbital sunrise, "an explosive arrival of dazzling brilliance . . . as if I were gazing at the sky through a crystal prism."

Gagarin had provided beautiful descriptions of his impressions from orbit, and even the pragmatic Gus Grissom had offered some vivid words about the brief glimpses he'd seen of the Earth. Unlike them, however, Titov's life was steeped in a profound love of literature, and his words have a power that other spacefarers have rarely matched, before or since. His recollections are both poetic and succinct. "I had the feeling that our Earth is a sand particle in the universe," he wrote, "comparable to a particle of sand on the shore of the ocean. It was strange to have a black dome above me and our earthly blue sky below. The Earth flashed as a multi-faceted gem, an extraordinary array of vivid hues that were strangely gentle in their play across the receding surface of the world . . . framed in a brilliant, radiant border. The colors were extraordinary—vivid, yet tender—and the light streaming through the cabin carried a strange shade as if it were filtered through stained glass."

Titov was not only there to observe, however—he was a test pilot with a demanding job to do. He took manual control of the spacecraft and successfully turned *Vostok 2* about its axis using the thrusters. He was pleased to discover how easy it was to perform this maneuver, and the experiment ran just as smoothly when he repeated it on the seventh orbit. "What a tremendous feeling," he later said, "to manipulate with just my hand the mass of a spaceship plunging through a vacuum at nearly eighteen thousand miles per hour . . . a vessel of space that responded to the flexing of my wrist and the pressure of my fingers like a well-trained animal. . . . I felt I was the master of the craft: it was obedient to my will."

During the mission the cabin temperature fell to chilly levels, reaching as low as fifty degrees Fahrenheit. It was later discovered that this was not a spacecraft malfunction; the system had been accidentally switched off at launch. Other equipment worked much better. Television pictures of the cosmonaut were transmitted to the ground, and Titov also became the first spacefarer ever to record movie images of the Earth passing below. He was entranced by the colors he saw outside, both on the Earth below and on the horizon during sunrises and sunsets. When talking to Earth, he used his call sign—*Orel*, meaning "eagle." "I had a fine call sign," he recalled; "it was like a second name."

Although he never felt hungry while in space, Titov ate on three separate occasions. Each time he consumed the specially prepared food without difficulty. Though much of what he ate was in tubes, he did have more appetizing choices such as bread and sausage. He even had cutlery and a small amount of coffee he could drink. He later recalled that when he drank black currant juice, some drops escaped and hung in front of his face. "It was fun watching them float in the air . . . with slight quivers, like small soap bubbles."

Titov ate and drank because it was part of his flight program, not by choice. In fact, he still felt particularly strange in the weightless environment, especially when he made a rapid movement of his head. When he loosened his straps and floated a little, the feeling grew worse. At times he almost felt like being physically sick, and in fact, drinking fruit juice and liquid chocolate did make him vomit on two different occasions. He even tried to shake off the strange feeling literally, by moving sharply in his seat, but it did not work. When he stayed absolutely still in his seat, however, he began to feel somewhat better. He even felt well enough to undertake some simple exercises, by pulling on rubber cords.

On the seventh orbit, it was time to attempt another unprecedented experiment—one whose outcome was decidedly unknown. Gherman Titov would try and sleep. It was a welcome respite, he remembered. "I was having moments of dizziness and nausea, so I was not displeased that the program now called for me to go to sleep." At first, it was hard for him to rest. Not only was he still feeling nauseous, but he also had to fix his hands under the harness straps to prevent them from floating. Eventually, he drifted off to sleep, but woke twice. "I awoke because of some kind of strange position of my body," he explained. "Weightlessness continued to play tricks with me, and for a long time I could not cope with my arms. I saw they had risen by themselves and were hanging in the air . . . I tucked them under the safety belt and went back to sleep." In fact, he fell into such a deep sleep that he dozed over half an hour longer than planned. Sleeping in weightlessness was very comfortable, Titov later related, a feeling as if he was on an ocean wave. For six orbits *Vostok 2* floated serenely around the Earth, carrying its sleeping passenger over ever-changing panoramas and through majestic sunrises and sunsets. Titov never saw them, nor did he dream of them. "I didn't dream at all," he later reported. "I had a good sleep, like a child."

When he woke he still felt tired, and the headache had not gone. Nor had the peculiar feelings of weightlessness completely disappeared. Yet he was pleased to note that his coordination was now sharper than earlier in the flight. "The nausea gradually went down," he recounted, "and things became much better." For the last five orbits of his flight, Titov felt perfectly fine. "I had become completely accustomed to weightlessness. I had grown a bit of a beard, and thought to myself it would be nice to have a shave before returning to Earth." His body may have finally been adapting to the space environment, but though this was the longest flight to date, there would not be enough time to find out.

Although profoundly moved by the views from orbit, Titov was more than ready to come home. The flight had lost a lot of its novelty, and he was ready to see other people again. "The first orbits," he recalled, "were rich in impressions. But the last ones? It was old stuff by then. It had become routine." The thought of returning to Earth was not something he relished completely, however. He was concerned about reentry, knowing it was one of the riskier parts of the flight. "Through the wonder, there intruded the very grim facts of life. As beautiful as the seventeenth dawn was, it was also dangerous . . . this is the moment that people in space fear. The slightest error at this stage could cause a great deal of trouble."

On the seventeenth orbit, Titov's program fully carried out, the retrorockets fired automatically, and Titov felt the g-loads rise as he began his journey home. He enjoyed the sensation, especially because he knew it meant the nausea would no longer return. Soon, however, he heard a soft banging noise. As during Gagarin's reentry, the instrument module had not fully separated from the cabin and was still attached by a cable. The spinning of reentry was less severe than for Gagarin, but Titov still found the g-forces tremendous. He wondered which would lose structural integrity first—the module he was trying to separate from or the module he was strapped inside. He switched on transmitters and recorders to document his descent in case he did not survive it. To his dismay, the clocks on his control panel were still running, which meant not only that the other module was still attached but that the attachment cable was still strong enough to send information between the modules. "The capsule rotated very fast," he recalled. "Then there was a huge shaking. Both compartments were hitting each other . . . it was rather horrifying to think that I was flying in the center of a furious blaze, and that the only thing protecting me . . . was a little thin layer of special material."

Fortunately, the connecting cable burned through, and Titov was finally free to make a smooth descent. Deciding not to cover the portholes of his spacecraft, he watched with fascination as flames enveloped him—"like the blazing maw of an erupting volcano . . . beautiful and rather awe-inspiring"—discoloring the glass and coating it with soot.

In the final stages of reentry Titov ejected smoothly from the spacecraft, just as Gagarin had done. "I felt a jolt and flew out of the cabin," he remembered, "and was blinded by the bright sun. Over my head the bright orange dome of the parachute blossomed." Below him, he saw swirling cumulus clouds, which he passed through before seeing the ground "covered with golden stubble; . . . swaying a bit, the chute carried me further and further down." Titov worried reflexively about his wrist as he drifted downward and prepared himself to avoid landing on that arm. He also had one last hazard to contend with. He was coming down about fifty yards from a railway line, and briefly feared he might hit a train that happened to be passing. Luckily, he missed the tracks, but about five yards from the ground a breeze turned him in midair. He hit the ground facing backward, and was rolled over three times before being dragged along the earth by his still-inflated chute. Titov ended his flight with a faceplate full of dirt.

Like many spacefarers who followed him in the decades to come, some of Titov's first impressions upon returning to Earth were how strikingly powerful the familiar smells of nature were after the comparatively sterile surroundings of the spacecraft. He grasped a handful of rich brown soil and smelled it deeply, appreciating it as he never had before. He felt tired but otherwise was in good shape, with no lingering aftereffects of the nausea he had suffered in space.

A car raced over to meet him, driven by an excited local woman in such a hurry to get there that she banged her head on the steering wheel while traversing the uneven ground. She "jumped out with blood on her forehead," Titov remembered. The cosmonaut found that his first task on returning to Earth was to apply first aid from his spacecraft medical kit to his would-be helper. He had landed close to Saratov, not far from where Gagarin had touched down months before. And just like Gagarin, he received a congratulatory telephone call from Nikita Khrushchev. True to form, Titov felt himself to be above such things, and although he did not disclose it at the time, he candidly admitted it years later. "Honestly speaking, I did not care. Party member or not, I was a specialist, doing my job . . . I really felt just how tired I was."

Moscow wanted to celebrate this latest space triumph, and a few days later Titov was at the epicenter of a huge parade in Red Square. He was understandably exultant. "Those wearisome, arduous months of training had paid off. *Vostok 2* and I were a success!" In the middle of the celebrations, Titov firmly embraced his colleague Gagarin for well over a minute. For the rest of that momentous year, they were the only two people ever to have orbited the Earth. In fact, as Gagarin had not quite completed an entire orbit, it could be argued that Titov was the first person to have achieved this honor. Gagarin would later say that he was glad to see his friend again. "Despite all the hardships of his flight, he was a picture of health. Only his expressive eyes betrayed how tired he was, something even his smile could not conceal."

As well as the political public relations duties, Titov had to undergo a rigorous debriefing. He was very honest about the feelings of sickness he'd experienced, and the information he provided would greatly assist flight planners in making decisions about future missions.

Gus Grissom had seen his spacecraft sink into the ocean only seventeen days earlier, while Titov's spacecraft was safely recovered from the field where it had landed. At the time, it was assumed that Titov's spacecraft would go on proud, permanent display in a museum, while Grissom's would be lost forever. In fact, fate would decide the exact opposite. Titov's heat-scarred Vostok spacecraft was ungraciously pulled out of retirement in 1964 and recycled to help test a proposed Voskhod parachute descent system. With callous disregard for its historical importance, the spherical craft was dropped from a great height. The prototype parachute did not work, and *Vostok 2* was shattered into tiny fragments. Conversely, Grissom's spacecraft was located decades later and recovered from the ocean depths, before going on a celebratory tour.

Titov's reputation for overconfidence and rule-flouting did not improve after his flight. In fact, it continued immediately after landing, when, on his way to the postflight debriefing, he opened and gulped down a bottle of beer. The doctors were appalled, but it was just a warning sign of more bad behavior to come. Late in 1961, both Titov and Gagarin were hauled in front of a Communist Party meeting and severely reprimanded for their womanizing, excessive drinking, and various other inappropriate behaviors. But Titov seemed not to care, and later said he enjoyed the company of his fellow reveler. "He was a good friend. He loved life, and liked having a merry time. It was easy to relax with him. . . . I like people with character." Over thirty years later, Titov confessed that he actually turned down most of the chances to womanize that came his way. "I had to limit myself," he admitted. "It would be one thing as a simple pilot, but not for a cosmonaut because of the reputation."

With official rebukes having little effect on him, Titov continued to enjoy his new opportunities and vices, which included a passion for racing fast cars. His newfound fame made it easier for him to acquire the luxuries he enjoyed, but heavy drinking and fast driving were always going to be a dangerous mix. The list of accidents he was involved in grew alarmingly, and all too often he found himself standing in front of a superior officer while being soundly berated. His father Stepan also expressed concern to Gherman about the effect that fame was having on him; he knew his son was still very young and felt the adulation might swell his ego too much.

Nevertheless, in 1961 Titov was sent on a world tour that didn't end until the following year. Unlike Gagarin, he did not always behave as he should have during official functions. While taking part in an official motorcade in Romania, for example, he chose to ride the motorcycle of an escorting policeman rather than remain in the official car. "They were not happy," he said with a laugh when recalling the incident. "It was a thing of youth." In April and May 1962 Titov and Tamara spent two weeks in the United States. If the Americans were expecting another Gagarin, they got a big surprise. Sightseeing in New York, Titov disparaged the famous skyscrapers, saying they "hid the sun from the people," and also commented that the city's neon signs "repulse and exhaust" rather than attract. To his hosts' discomfort, he even ridiculed the paintings he saw in the Museum of Modern Art.

During this tour, John Glenn had only recently returned from his own orbital flight, whose schedule, he later admitted, had been moved up largely because of Titov's successful mission. Because he and Titov were both due to speak at a space research symposium in Washington, Glenn was asked to

Cosmonaut Gherman Titov meets President John F. Kennedy and astronaut John Glenn. Courtesy NASA.

be Titov's social host. The Mercury astronaut took him around the famous sights of the nation's capital city, including a brief visit to meet President Kennedy at the White House. He was also shown Alan Shepard's spacecraft, on display at the Smithsonian Institution. As they traveled around, the two spacefarers discussed the feeling of weightlessness, their training regimes, and other things they had in common. However, as Glenn related in his memoirs, security restrictions prevented Titov from speaking in more than generalities about most subjects.

Glenn remembered Titov as "cordial but forceful." Paul Haney, present to assist Glenn during the official visit, could afford to be less diplomatic. He remembered that Titov was disparaging of almost everything he was shown:

The first thing we noticed was what good English Titov spoke when there were no other Russians around. When you visit another country, you are supposed to at least appear interested in everything you are shown. Everything we showed him, he would use a word that I was to hear often in East Texas, and the word is, "Shee-ut." It's a negative term. Everything we'd show him, such as the Washington Monument, 555 feet and five inches tall, he'd say, "Shee-ut! We got obelisk in South Moscow 1,500 meters." It didn't matter what we showed him. We took him to the biggest rolling steel mill in the United States, in the nearby suburbs of Baltimore. "Shee-ut! We got one in Novorossiysk that makes three times that much steel."

I couldn't resist. The second day we were going down Pennsylvania Avenue, and we had a little time to kill before going up to Capitol Hill, so I told the driver to stop at the National Archives. Inside, I showed him an original copy of the Declaration of Independence and the Constitution. He just looked at it, and said merely that he had never seen those documents before.

The icebreaker came when Glenn asked Gherman and Tamara if they wanted to come to dinner the next night at his house. According to Glenn, Titov said he'd like to, but he'd have to check with his superiors. Diplomats on both sides rejected the idea, choosing a formal embassy reception instead. However, Haney later recalled that Titov had other plans:

All through the two days, he kept quizzing us about "barbecue." He was terribly interested in it—how to pronounce it, how to fix one, how to serve one. For a day or so, Glenn and I could hear him saying "bar-bee-cue" under his breath in the car; it was like a word he'd just discovered. Glenn jokingly explained it to him: a barbecue is an excuse for neighbors to get together and burn some steaks.

The second night the Russians, following diplomatic protocol, entertained us—Glenn, Al Shepard, and me—with a reception at the Soviet Embassy. We were going down the handshake-introduction line, and I was about two people behind Glenn. A grinning Titov put out his hand and introduced John to the ambassador. Then he pointed at John and said, "I come your house tonight six o'clock for barbecue." He knew John lived across the river in the Virginia suburbs of Washington. Awestruck, John said, "You will? Tonight?" It was already five o'clock.

Haney remembers stalling the Russians as long as he could at the embassy, drinking as much as he could to delay them, while Glenn asked Haney to take as long as possible before delivering Titov and friends at his home—if necessary, they should make a wrong turn or two. Accordingly, on the way,

Haney took the entire party twice around the Pentagon. Since he was the only one who knew where Glenn lived, there was little the Russians could do but follow. "I considered taking them by George Washington's place," he said, "but I was afraid the Russian ambassador would know better."

John and Annie Glenn, in the meantime, had raced home with a police escort. Glenn got all of his Marine Corps neighbors to raid their freezers for steaks then get some braziers and charcoal going. The police escort was deputized to go and buy more groceries. Glenn frantically lit the charcoal in the middle of his carport, fanning the flames to get them going fast. However, the flames grew so hot the grill's center post failed, the steaks fell on the charcoal, and as the fat caught fire the flames flared up into the roof. At this moment, Haney arrived with the large Russian delegation in five limousines, trailed by photographers and reporters. "As we pulled up in front of John's house," Haney narrated, "one of the braziers had sparked off the paint on his carport under the garage. The paint was on fire: you didn't need any interpreters to know what was going on. These Russians poured out of their limousines, and grabbed garden hoses and buckets of water. The fire was out in about three minutes. It really didn't burn much of anything except the steaks. But everybody said, wow! The photographers also got a great bunch of pictures."

The steaks were saved, and Glenn and Titov finished cooking them together. "Later Titov walked up to John," Haney recounted, "and in excellent English said, 'Tell me. Every time you have barbecue, you burn down house?' One of the greatest lines I think I've ever heard. It was a long evening, filled with burned steaks and superb Russian vodka." Titov told Glenn that the evening was the most fun he'd had while in Washington. Despite his boasting and disparaging remarks, Titov had ended up making a great impression.

Unlike Gagarin, who would forever be celebrated as the first to fly, Titov's fame gradually diminished as the years passed and other cosmonauts flew. This allowed him to live a far more normal existence than Gagarin ever could. He remained ambivalent about the fame that came his way. "You never get used to it," he later admitted. "Of course, it is pleasant when they say nice things, such as you are so smart, good, famous, and achieved such a great thing. But when it disturbs your work, you grow tired of it."

Yet even after the touring came to an end, Titov's bad behavior continued. May 1963 seems to have been a particularly bad time for the outspoken young man. During a trip to Kiev that month, he purposefully insulted an officer with the scornful remark: "I am Titov, who are you?" He escaped a serious reprimand only by asking the wives of the generals in attendance to help him. In the same month he was traveling with a journalist and carelessly left in his car a folder full of sensitive classified material, including secret state decrees and security passes allowing access to many of the Soviet Union's top secret space facilities. The folder was stolen from the vehicle, and this time Titov did not escape an official rebuke. "Korolev told me what he thought of me," Titov recalled, "how I was slovenly and not serious enough. He said, 'I'll write you new passes, but if you lose them you'll go to prison.'"

September 1963 proved to be a happier month for Titov. He and Tamara had a daughter, the first child born to a person after they had been into space. Unsurprisingly, the couple named her Tatyana—the name of a character in Pushkin's *Onegin*. In 1965, they celebrated the birth of another daughter whom they named Galina. In September 1961, Titov had begun studying spaceflight engineering at the Zhukovsky Air Force Engineering Academy. It seems he did take this aspect of his work very seriously, as he graduated from the demanding courses in 1968. The year before, he also began test-pilot training under famed aircraft designer Vladimir Ilyushin. "He is teaching me to master my job as a test pilot," Titov said at the time. "Ilyushin is a great master. I fly a lot; I want to get acquainted with every type of aircraft, including the most modern. I fly in supersonic planes of various types. At times you go to such heights that it seems as if you are on a space flight." He was eventually awarded the title of Test Pilot, Third Class.

There was a specific reason that Titov was apprenticed to an experienced test pilot: he planned to fly the proposed Spiral space plane. Though he had the opportunity to continue training for the more mainstream space missions, he chose instead to stake his future on a program that promised more real flying. In 1965, he was named head of a cosmonaut team of five dedicated to this ambitious program to create by the 1970s a two-man, reusable Earth-orbiting spacecraft incorporating an ingenious movable-wing design. Titov became absorbed in the project, saying at the time that "new types of flying machines are needed if not only for the fact that the higher reaches of the atmosphere cannot remain unconquered. It is not enough to pass through it at the time of the spacecraft's exit into orbit and descent to the Earth. Controlled flight must take place in it . . . the further progress of astronautics is unthinkable without its merging with aviation." Development work on the craft began in earnest in 1966, but though it reached the stage of wind tunnel tests, the program was eventually cancelled. Titov had spent several years working on a spacecraft that could conceivably have returned him to orbit, but his luck had finally run out.

In March 1968, after his close friend Yuri Gagarin was killed in an air crash, Titov was barred from flying jets. His superiors, worried about losing their country's second cosmonaut on top of their first, put an end to his flying ambitions. In truth, Titov had also never quite escaped the unruly reputation he had earned among his superiors. Kamanin had protected Titov, hoping to guide his career, and, encouragingly, he saw Titov evolve not only into a more experienced pilot but also a much more mature person. Kamanin's diary entries for 1967 frequently state his pleasure at this development. He even hoped to steer Titov toward a key role in the lunar landing program, and so was surprised when Titov wanted to stay with the faltering Spiral program until the end.

But Titov's apparent turnaround did not last. By 1969, he was again in serious trouble after a never publicly disclosed but apparently very embarrassing incident involving a movie crew. He was banned from driving cars for two years as well as from the most prestigious of the cosmonauts' foreign tours. When Kamanin was also reassigned at the end of the year, Titov lost one of his key protectors and helpers. In 1970, after two years of frustration, Titov left not only the Spiral team but the entire cosmonaut corps. "We thought that our careers as cosmonauts would end with a flight to Mars," he mused. "But you see, life made some course corrections." He turned down the opportunity to stay on as an administrator. The idea of watching while others flew jets was too galling for a man who loved flying as much as Titov.

Titov began a new phase in his life by earning a military sciences degree from the Voroshilov Military Staff Academy in 1972. He then went to work for GUKOS, the main directorate for space facilities of the Ministry of Defense. Serving as deputy director of the ministry's command and control center, he oversaw 1974's *Salyut 3* military mission. Until 1979 he worked on military spacecraft and launch vehicle research and development, at which point he was appointed GUKOS's deputy director. While at GUKOS, he also headed the state commission for the Zenit medium-lift booster, considered by many to be the most advanced launch vehicle the Soviet Union ever produced. Titov left the military in 1991, retiring with the high rank of colonel general. The GUKOS center is now named after him, and his office is maintained as a memorial to him—a great honor in Russia.

After private work in the space business field, Titov was elected to the Russian parliament in 1995, representing the city of Kolomna. It was apparently the fulfillment of a long-held ambition. Paul Haney remembered Titov saying in 1962 that he was interested in getting into the Soviet political system, and perhaps running for office. Haney recalls Glenn, the future senator, finding that very interesting. Despite the faltering fortunes of the Communist Party, Titov remained a loyal adherent to its principles and was twice reelected.

Over the years, Titov had become well known in the West as one of the most accessible cosmonauts and was often the one selected to represent the Soviet space effort to the Western press. As late as 1988 he was leading visiting journalists around Soviet space facilities. In these later years, he was still wistful about the fact that he had never flown again. "Space is like a drug," he explained. "Once you experience it, you can't think of anything else."

When Colin Burgess met Gherman Titov in October 1993, the former cosmonaut was taking part in the Association of Space Explorers' annual congress. Attending along with dozens of spacefarers who flew after him, Titov could have basked in the attention and glory of his past. Instead, Titov seemed to have changed. Not only did he look different—the once lean young man was now a silver-haired, heavier-set fifty-eight-year-old—but he had matured. To Burgess, he now seemed remarkably shy and modest. When he was frequently hailed throughout the conference as Russia's "senior citizen" of spaceflight, he gave only a humble smile and small salute in acknowledgment. He seemed happy to stand in the background, observe others, and offer a warm handshake to those who wanted to meet him. Burgess took away the impression of a gentle, unassuming man who seemed bemused but tolerant of the extraordinary fame that his historic space flight had thrust upon him.

When the year 2000 began, Gherman Titov was the world's senior living space explorer; he had outlived Gagarin, Shepard, and Grissom. He'd lost the chance to fly first all those decades ago, but for a time—all too brief a time—he headed the list of surviving spaceflight pioneers. He would not, however, live to see out the year. On 20 September, not long after Titov's sixty-fifth birthday, he was found dead in the sauna of his Moscow home. Initially, it was believed that he had died of carbon monoxide poisoning, but this was quickly discounted. Titov had been suffering from heart problems, and a heart attack was the actual cause of death. He was buried in Moscow's Novodevichye Cemetery with full military honors. It was an appropriate burial site, as he was laid to rest near the graves of many people who had proved influential in his life. Among them were fellow cosmonauts Pavel Belyayev and Georgi Beregovoi, former premier Nikita Khrushchev, aircraft designers Sergei Ilyushin and Andrei Tupolev, and a large number of Titov's favorite writers and composers. His funeral was attended by many of his fellow cosmonauts. Andrian Nikolayev wept openly as he gazed at his colleague's coffin. Pavel Popovich remarked that "We have not just lost a comrade in arms, but a true friend."

The day Titov died, two Russian cosmonauts were returning to Earth in the space shuttle *Atlantis*, having worked with American astronauts aboard the International Space Station. It was a sign of how much things had changed since Titov's flight. The Russian duo, Yuri Malenchenko and Boris Morukov, first learned of the death of the pioneering cosmonaut at their postflight press conference, and both were deeply shocked. Morukov, a medical doctor, felt it important to acknowledge the legacy Titov had left behind. "He was a very interesting man," he told the press. "We got maybe the first interesting medical data after his flight because Titov explained that microgravity may cause trouble with the vesicular system. So it was very important for the next missions because we had to create countermeasures to ward off the potentially debilitating effects of long stays in weightlessness. In the last several years of his life he was our congressman, and he supported the Russian space program very well. It's very bad news for us."

Gherman Titov was not a simple man. After interviewing him at length for a book on his life and flight, Western journalists Wilfred Burchett and Anthony Purdy concluded that "like so many great men, his character is full of anomalies and paradoxes." Yet Titov's outspokenness, particularly the open, frank way in which he discussed his episodes of space sickness enabled space exploration to advance at a critical time. Titov himself believed that being a normal person, flaws and all, was actually a strength in his profession. When discussing future flights to Mars, he dismissed plans calling for unemotional individuals who would not be overwhelmed by the long voyage through the void. Titov believed sending such "narrow-minded" people would be profoundly wrong. "In that case, why send a human up into the cosmos at all? What will they see or learn there? What will they bring back to people when they return to the Earth?"

Given Titov's strong personality he was rarely unemotional, which at times made for an eventful and turbulent career. Yet, as he put it: "We cosmonauts are anything but supermen. Human reason . . . attended our flights. Reason is inquisitive, able to analyze, observe and decide." Stepan Titov once said that, though his son was somewhat hot-tempered, "perhaps his strongest point was tenacious willpower. He was very ardent, perhaps too much so at times, in going after the things he believed in." Considering the political nature of the time, Gherman Titov was perhaps an odd choice to become the second cosmonaut, but in the end he proved to be a wise one. The man Yuri Gagarin once described as "a very good flier, an intelligent man and a wonderful friend" carried out his spaceflight assignment perfectly, and opened the door to the longer, more difficult flights ahead.

5. To Rise Above

Man must rise above the Earth,
to the top of the clouds and beyond,
for only thus will he fully understand the world in which he lives.

Socrates

On 24 May 1962, a small spacecraft named *Aurora 7* approached the west coast of Australia. Weightless inside the snug cabin, strapped to a custom-made contour couch, Lt. Cmdr. Scott Carpenter (USN) soared over the Muchea tracking station 146 miles below. The Mercury astronaut had already witnessed his first sunset. Below him it was already the middle of the night. Traveling at almost three hundred miles a minute, circling the planet every ninety minutes, he was truly on the ride of a lifetime. After three circumnavigations of the planet, Carpenter would return, safely, to a landing in the Atlantic waters northwest of Puerto Rico.

But that was about four hours away. As he approached Muchea, the astronaut spoke with CapCom (and fellow astronaut) Deke Slayton, who, after a greeting, prompted Carpenter for a status report. The second American to orbit the Earth obliged, and added: "Tell John Whettler to saddle up Butch."

Despite a busy flight plan, Carpenter had taken the time to honor a friendship forged the previous year in a small, isolated town in the Australian outback.

Before it fell victim to improved technology and ceased operations in 1963, the tracking facility at Muchea was a key command station during the Mercury program, for which it had been developed. Located less than forty dusty miles up the Great Northern Highway, north of Perth, the station sat halfway around the world from Cape Canaveral and, more importantly, beneath the highest point of a Mercury spacecraft's orbital path.

A typical outback township, Muchea offered little more to the visitor than a small general store, which was run by Blanche Peters, who also tended the town's sole means of outside communication, a three-line telephone switchboard. A small primary school and even a railway station had once graced the town. By the time the space age arrived in the outback, however, both had fallen into disuse and been demolished. For visiting NASA technicians accustomed to the relative sophistication of American tracking stations, Muchea was an eyeopener—and inadvertent adventure. They delighted in the region's unspoiled beauty, and like the part-time tourists they were, lunged for their cameras whenever kangaroos and wallabies strayed into the compound.

An assignment to Muchea in September 1961 had brought Mercury astronaut Scott Carpenter to the vast spaces of western Australia, and he was an immediate convert. A test pilot with a natural love of fast cars, he saw the road sign on the way to Muchea—"End of Speed Limit"—as an irresistible invitation to the region. As a member of NASA's team of technicians and scientists, Carpenter was present to assist in a global tracking effort supporting the upcoming launch of *Mercury Atlas 4*. Although unmanned, MA-4 (a designation that meant the fourth launch of the Mercury-Atlas rocket) was crucial preparation for what would eventually be MA-6, the first, epic orbital mission for the fledgling U.S. space program. John Glenn had already been assigned as pilot for that flight.

After a stopover in Tahiti, the team arrived in Perth and took up accommodations at the Morley Park Hotel, managed by the amenable but dour John Whettler. Once Carpenter learned that the hotelier had flown Spitfires during World War II he naturally pressed for stories. But Whettler, like most combat veterans, was reticent about speaking of his wartime service. He did let slip, however, that after the war, through some fluke of opportunity, he had taught at the famed riding school in Vienna. In fact, he owned two retired thoroughbred racehorses. Carpenter, an accomplished horseman who had spent a good part of his childhood on horseback, was delighted to hear this. The astronaut had soon elicited a promise from Whettler to take him riding during any free time he could muster.

Given the time difference with Cape Canaveral, where Mercury Control was then headquartered, Carpenter's mornings were free. The next day the two men saddled up for a tour of the local countryside. In this way the astronaut was introduced to a large, powerful ten-year-old chestnut gelding named Butch. Years before, under the track name of Time Gentlemen, the horse had won four races. Carpenter could think of no better introduction to the region, as he later recalled for the authors: "It was a marvelous opportunity to see country that could have been on a different planet. I was just astounded by the flora and fauna. It was there I learned the demanding sport of kangaroo hunting from horseback, and witnessed some of the greatest displays of horsemanship ever. There was joking concern that NASA might never get me back from Australia!"

On one occasion, the regal Butch was unavailable, so Carpenter rode a neighbor's black filly—with wholly unexpected results. "We were out on a trail," he recounted, "and she started acting up." Whettler was by then well out in front, which distressed Carpenter's mount. Urging her to follow, Carpenter found that *she* had to take the lead. The more he tried to control her, the more stubborn she became, to the point of walking backward—a defiant stratagem that eventually placed her hindquarters squarely, and painfully, in a thorn bush. "Well, then she took off!" Carpenter remembered. "Straight through the brush and woods and everything, and I was *totally* out of control." After spying a good-sized tree, he recalled, "my horse figured that its shoulder-high branch might knock me off her back." Sure enough, Carpenter was neatly unhorsed. Whettler didn't find the filly for a couple of days until she finally wandered home.

The astronaut limped back, nursing a badly bruised shoulder, in search of aspirin and a heating pad. Whettler had plenty of aspirin, but instead of a heating pad the injured astronaut had to make do with an electric blanket. Later that night, after folding the blanket into a compact square and clicking it on, he placed it beneath his shoulder for warm relief. Carpenter woke a few hours later to find the hotel room filled with acrid-smelling smoke. His makeshift heating pad was smoldering on the floor, ready to burst into flames. "I crawled to the bathroom, stuck it in the bathtub and ran some water on it," he recalled. He may have ruined Whettler's electric blanket, Carpenter reasoned, but he had prevented the hotel from burning down.

Though the Redstone rocket was capable of lofting manned Mercury capsules on ballistic flights, the forthcoming orbital missions required a more powerful booster. Originally a prototype of the operational Atlas intercontinental ballistic missile (ICBM), the Atlas D variant had three liquid-fuel engines that generated a combined thrust of 367,000 pounds—more than four times that of the Redstone. But it would have to be tested.

The fourth test launch of the unmanned Mercury spacecraft finally took place, after innumerable delays, on 13 September 1961. Separation from the Atlas rocket took place as scheduled, and the Mercury craft, with a mechanical crewman simulator aboard, became the first to attain Earth orbit, achieving a peak altitude of 158 miles. After its sole circumnavigation of the planet, mission controllers fired the spacecraft's retrorockets with a remote signal. The spacecraft began its fiery return, plummeting toward the Atlantic. A six-foot-diameter drogue parachute automatically deployed at 41,750 feet, followed by the main parachutes at 10,050 feet, slowing the spacecraft's descent to an uneventful splashdown. The destroyer USS *Decatur* steamed in to pluck it from the ocean. Despite an abrupt drop in the craft's oxygen usage rate, produced when excess vibration caused a valve to crack open, operations director Walt Williams concluded that an astronaut would have survived the flight by taking corrective action.

Just eight months later, strapped into a spacecraft atop another Atlas rocket, one of those astronauts—Scott Carpenter—was blasted into orbit from Cape Canaveral. Planners hoped his would not only duplicate but also exceed the hugely successful mission flown three months earlier by Lt. Col. John H. Glenn Jr. (USMC). Toward this end, Carpenter was assigned a host of in-flight tasks to accomplish. As it happened, however, Carpenter's flight aboard *Aurora 7* would be the most perilous and controversial mission in the entire Mercury program.

Before they became Mercury astronauts, Scott Carpenter and John Glenn were test pilots at the navy's Test Pilot School at Patuxent River, Maryland, where they met for the first time in the mid-1950s. Their paths crossed again, early 1959, in even more rarefied aviation circles. They and about seventy other jet-qualified military test pilots had been ordered to report to the Department of Defense for a top-secret government program: Project Mercury. The country required some very special pilots—ones who would fly into space. Mature, highly competitive, articulate, intelligent (with IQs above 130), extraordinarily fit, the two candidates soon caught the attention of researchers and clinicians assessing them for phase two of the selection process. By the time they advanced to the Lovelace Clinic in New Mexico and to Wright-Patterson Air Force Base in Dayton, Ohio, for phases three and four, Glenn and Carpenter had been identified as the pacesetters in the grueling psychophysiological contest. Their mutual love of flying, forged as young men in wartime service, had brought them together in the most spectacular technological undertaking of the century.

When NASA officially introduced the Project Mercury astronauts to the world in April 1959, John Herschel Glenn Jr. was, at thirty-seven years of age, the oldest of the seven men selected. Born in Cambridge, Ohio, on 18 July 1921, Glenn grew up in nearby New Concord. An honor student in high school, Glenn excelled at everything. In sports, this included basketball, football, and tennis. He also performed in school plays and was a reporter for the school newspaper. At summer camp, he was a lifeguard, and at church the devout Presbyterian sang in the choir. He made two fateful determinations in high school. The first was to fly airplanes. Marrying pretty Anna ("Annie") Castor, his childhood sweetheart, was the second—although "childhood sweetheart" doesn't quite capture the depth of their history. Their parents belonged to a group dubbed the Twice Five Club. Each month one couple would host dinner for the other four—along with their babies. "They put us in a playpen together," he revealed in his autobiography, *John Glenn: A Memoir*, "and she was part of my life from the time of my first memory."

In 1939 Glenn enrolled at university and signed up for a civil pilot training program at New Philadelphia (Ohio) Airport. In March 1942, three months after the attack on Pearl Harbor, he was accepted into the U.S. Navy's elite V-5 program, which offered a college education along with officer and flight training. Glenn received his navy wings of gold a year later and promptly transferred his commission to the Marine Corps Reserve. A week later, on 6 April, John Glenn married Annie Castor in New Concord, Ohio. He had achieved his two principal goals. Next on his list was to survive the war—and to be the best marine pilot there was.

After a fleeting honeymoon and some advanced fighter training, Glenn shipped out to the Pacific theater, where he was soon promoted to first lieutenant and served in the Marshall Islands with Marine Fighter Squadron 155. Flying fifty-nine combat missions aboard F4U Corsair fighters on his first tour of duty, Glenn earned two Distinguished Flying Crosses and ten Air Medals. He was promoted to captain in July 1945 after returning stateside for service at Cherry Point, North Carolina, and Patuxent River. Two years later he was patrolling the Chinese coast with VMF-218, a Corsair squadron of Marine Aircraft Group 24, First Marine Aircraft Wing. His unit moved to the mid-Pacific island of Guam the following year.

By now wearing the oak leaves of a major, Glenn flew ninety wartime missions during the Korean War, in F-86 Sabres with the 25th Fighter-Interceptor Squadron. His combat missions over the notorious "MiG Alley," along the Yalu River, earned the marine his third and fourth Distinguished Flying Crosses and eight additional Air Medals. In the last nine days of fighting, he downed three MiGs, the fearsome Soviet-made jet fighter.

After Korea, Glenn won an appointment to the navy's Test Pilot School at Patuxent Naval Air Station, where he stayed for two years working on armament systems. He was then transferred to Washington DC to work as a project officer in the Fighter Design Branch of the Navy Department's Bureau of Aeronautics. His main work focused on further developing the F-8U-1 Crusader—a sleek, swept-wing fighter capable of sustained supersonic flight. On 16 July 1957, flying from Los Angeles to Floyd Bennett Field, New York, Glenn set a speed record in the Crusader (averaging 723 miles an hour)—the first nonstop supersonic transcontinental flight. It took him just over three hours and twenty-three minutes and earned him his fifth Distinguished Flying Cross. He made lieutenant colonel less than two years later—and also survived the Project Mercury selection process. Eight days later, he and six other military pilots were named Mercury astronauts.

Unlike Glenn, whose college career was interrupted by World War II, Carpenter was only sixteen and a junior in high school when Pearl Harbor was attacked and the United States entered the war. With the world war, a new round of heroes came to the attention of every boy—combat pilots flying in the navy or marines.

For young Carpenter, born in Boulder, Colorado, on 1 May 1925, it also brought a compelling new focus to his life, which had previously been dominated by dreams of Olympic glory in downhill skiing. Like Glenn, he was a standout athlete in high school, lettering in gymnastics and serving as president of the ski club. Carpenter also sang in the glee club, acted in school plays, and served as acolyte at the Episcopal church where he had been christened and confirmed. (Interestingly, both Glenn and Carpenter were known by identical nicknames—"Bud"—in their hometowns.) Carpenter eagerly made plans to enter the Navy Air Corps before he turned eighteen, journeying to San Francisco and Twelfth Naval Headquarters to qualify for the V-12a program—quite an achievement for a seventeen-year-old. Carpenter returned to his high school a minor celebrity. His friends had all entered the U.S. Army Air Corps; he alone among his classmates was a U.S. naval aviation cadet.

That fall, in September 1943, the eighteen-year-old began basic training in the V-12 a program at Colorado College. By January 1945 he had advanced to preflight at St. Mary's College, California. That summer he moved on to primary flight training in Ottumwa, Iowa. But combat flying, his goal since boyhood, would elude him. In August 1945, Japan capitulated shortly after U.S. B-29 bombers dropped atomic bombs on the cities of Hiroshima and Nagasaki. "The war was over before I learned to fly," Carpenter explains. "I only had eight hours in the Stearman N2S. So I returned to Boulder, reentered school at the University of Colorado, and nearly got myself killed in a car accident."

In 1949, after a long convalescence in which he married Boulder coed Rene Price, Carpenter rejoined the U.S. Navy. The Korean War began just as he was completing flight training at Pensacola Naval Air Station in Florida. Following a tour of duty flying P2Vs, a lumbering maritime patrol plane, and a string of outstanding fitness reports, Carpenter won an appointment to the U.S. Navy's Test Pilot School at Patuxent River. There he would finally fulfill his boyhood dream of flying the fastest, newest, and finest airplanes in the world. The credential would also qualify him for another, even more elite flying fraternity.

The United States Navy trains and educates aviators. But it also demands that they be well-rounded officers. There is schooling and ship duty. There are nonflying billets, during which aviators must maintain their flying hours. There are moves, with young families, that require among other things cross-country driving marathons at the navy's whim. Worst of all for these born-to-fly creatures, navy life sometimes means time away from their beloved airplanes.

In October 1957 the navy ordered Carpenter on one such cross-country trip. The thirty-two-year-old engineering test pilot had just completed a year's training in electronic intelligence (ELINT) at the navy's Postgraduate School in Monterey, California. Next stop was the navy's Air Intelligence School in Washington DC. The lieutenant could then expect a three-year tour of sea duty as an intelligence officer. Carpenter made use of his travel orders to do a drive-and-camp adventure with his sons, Scotty and Jay. Rene and their daughters, Kristen and Candace, had flown ahead to Boulder, where they planned to meet before he and the boys pressed on alone to Virginia.

One clear autumn night, while camped out beside the car in Nebraska, Carpenter caught sight of the upper stage of the Russian satellite *Sputnik* gliding serenely across the night sky. News of the satellite had been announced on the radio, so he was looking for it and pointed out the bright, history-changing dot of light to his sons, explaining how a brand-new age had just dawned. Pondering the starry sky, however, he had no inkling of how *Sputnik* would shape his future. He knew only that he was to report to the Air Intelligence School by 31 October.

The space race was on. The United States had to develop the capability to launch men into space—without it the country had no hope of maintaining either technological parity or its national security. So a civilian government agency, called NASA, was created. After a lot of to-and-fro among experts about what sort of candidate was best suited for spaceflight President Dwight D. Eisenhower intervened. Candidates, he decreed, would come from the ranks of jet-qualified military test pilots—men who had completed test-pilot schooling at Edwards (Air Force) or Patuxent (Navy). A NASA working group quickly established several other criteria, seven in all, specifying age, height, education, intelligence, temperament, and physical ability.

Five hundred and eight such men were eligible. A U.S. Army flight surgeon worked methodically through their military personnel files, eliminating hundreds of candidates for medical or other reasons. When he was done, 110 pilots were considered fully qualified according to the seven criteria. Phase one of the selection process was complete. These men (no women were military, let alone jet-qualified, test pilots in 1959) then advanced to phase two, beginning in February 1959 at the Pentagon.

By 9 February, 69 of the 110 candidates had reported to Washington for initial evaluation. Of these, 32 were said to have passed all the tests with "flying colors." Thirty-two candidates were more than enough for the selection needs of the working group; it scrapped plans to bring in the third group. After a week of top-secret briefings, interviews, and evaluations, tight-lipped candidates returned to duty to await news, they hoped, of their advancement to the next selection phase. For Carpenter, news of his advancement arrived via civilian channels—namely, the U.S. Postal Service. But he was aboard the *Hornet*, then undergoing sea trials. As he recalls in his memoirs, *For Spacious Skies*,

Rene immediately opened the envelope, marked special-delivery, registered mail, and basked in the historic language: "You have been chosen to proceed with further interviews and tests in connection with project Mercury. Please call —— in Washington DC by noon Monday, if you wish to continue."

Rene thought, I knew he'd make it! How incredible! She reread the happy letter in a rush until her eyes snagged on the phrase "by noon Monday." It was Tuesday afternoon! She screamed and dialed the telephone number. "We volunteer!" she told a startled Dr. Allen Gamble, whose office she had rung.

Years later, Carpenter smiled at the memory: "I gained a lot of print because of the lady who volunteered her husband for space flight."

By the end of March 1959, the NASA working group had winnowed the group of thirty-two candidates down to eighteen, ranked numerically, and then forwarded the names and service jackets to the Space Task Group Selection Committee. Three air force pilots, three naval aviators, and one marine colonel made the final cut. On 3 April NASA's new administrator, T. Keith Glennan, approved the recommendations. The task of notifying the seven victors fell to Charles J. Donlan, assistant director of the Space Task Group.

After learning from Rene that Carpenter was aboard the *Hornet*, Donlan placed a call to the ship. Someone handed Carpenter a slip of paper showing a five-digit telephone number. "Call that number," he was told. He rang from a payphone on the wharf and was soon speaking with Charlie Donlan, who gave him the triumphant news. Getting off the *Hornet*, however, would involve a less than triumphant encounter with his skipper, Capt. Marshall White. Carpenter explained:

I had already left the ship twice . . . and now I was leaving again while we were training for deployment. He was getting very angry . . . and it took the CNO on

the telephone to call White and say, "I know you need an air intelligence officer for your next deployment, but right now the nation needs yours . . . !"

Everyone else was walking aboard with their gear, and I was walking off. Finally Captain White, in reasonable and understandable impatience (remember, this is a trained, dedicated naval officer) said, "Would you mind telling me what this is all about?" Of course I was bound to secrecy by these orders, but his disappointment and frustration were evident. I said to this fine navy captain, "I am not supposed to tell you, but I am going to Washington to report to NASA, *and I am going to ride the nosecone of a rocket around the Earth three times."*

The skipper's reply? A single, skeptical, vivid "*Bullshit*!"

In the fall of 1961—in fact, on the fourth anniversary of *Sputnik*—the seven Mercury astronauts were locked in a meeting with Bob Gilruth, director of the Manned Spacecraft Center, and Walt Williams, head of Space Task Group operations. Earlier that year Alan Shepard had become the first American in space, aboard MR-3 (meaning the third launch of the Mercury Redstone); Gus Grissom followed two months later with MR-4. Both were ballistic flights, both were successes, and both led to the meeting that day during which Gilruth would name the prime-alternate teams for the first two orbital missions.

The news was grand for some. Glenn, backup for both the Mercury-Redstone flights, learned he would fly the MA-6 mission then scheduled for December. Carpenter was named John's backup. But there was also disappointment. Like all pilots, and astronauts, Carpenter had wanted a flight of his own. Instead, the follow-on mission, MA-7, went to Deke Slayton, with Wally Schirra named backup. Shepard and Grissom would have to wait for their second flights; Gordon Cooper had missed out altogether.

With the selection made, however, and the sharp rivalry for the moment resolved, the seven men reverted to another accustomed role: that of team player. In fact, one underreported aspect of Project Mercury was the largely successful effort of the seven astronauts to balance their intense individual rivalries with an equally intense need to work as members of a team. Strong bonds of camaraderie were forged in the process, as Carpenter confirmed: "Everybody wanted to take the next flight, whatever it was. We were all in

John Glenn and Scott Carpenter strike a pose for photographers at Cape Canaveral. Courtesy NASA.

a heated competition with each other. But the Three Musketeers came to my mind—we were all for one, and one for all. Truly we were. Although we struggled sometimes with petty difficulties, we were all one team where the program was concerned." Asked if he believed Gilruth weighed the seven's peer ratings—an evaluation tool often used in the military—in the selection process, Carpenter said: "Sure. Bob made the decision [and] he probably took the ratings into consideration—he never told us what his decision-making process was. But I remember saying, in my peer review, that John was my first choice as backup, and conversely that if I had to be anyone's backup then I'd prefer being John's. Of course, in the end, John got the flight. I have an idea he named me as his backup of choice. So John got *two* things he wanted, and I only got one of mine!"

Glenn also got to name his spacecraft in a family-centered process he described in *Life* magazine. Shepard's choice, *Freedom 7*, had struck the right note, Glenn thought. And he liked Grissom's patriotic-sounding *Liberty Bell*

7, which also suggested the shape of the Mercury capsule. Glenn wanted to involve his children, David and Lyn, so one night he sat them down with a thesaurus and a notebook. *Friendship* was their choice, sounding a warm note at a chilly moment in the Cold War.

As training for the first orbital mission began, Carpenter was determined to keep pace with the dogged marine from the start. But the effort took a toll, as he explained: "You know, John never quit. I could stay with him in all of the long hours in the Sim [simulator], training, studying, and everything else we had to do, but he would not relax. Maybe he couldn't relax. He was single-minded, and that's one of his virtues. What really wore me out, though, was his passion for running. I don't care that much for running—never have. There are other exercises that I'd prefer to just pounding the pavement. He did wear me out, though, because of his dedication."

Another motivator besides the astronauts' competition with each other was the Cold War competition with the Soviets. It "was a powerful driver" that energized *all* the astronauts, Carpenter recalled:

Both countries knew that preeminence in space was a condition of their national security. That conviction gave both countries a powerful incentive to strive and compete. The Soviets accomplished many important space firsts, and this gave us a great incentive to try harder.

The space program also accomplished another vital function in that it kept us out of a hot war. It gave us a way to compete technologically, compete as a matter of national will. It may even have prevented World War III, with all the conflict and fighting focused on getting to the moon first, instead of annihilating each other. There's no evidence of that, but as eyewitness to those events, I think that's what happened.

With the focus well and truly on Glenn during the long months of training and the several launch delays leading up to 21 February 1962, Carpenter was philosophical. "The fact that he got all the attention didn't bother me at all," the soft-spoken astronaut recalled forty-one years later. The two men had developed a deep friendship and durable working relationship. "That lasted throughout the program," he added. "The five others often shifted with the issues, but ours did not. We always seemed to agree on the issues. And our friendship, and the trust, was unshakeable."

Nevertheless, waiting for MA-6, the country's first orbital mission, was a trial. The launch, initially scheduled for 20 December 1961, was plagued by technical and weather delays. On 15 February NASA scrubbed yet another attempted launch, the ninth, citing vicious storms lashing the Atlantic, where Glenn would splash down at the end of his three-orbit flight. Two weeks earlier, Glenn had spent five hours and thirteen minutes waiting atop an Atlas rocket booster loaded with close to 250,000 pounds of liquid oxygen and kerosene. That launch too was scrubbed. A tenth attempt was set for the next morning in the now-daily effort to exploit any break in the weather.

Glenn remained calm throughout the fifty-seven days of delays. In a meeting with reporters after the 15 February scrub, Lt. Robert B. Voas reported that the astronaut was still "taking it very well." The newsmen were not, however, and in fact were increasingly vexed by the mounting delay. Their repeated question—how was Glenn handling the pressure?—brought the repeated answer: "There is no evidence," Dr. Voas responded, "that he is building up any frustrations or annoyance." Although he conceded that Glenn was "anxious to get in three full orbits," Voas added that he "would prefer to wait rather than not have everything right for the full plan."

The next launch attempt was set for 20 February. Two days before then, a group of reporters stumbled across the forty-year-old astronaut relaxing in a barbershop. Fielding the customary question, he replied simply: "I have been training and waiting for three years, and a few more days won't matter."

Years later, Carpenter agreed. "The delays didn't bother either of us much. It gave us both more time to get ready. And you figure it's just part of the job—if something fails, you wait for it to be fixed and try again."

Gene Kranz, then assistant flight director, explained their patience:

We had no orbital experience. We had about thirty minutes of experience with Alan Shepard and Gus Grissom's missions. The preceding orbital, unmanned mission was targeted for three orbits but actually completed only two. There was risk with that, but the real question with me was the integrity of the Atlas booster. Two of the previous five missions had blown up or had to be destroyed. When we were flying John Glenn, we did not know if we were going to have the third bad rocket, or the fourth good one! It was 60/40, if you look at the odds. Unheard of—in American society at least.

Before retiring on the evening of 19 February, Glenn phoned Annie at home and also spoke with his two teen-aged children. Everything was fine, he reported; he was looking forward to the launch attempt next morning. By the time he climbed into his double-bunk bed in crew quarters at the Cape, the weathermen were being coy. It seemed a launch the next morning was only a fifty-fifty proposition. Conditions in the recovery area were nearly perfect, but at the Cape they were marginal at best, with overcast skies.

At 1:30 a.m. Glenn woke without assistance, lying quietly in the dark until team physician Dr. Bill Douglas arrived at 2:00 a.m. to rouse him. Carpenter, awake since midnight, had already reported in from the gantry's astronaut insertion area: *Friendship 7* was checked and ready. Glenn rose, shaved, and showered. Then, clad in a white bathrobe, the astronaut sat down for his prescribed low-residue breakfast—steak, scrambled eggs, jellied toast, orange juice, and Postum, a decaffeinated coffee substitute. After a brief medical exam, he stood patiently while technicians attached biomedical sensors to his body. These were then checked to make sure they were telemetering correctly. Then, with the aid of suit technician Joe Schmitt, Glenn pulled on his spacesuit—a procedure they had practiced many, many times before. Nothing seemed out of the ordinary that morning, he would later report.

Launch crews confirmed to the Mercury Control Center (MCC) that they were on schedule, and Glenn boarded the transfer van for yet another ride to Launch Complex 14. Through a van window, he took in the incredible sight. Framed by the night sky, huge arc lights illuminated the gleaming Atlas D booster, poised beside the orange 145-foot gantry. By itself, the Atlas was 65 feet tall, but with a 2.5-foot adapter collar, the 9.5-foot spacecraft, and a 16-foot launch escape tower perched on top, the entire assemblage stood at 93 imposing feet. A remarkable machine, the Atlas D was basically a stainless steel balloon, with skin thinner than a dime. Unable to bear its own weight, the rocket relied on pneumatics to stay erect. Once loaded with supercold liquid oxygen, however, the rocket grew stronger and more rigid.

Before Glenn could disembark from the van, a hold in the countdown was announced, first because technicians had to replace a radar-tracking device in the Atlas booster's guidance system and then when a faulty respiration sensor in Glenn's helmet was discovered. Finally, at 5:59 a.m., some encouraging news was announced: with the sensor replaced, Mercury Control had resumed the countdown. Cleared to leave the van, Glenn entered a rickety elevator at the foot of the gantry and rode up to the eleventh deck. There at the end of the crew access arm stood a canvas-clad, environmentally controlled chamber known as the "White Room," where Guenter Wendt and his team would make final checks. The scene was one of orderly tumult when Glenn finally arrived because the weather forecasters were suggesting that a break in the cloud cover might occur by midmorning. "Everyone seemed to sense," Glenn later wrote, "that we were going for real this time."

Glenn noticed his friend and alternate standing off to one side and approached him with some last-minute words. "I know something about being a backup pilot," he later wrote in *We Seven*,

having done the job twice myself. It is hard work and the personal satisfaction is limited to helping someone else. Scott had pitched in from the beginning as if this were going to be his own flight. He took an enormous load of detail off my mind and left me free to concentrate on areas where I thought my attention was most needed. All this time he had to keep himself in shape and trained to a fine edge so that he would be ready if I was unable to go at the last minute. But Scott is much more to me than a willing colleague, and we both felt emotions we didn't express when we shook hands and he wished me good luck.

Glenn squeezed into his spacecraft. Technicians hooked him up to the environmental control system and then closed and bolted the hatch. He settled into quiet, watchful waiting. Vigilance, among other things, had helped the U.S. Marine survive aerial combat in two wars. His pioneering spaceflight would require these same skills, and more. He systematically ran through his checks, looking for any fault, any potential problems. Below him, the temperamental Atlas rocket creaked and fumed. He was ready for all these sounds from "a squat, ugly brute"—Tom Wolfe's unforgettable description of the adapted military missile Glenn would ride into space.

At 7:25 a.m. there was another delay. A hatch bolt had broken during the initial closure and would have to be replaced. Once the hatch had been repaired and closed again, the countdown resumed. Just after 8:00 a.m., claxons began to sound outside the White Room, warning personnel to leave the gantry area. Carpenter made his way to the blockhouse, where he spoke with Glenn by radio and, by prearrangement, patched a call through to Annie. The couple spoke as the rust-red gantry was rolled back. "The initial unusual experience of the mission is that of being on top of the Atlas launch vehicle after the gantry has been pulled back," Glenn said during his later pilot's flight report at the Manned Spacecraft Center: "Through the periscope, much of Cape Canaveral can be seen. If you move back and forth in the couch, you can feel the entire vehicle moving very slightly. When the engines are gimbaled, you can feel the vibration. When the tank is filled with liquid oxygen, the spacecraft vibrates and shudders as the metal skin flexes. Through the window and periscope the white plume of the lox [liquid oxygen] venting is visible."

At 8:57 a.m., Mercury Control announced that all systems were in a ready, or "go," condition. The weather was nominal for launch. On nearby beaches tens of thousands of spectators gathered, as reports of an imminent launch were broadcast over tiny portable radios. At homes across the country, black-and-white images of the waiting rocket shimmered on television sets. In New York's cavernous Grand Central Station, commuters stood riveted by huge television screens showing live images from Cape Canaveral.

Glenn heard ground controllers report in.

"Communications?"

"*Go!*"

"ASCS? [Automatic Stabilization and Control System]"

"*Go!*"

"Aeromed?"

"*Go!*"

Voices tumbled over each other as the clock ticked down. It sounded chaotic, but was in fact well rehearsed and orderly.

"*Minus forty!*"

"Status check: pressurization?"

"*Go!*"

"LOX tanking?"

"I have a blinking, high-level light."

"You are *Go!*"

"Range operations?"

"*All are clear to launch!*"

"Mercury capsule?"

Glenn's transmission was clear and calm: "*Go!*"

"All prestart panel lights are correct. The ready light is on. Eject Mercury umbilical. All evacuate!"

"*Mercury umbilical clear.*"

"All recorders to fast: T minus eighteen seconds and counting. Engine start!"

"May the wee ones be with you, Thomas," came the quiet voice of General Dynamics' astronautics base manager, Byron MacNabb, seated in Mercury Control.

"The good Lord ride all the way," responded test conductor Tom O'Malley, also from General Dynamics. He was the person who would press the black button to send the mighty Atlas on its way.

Stationed in the blockhouse, just yards from Pad 14, Carpenter intoned: "*Godspeed, John Glenn.*" It was a spontaneous remark, he later explained: "The two previous Mercury flights were powered by the Redstone, a tiny rocket that couldn't provide enough power to give John the speed he required for orbital flight. What he needed, and what everyone hoped the Atlas would provide, was speed. I had not pondered this, just as Neil Armstrong maintains he'd never pondered the phrase, 'One small step for man, one giant leap for mankind.' It just popped out of thin air. He needed speed, his name was John Glenn, and it was sort of a salute to a friend, and a plea to the higher power. *Godspeed.*"

Glenn heard none of the prayerful transmissions—only Shepard continuing the steady countdown:

" . . . *eight, seven, six, five, four, three, two, one, zero!*"

"*Ignition!*"

"*Liftoff!*"

At 9:47 a.m., two small stabilizing rockets at the foot of the frost-covered Atlas shrieked into life. Then the three main engines ignited. Dense clouds of smoke billowed from the launch pad, and for a heart-stopping moment nothing seemed to happen. Amid this conflagration the Atlas, still restrained, gathered its strength until the explosive clamps fell away. Ponderously at first, then quickly gathering momentum, the rocket surged higher and higher. Beneath it smoldered a blackened, smoking launch pad.

Crackling, man-made thunder swept across the Cape. Cheering onlookers watched the Atlas rise, carrying *Friendship 7* and Glenn into space and history. He had always imagined that it would seem slow, even smooth, like an elevator rising. But as he explained in his postflight pilot's report, it was not like that at all. "When the countdown reached zero, I could feel the engines start. The spacecraft shook, not violently but very solidly. There was no doubt when lift-off occurred. When the Atlas was released there was an immediate gentle surge that let you know that you were on your way."

Immediately after liftoff, Alan Shepard began checking Glenn's progress over the radio. The launch was the first of several hurdles the astronaut had to overcome in order to achieve orbit, and as he said, "it was a big one." His Atlas booster had to perform flawlessly. The planned roll to the correct azimuth took place, and he could see the ground moving away beneath him. He also glanced up through his window and could see the horizon turning. There had been some vibration after liftoff, but this, he said in his report, "smoothed out after about ten to fifteen seconds of flight but never completely stopped. There was still a noticeable amount of vibration that continued up to the time the spacecraft passed through the maximum aerodynamic pressure or maximum q at approximately T+1 minute. The approach of maximum q is signaled by more intense vibration. Force on the outside of the spacecraft was calculated at about 982 pounds per square foot at this time."

One minute after launch, controllers relayed to a worldwide audience that Glenn had reported that all systems were "go," and his flight was proceeding smoothly. Following the successful launch, which they had watched on TV monitors, most of the blockhouse crew had no further duties to perform. For his part, Carpenter called it a day: "There was nothing for me to do," he recalls. "The blockhouse was totally out of the loop, and I was totally out of the loop, so I went to the *Life* house—a kind of retreat for the astronauts—where I watched the flight on television." The astronaut had sometimes repaired to the magazine's private beach house in the months leading up to Glenn's flight because he explained, "These few quiet moments, alone with the setting sun, a scotch and ice, and complete quiet and isolation were the most relaxing times I have spent since I was in the Australian outback."

Glenn's flight, scheduled for three orbits, proceeded smoothly despite some stabilization and control system problems. This was not a serious difficulty, as the pilot could disable the autopilot and assume manual control. But on Glenn's first circumnavigation of the Earth, a warning light flashed in Mercury Control. While routinely scanning the bank of meters in front of him, technician Bill Saunders noticed meter number 51 flashing. Immediately, he informed flight director Chris Kraft and Cape CapCom Alan Shepard:

"I've got a valid signal on segment 51!" Segment 51 was a reference to *Friendship 7*'s heat shield; the telemetry showed it had come loose—a dangerous development. With a blunt-end reentry, the Mercury capsule was designed to withstand temperatures of up to four thousand degrees Fahrenheit, thanks to an ablative fiberglass heat shield. Without this heat shield, a returning astronaut would be incinerated. It was as simple as that. As a precaution, Mercury Control decided that the retropack and its restraint straps would remain in place to help hold the heat shield in place during reentry. They also elected not to tell Glenn about the suspected malfunction.

But as retrofire approached, the astronaut sensed trouble. Hawaii CapCom asked him to check the landing bag indicator light, which might confirm a loose heat shield. Glenn obliged, flipping the switch on and off. No light, he reported, and returned to his retro-sequence checklist. Although Glenn's transmission provided some reassurance at the Cape, the problem became how to tell the famously self-possessed marine *what* to do with his spacecraft without telling him *why* he had to do it. Seventeen minutes before splashdown, they decided to frame their decision as a recommendation. "We are recommending," ground control said; "that you leave the retropackage on throughout the entire reentry."

"This is *Friendship Seven*," Glenn replied, "What is the reason for this? Do you have any reason? Over."

"Not at this time," came the implacable response. "This is the judgment of Cape Flight. . . . Cape Flight will give you the reasons for this action when you are in view." Thirty seconds later the astronaut was in range. Shepard relayed instructions for retracting the periscope. "While you are doing that," Shepard added, with a forced nonchalance, "we are not sure whether or not your landing bag has deployed." He added, by way of explanation: "We feel it is possible to reenter with the retro-package on. We see no difficulty at this time in that type of reentry."

"Roger, understand," Glenn replied, knowing he could save any criticism for the debriefing.

The retrorockets fired, with Glenn reporting back to Hawaii that he felt "a big boot." Ten minutes later *Friendship 7* was over Florida, hurtling toward a landing in the cold waters of the Atlantic. Glenn would later state in his pilot's report that as he watched, incandescent pieces of the retropack flew past his window:

I could feel something let go on the blunt end of the capsule behind me. There was a considerable thump. I saw one of the three metal straps that hold the retropack in place start flapping around loose in front of the window. This was not abnormal. The loose strap burned off at this point and dropped away. Right away I could see flaming chunks go flying by the window. Some of them were as big as six to eight inches across. I could hear them bump against the capsule behind me before they took off, and I thought that the heat shield might be tearing apart. This was a bad moment. But I knew that if that were really happening, it would all be over shortly and there was nothing I could do about it. So I kept on with what I'd been doing—trying to keep the capsule under control—and sweated it out.

As he "sweated it out," Glenn was also enduring a communications blackout. This occurs when the atmosphere becomes so hot from friction against the returning spacecraft that it becomes electrically charged, or ionized, which prevents all transmissions. Glenn could make voice reports, but only to a tape recorder. Ground control could hear nothing. As the heat outside gradually faded, Glenn felt the g-forces increase to around eight times the standard gravitational load—all entirely normal. With voice communications restored, Glenn told an anxious Shepard: "My condition is good, but that was a real fireball, boy!"

Twelve minutes later, floating beneath a huge red-and-white parachute, *Friendship 7* splashed down only six miles from the recovery ship, USS *Noa*. Mindful of Grissom's near-disaster just months before, Glenn elected to remain inside his spacecraft. The spacecraft was hoisted aboard the navy destroyer, where Glenn blew the hatch. "It was hot in there," he declared as he stepped onto the *Noa*'s deck. A welcome glass of iced tea was soon in his hands. Postflight inspection disclosed to technicians the cause of the erroneous Segment 51 signal—an improperly rigged switch. The heat shield, and the astronaut, had never been in jeopardy.

The U.S. Navy ferried Glenn to the Grand Turk airstrip aboard an S2F patrol plane, arriving about 9:30 p.m. Seven hundred miles southeast of Cape

Canaveral, Grand Turk is a tiny British island at the southeastern end of the Bahamian archipelago. It was also occasional home to Project Mercury debriefers, a missile tracking station, and an auxiliary U.S. Air Force base. The sun-drenched, seven-mile-long island was, in short, an ideal setting for debriefing America's first person to orbit the Earth.

A small crowd was there to greet Glenn, including Deke Slayton, scheduled to fly MA-7, the follow-on mission, and Scott Carpenter. After shaking Slayton's hand and sustaining innumerable backslaps, Glenn embraced Carpenter, both men exulting in the extraordinary success of the mission. A cluster of pool reporters fired questions at Glenn. "Fine, wonderful, I couldn't feel better," he replied to an obvious question. Another reporter asked for a comment: "Well, it's been a long day," he offered, "and an interesting one too, I might add."

After enjoying a steak dinner, the astronaut was taken to a nearby hospital specially built for the historic occasion; postflight medical exams and debriefings began there. Later, after a weary Glenn had been allowed to retire, Mercury team physician Bill Douglas told the reporters what they'd been waiting all day to hear: "John is in excellent condition."

The Turk and Caicos islands boast beautifully preserved reefs—perfect for the scuba diving the astronauts had learned to enjoy as part of their physical training. On Wednesday, 22 February, John and Scott grabbed some recreation time to go spearfishing from a couple of glass-bottom boats. Along for the dive on Carpenter's boat were Dr. Douglas and Wes Vickery, a technician attached to the base's Timing Section. Once at anchor, the boat's crew dropped a measured descending line over the side to help them gauge the water's depth in their dive and ascent. Then Vickery asked for Carpenter's help in setting a personal breath-hold dive record, and the astronaut agreed. "He wanted to see if he could get to a hundred feet and go back up," Carpenter recalled. "I was there as his safety diver."

Carpenter turned on the air in his tanks, adjusted his mask, and slipped over the side. Beneath him coral reefs gleamed among spits of white Bahamian sand. He was soon floating near the hundred-foot mark and looking up at the glass-bottom boat. Vickery did some deep-breathing exercises, slid over the side, and began treading water. Then, pulling in a last gulp of air, he plunged beneath the surface. Holding his post below, Carpenter watched from far beneath the boat's shimmering hull. "I saw Vickery coming down the line. Then, at about the fifty- or sixty-foot mark, he just stopped moving. So I swam up." Finding Vickery unconscious, Carpenter tried placing his tank's mouthpiece into Vickery's mouth, but there was no response. With precious seconds ticking by, he quickly took hold of the inert diver and hauled him to the surface. Bill Douglas and the skipper reacted quickly when they heard Carpenter's cries for help, grasping the unconscious man and dragging him into the boat.

Vickery, blue around the lips, was not breathing, but Douglas found a pulse and immediately began resuscitation. The diver began responding within seconds. "He came around, and he was okay," Carpenter later said of the incident, still indelibly clear in his mind. He then recalled with quiet satisfaction: "Only time I know of that I saved somebody's life!"

Universal lionization being what it is and Glenn being the man of the moment, newspapers were soon trumpeting the news: "John Glenn Saves Diver's Life!" The astronaut was in the vicinity, of course, but Carpenter was the real hero that day. Given Glenn's status, however, no one thought to write the headlines any other way.

Later that same day the USS *Randolph* and two smaller navy ships positioned themselves off Grand Turk. Sailors off-loaded *Friendship 7* at the island's dock and brought it ashore. Once secured, the spacecraft was transported to the airstrip, where workers reverently loaded it onto a military aircraft for conveyance to Patrick Air Force Base in Florida. Joining Glenn, Carpenter, and Slayton that day were fellow astronauts Alan Shepard, Gus Grissom, and Wally Schirra, all in high spirits after the historic mission. Gordon Cooper was still half a world away in Muchea, busily packing after his tracking duties as station CapCom in the Australian outback.

That evening, most of Grand Turk's citizenry, a couple of hundred people, celebrated Glenn's flight at the beachside Conch Club. Amid the congratulations, the shouting, the drinking, and the happiness, Grand Turk islanders demanded their due—they wanted to meet the free world's new hero, to shake his hand and get his autograph. He willingly obliged. Carpenter too was swept up in the festivities that balmy evening. But he was also quietly celebrating some good news from Bill Douglas. A chagrined Wes Vickery was resting comfortably in the hospital and would make a full recovery.

At a press conference held the day after Glenn's flight, a jubilant Brainerd Holmes, director of the Office of Manned Space Flight, repeated that Slayton would fly the next Mercury mission. Reporters wanted to know when this would happen. "As soon as we feel we are ready to go," Holmes replied, adding: "We will now fly more extensive and complicated missions. Present plans are for four more flights this year, identical to Glenn's, with one of eighteen orbits at the end of the year." There would, in fact, be just two more missions that year: one of three orbits, the other, six.

NASA's deputy administrator, Dr. Hugh L. Dryden, responded with a scientist's caution: "This is just the beginning—the first step," he told reporters. Dryden had been with the agency since its inception in 1958. He had overseen its momentous transformation from the National Advisory Committee for Aeronautics (NACA) into a young space agency. He also had a strong sense of history. "A few years from now," he explained, "we will look back on the Mercury capsule and think of it as we do now of the Wright Brothers' airplane."

During postflight briefings for MA-6, Glenn weighed in on his flight and, among other things, voice communication procedures. Astronauts should be informed, he said pointedly, about every possible malfunction affecting their spacecraft. Years later, Gene Kranz, then assistant flight director, agreed: "We didn't know what we were doing."

We were still in the process of trying to sort out the relationship between ground team and the crew. Until we knew we had a problem, and decided what we were going to do about it, was it worthwhile to distract the pilot? In retrospect, I believe that was a bad call. It is my belief as a fellow pilot that you want to keep the guy onboard as fully knowledgeable of everything that is going on, even if he can't help you. Later on, anytime we suspected something was going wrong, we'd advise the crew; it turned into a much more free and open exchange. But we were just utterly naïve in the early days of Mercury. We were still carrying over some of the traditional ground role separate from the flight role we used in [airplane] flight test.

Back home, the jubilation awaiting Glenn was overwhelming—national joy not seen since 1927, when America's "Lone Eagle," Charles Lindbergh, returned to the States after his solo flight across the Atlantic. Senators wept openly during Glenn's address to the U.S. Congress. A whistle-stop tour of several major U.S. cities culminated in a tumultuous parade through the packed streets of New York on 1 March 1962. In a powerful display of emerging "Astro Power," Glenn insisted on the presence of his six Project Mercury colleagues and their wives, prevailing over the objections of NASA administrator James Webb.

Photographs from the day show a storm of ticker tape—detritus from a Wall Street that still used stock price ticker machines—obscuring a convoy of VIP limousines carrying astronauts and other dignitaries. The New York City sanitation department, responsible for cleaning up after the great street party, estimated that New Yorkers had tossed some thirty-five hundred tons of paper out of Wall Street windows. The emotional force of these and other displays caught nearly everyone by surprise. Shepard and Grissom had been feted well enough after their flights, but that was nothing compared with the hoopla now attending Glenn's achievement. Slightly bewildered, they pondered their new status as celebrity-astronaut Cold Warriors. The prospect of such relentless adoration was enough to give anyone a shuddering dose of cold sweats. Next in line for spaceflight, Slayton could only shake his head. He would later recall: "It was a madhouse."

When NASA first announced that Glenn would pilot the first orbital mission, it also named Slayton for the follow-on mission, MA-7. But in mid-March 1962, barely three weeks after Glenn's flight, the gruff air force pilot received devastating news. He was grounded—scratched from the roster. It was a complex story. Slayton had been quietly diagnosed, back in 1959, with idiopathic atrial fibrillation, a rare cardiac disorder. On the ground, he regarded his primary symptom—an occasional racing heartbeat—as annoying but nothing worse. He adjusted his diet, quit smoking and drinking, and increased his running schedule. Slayton's six colleagues had known about his condition ever since Bill Douglas pulled Deke off centrifuge training after hearing the telltale racing pulse through his stethoscope. It was never discussed—among military pilots, ill health is not a topic of conversation.

Etiquette wasn't Jim Webb's concern, however. He was worried about the wider operational implications of subpar astronaut health and about the cardiac stresses caused by a high-g launch and reentry. But mostly he was worried about an astronaut losing consciousness, or dying, during spaceflight. He also had a letter from an air force cardiologist. Unknown to Slayton, the physician had written to Webb in 1959 recommending that the air force pilot *not* be given a flight. After a lot of to-and-fro within NASA, Webb asked three "nationally eminent" cardiac specialists to render an opinion. "If NASA had an available astronaut who did not 'fibrillate,'" they wrote in a consensus recommendation, "then he should be used rather than Slayton." Worse, NASA then ordered the distraught astronaut to preside over a curt press conference announcing his grounding. He may have appeared stoic in public, but in private Slayton was in a white-hot rage.

The ensuing media glare focused first on Slayton and his physical issues, then latched onto Carpenter. Acting on the advice of Walt Williams, head of the Space Task Group, Bob Gilruth had named Glenn's backup, Carpenter—rather than Slayton's backup, Wally Schirra—to pilot MA-7. Schirra argued, with justification, that as Slayton's alternate *he* should have gotten the flight. No, Gilruth explained, he would remain backup. Serving as Glenn's backup for more than five months, Gilruth reasoned, Carpenter had logged more training hours for an orbital mission than either Slayton or Schirra. As it turned out, the decision to ground Slayton gave Schirra an assignment to MA-8, planned as a six-orbit mission later that year, with Gordon Cooper as his backup.

Although a gratifying endorsement of his hard work, the news caught Carpenter by surprise. He had just eleven weeks to train—and a spacecraft to name. Carpenter settled on *Aurora*. At the time he explained, "I think of Project Mercury and the open manner in which we are conducting it for the benefit of all as a light in the sky. Aurora also means *dawn*—in this case the dawn of a new age." Hometown reporters later made the connection with Carpenter's boyhood home, a stately brick house on the corner of Aurora and Seventh Streets in Boulder, Colorado. It is a link with which his daughter and biographer Kris Stoever is very familiar:

The childhood connection, for my father, with Aurora *and* Seven *is indelible and indisputable. Even more indelible for my father was the sight of celestial or other mysterious visual phenomena—he loves the numinous. It fires his considerable imagination. As it happens, this love, in addition to his twenty-ten eyesight and scientific temperament, suited him ideally for observation-rich duties required in particular of the flight he was given. He appreciates such things, and ponders them, more than I can possibly convey in words.*

Also, recall that his flights near the North Pole, during the Korean War, brought him time and time again in view of the aurora borealis. He wrote home about them to my mother. Remember, too, that many of the early flights and rockets and programs were taking their names from classical or mythological themes or persons. Aurora was the goddess of the dawn, and Dad said at the time that he felt we were at the dawn of a new age, space exploration, and Aurora, *the name, was one way to declare that belief. Interesting, too, that among the manly names being chosen and applied—Atlas, Mercury, Thor—my father wasn't afraid of honoring a gentle, light-bringing female deity.*

Brainerd Holmes had been right: the Mercury missions were growing more complicated. They had to, since President Kennedy had committed NASA to landing men on the moon and (the hard part) bringing them safely home to Earth. There were engineering, physics, and navigational mysteries to resolve. Could liquid fuel be pumped in a zero-g environment? Could humans distinguish colors in space? If so, what spacecraft colors were best for the rendezvous and docking missions then being planned? What of turbulence in the slipstream, the airglow layer, weather patterns, photometer readings of stars? There were also photographs of celestial and weather phenomena to take.

So the MA-7 mission planners chose five experiments for the five-hour flight. Slayton had complained rather loudly about the science mission when MA-7 was his to fly, but Carpenter was different. As the son of a research scientist, he appreciated the merits of experiments, particularly on the new frontier of space. But there was more: with Carpenter's extensive training as a patrol plane pilot, whose primary job, aside from flying the plane, is to make external observations, MA-7's observational duties were for him second nature. He was, in fact, the perfect astronaut for a pioneering science mission.

Carpenter and Schirra put the politics of crew selection to one side. They were professionals. But the close working relationship Carpenter had enjoyed with Glenn would not be replicated, as he later explained:

It was a question of friendship and priorities: Wally was a valued colleague, but he and I didn't have anything near the friendship John and I had. Wally also had a competing interest at the Cape. He'd already been assigned to the next flight, MA-8, and he had his own preparation to do for his own flight while he was backing me up. It was different when I was John's alternate. All that time training for MA-6 was focused entirely on MA-6. I had no competing interests. No flight waiting for me (I thought) after John flew. But Wally had a lot of his own stuff to do, as well as doing stuff for me.

The final weeks of training had gone well; Carpenter felt calm, confident, and ready to carry out his mission to the full. He had some lingering concerns, however, in addition to some sobering setbacks. On 9 April an Atlas F rocket exploded in a massive fireball a few feet off the launch pad—a mishap that delayed the flight while everyone pitched in to troubleshoot. Had Carpenter discussed the implications of the explosion with Rene and their four children? "Certainly," he explained. "You look at all the data and figure out what went wrong and you fix it and press on. But it didn't affect my resolve one bit. Nor would it have affected anybody else's. Did I discuss it with my family? I'm sure. They're practiced at all of this too; they understand what's going on as well as I do. I mean you have failures and you fix them and you keep going, and you just don't pay attention to it anymore. It happened, but it's going to be fixed."

Originally scheduled for an April launch, MA-7, like its predecessor, endured its own share of scrubs. But Carpenter welcomed the delays; they gave him precious extra training with the flight equipment and time to resolve other matters that concerned him. As the next launch date was set, the pilot of *Aurora 7* was ready. Slayton flew to Muchea, a world away, for CapCom duties.

Early on 24 May 1962, Dr. Howard Minners gave Carpenter a gentle shake. The astronaut, immediately alert, remarked that he had slept soundly. Thirty minutes later he was having breakfast with Glenn and Drs. Douglas and Minners, who gave the astronaut his final preflight medical check and declared him in perfect health.

Having raced through his preflight checks, fitting of body sensors, and spacesuit donning, Carpenter emerged from Hangar S and entered the transfer van a little after 3:40 a.m. Glenn had not left the hangar until after 5:00 a.m. for his flight. For all his alacrity, however, Carpenter had to cool his heels for nearly an hour after the four-mile journey to Pad 14:

It was dark when I rode the elevator up to the gantry to be inserted into the capsule. Wally was waiting to greet me after having checked out the capsule through the preliminary stage of the count. "It's ready, Scott," he said. "It's all yours."

Through the small window above me, I watched the sky lighten. At T-6 *minutes, visibility was still impaired by the usual morning haze, now thickened by smoke from distant swamp fires. We knew the sun would soon burn away the haze, and Walter Williams, Project Mercury's operations director, decided that with so many things going right it would be best to wait a while. Consecutively, he ordered three fifteen-minute holds. I used the time to put a telephone call through to my wife, Rene.*

The ground haze quickly dispersed as the sun rose, and the belt of dark scrubland smoke was soon swept away as winds swung to the east. Launch was set for 7:45 a.m. Carpenter recalled the moment:

The feeling, when you hear the countdown approach zero, and you realize you are really going to launch, is one of relief. Finally, you're not engaged in make-believe. This time it's real, and you can't wait—you think this in your head—you can't wait *to get started.*

I am frequently asked if I was afraid. Yes, I was afraid. But fear is neither shameful nor to be feared, it is to be respected; it does a lot of good things for you when things get dicey. It improves your vision, extends your peripheral vision, reduces reaction time, makes you stronger, and improves your endurance. It is a very valuable cockpit companion. What people, I think, are chary about here is that fear is the same as panic. But panic is uncontrolled *fear—and that is deadly. But if you* use *fear, and control it, it can save your life. It can be a pistol!*

It was a flawless launch. Right on schedule, two brilliant arrows of fire discharged from the flanks of the Atlas rocket, piercing the air with a shrill, waspish scream. Then the main engines thundered into life. Eighty feet above, strapped tightly into his contoured couch, Carpenter could hear the swelling rumble of ignition. He felt Atlas 107-D respond, shivering into life. As part of the launch procedure, he was holding the abort handle (known in the astronaut lexicon as the "chicken switch") in his left hand. In his right was the control stick. In the event of a launch catastrophe, he could yank the abort handle, and in an instant the seventeen-foot escape tower strapped to the top of *Aurora 7*, loaded with 285 pounds of solid fuel, would ignite and blast the spacecraft clear of the booster. A parachute would then deploy, bringing the spacecraft back to a landing well away from the conflagration.

But that day all went well. Steel shackles fell away, and the Atlas was unleashed. Carpenter, feeling the sudden surge of acceleration, calmly reported "I feel the liftoff. The clock has started." Grissom, watching the ascent on a TV monitor, acknowledged the transmission from his post as Cape CapCom. At around thirty thousand feet, accelerating through the zone of maximum vibration, Carpenter could feel the Atlas shaking but not as much as he had anticipated. Maximum acceleration during his launch translated to just over eight g's. Within a minute and a half he had passed through most of the atmosphere and was watching the blue sky become black space. The booster engine cut off right on time, and unneeded propulsion hardware was jettisoned, while the sustainer and vernier engines continued burning. Twenty seconds later, the emergency escape tower was also jettisoned. "My status is good," Carpenter reported, as he watched the tower recede toward the horizon.

Three hundred seconds after launch, the sustainer engine shut down. Explosive bolts detonated, clamp rings and metal connections holding *Aurora 7* to the booster were severed, and three solid-rocket posigrade thrusters at the capsule's blunt end fired briefly. A one-second burst propelled *Aurora 7* away from the inert, hollow Atlas rocket at twenty-four feet per second.

Carpenter's rocket had found the "keyhole"—an exact point in space five hundred miles from the launch pad that would place *Aurora 7* into orbit at precisely the desired angle. Carpenter was traveling at close to 17,500 miles an hour, and would later report that those had been the fastest five minutes of his life.

His first maneuver was to turn his spacecraft around so its blunt end was heading into the line of flight and he was facing backward. Against the spectacular backdrop of the Atlantic, Carpenter could see the spent Atlas booster tumbling slowly. Then Grissom gave him excellent news: "We have a Go, with a seven-orbit capacity." The flight could continue. "Roger," an exultant Carpenter replied. "Sweet words!" He summed up his initial reaction to space flight in *We Seven*: "The first thing that impressed me when I got into orbit was the absolute silence. One reason for this, I suppose, was that the noisy booster had just separated and fallen away, leaving me suddenly on my own. But it was also a result, I think, of the sensation of floating that I experienced as soon as I became weightless."

Carpenter could no longer feel the pressure of his body against the couch, and his pressure suit, which on the ground was noticeably constricting and uncomfortable, had suddenly become entirely comfortable. As he later recalled, part of his routine was to report these sensations to the ground: "It was such an exhilarating feeling . . . that my report was a spontaneous and joyful exclamation: 'I am weightless!' Now the supreme experience of my life had really begun." As Africa loomed beneath him, barely minutes into the flight, Carpenter reported among other things that his pitch altitude "did not agree" with his onboard navigational readings. Then he turned to his duties—among them, a hefty load of experiments. So busy an activity schedule would not be duplicated until well into the Gemini program.

In spite of the workload, Carpenter recognized the splendor outside his cockpit window through his periscope. He conveyed vivid, information-packed descriptions to the ground—and to the voice recorder when he was out of range. While shooting across the Indian Ocean for the first time, he witnessed the setting sun—the first of three sunrises and three sunsets he would see from his exclusive perch in the heavens. Carpenter would later report that they were the most awesome phenomena of his flight, beggaring his powers of description. As the sun dipped below the horizon, he saw a lingering, intensely colored glow. The rare sunset from space might be captured on film—but only a spacefarer could glimpse the thing itself.

Glenn had described the sights from space to him, so Carpenter was prepared. But he expected to see stars in greater numbers than ever before. Instead, it was more like the night sky on a clear desert evening, with one difference: with no interceding atmosphere, the star shine was constant and unblinking—and the stars themselves, he saw, slipped below the horizon at nearly eighteen times normal speed.

During the first orbit, Carpenter released a tethered thirty-inch Mylar balloon painted five different colors. The multicolored balloon bumped gaily along after *Aurora 7* at the end of a 100-foot nylon line. It had inflated only partially, so the experiment to measure the amount of drag in the near vacuum of space was a failure. But Day-Glo orange, he could report, was the most visible of the balloon's five colors—a useful observation for later space-rendezvous flights. Carpenter was also the first astronaut to eat and digest solid food in space. Glenn had made do with pap squeezed from toothpaste-tubelike containers.

Above Carpenter's right ear hung a fluid-filled glass globe, designed by NASA's Lewis Research Center. He was to observe and note any movement of fluids in the zero-g environment. Using special filters, and an extraordinary camera, he also snapped sixty photographs of the Earth's curvature for the Massachusetts Institute of Technology, which was developing a navigation system for the lunar flights.

Some of Carpenter's most valuable experiments were his measurements of Earth's luminous layer, or airglow. First reported by Glenn, the airglow layer extends all the way around the planet. Using a photometer, Carpenter was able to measure its distance from the horizon as well as the light intensity of stars observed through the layer. He also inadvertently solved the mystery of the "fireflies" that a puzzled Glenn had repeatedly observed during MA-6. They turned out to be nothing more sinister than ice particles clinging to the spacecraft's cold exterior, as he discovered when he accidentally rapped the hatch while reaching for his camera. Carpenter then found he could produce flocks of them simply by banging hard on the cabin walls.

He also discovered that his state-of-the-art three-axis control stick had a design flaw: it only took a "wrist flick" to activate the larger, or "high," thrusters and compensatory countermovement to activate them once more. Yet, as Carpenter would later explain to debriefers, the thrusters offered a maneuvering ability that was not needed until reentry. The smaller thrusters were entirely adequate for spaceflight itself, and yet it was difficult to move the stick subtly enough to activate them alone. Mercury spacecraft had eighteen thrusters in all, powered by hydrogen peroxide that was forced through exhaust jets in the form of decomposed steam and oxygen. There were twelve one-pound thrusters, which used just a little fuel, and six twenty-four-pound thrusters, which used a lot. To pilot his craft, the astronaut could fire the thrusters in all three navigational axes: roll (rotation around the long axis), pitch (up-and-down navigational axis), and yaw (side-to-side axis). But the spacecraft could also be flown on autopilot (or the "Attitude Stabilization and Control System" [ASCS]), which controlled yaw, pitch, and roll attitudes according to data supplied by automatic attitude sensors. Given the many observational duties specified in the flight plan, the MA-7's control stick and navigational system got an extensive workout. Carpenter's postflight critique led to a useful redesign that enabled later astronauts to lock out the high thrusters, thus reducing fuel usage, and allowing them to reenter with a safe margin of maneuvering fuel in their tanks.

Carpenter had a heavy schedule of experiments to carry out, sometimes performed with the spacecraft in manual mode, sometimes on autopilot, with frequent maneuvers and mode changes. This meant there was no time for the astronaut to actively search for any other problems with the spacecraft. Worse, even on ASCS, symptoms of trouble were intermittent and obscure, without discernible pattern. On *Aurora 7*'s second orbit, both Carpenter and controllers noticed mysteriously low fuel readings—not dangerous this early in the mission but still a problem. Lacking the option to lock out the high thrusters, Carpenter had spent too much fuel on the radical maneuvers called for in the flight plan. As Carpenter soared over Zanzibar, through the CapCom, Mercury Control proposed that the thrusters were causing the low readings. Carpenter, however, had already checked them thoroughly and reported that the thrusters were *not* malfunctioning. No one suspected a malfunctioning attitude sensor. Indeed, throughout the flight, the pitch reading was sometimes correct, sometimes not.

Only postflight inspection determined the prime cause of low-fuel trouble: an intermittently malfunctioning pitch horizon scanner (PHS). When the ASCS or autopilot was in control, the PHS fed erroneous data about *Aurora 7*'s pitch attitude to the thrusters, which then automatically fired to "correct" its attitude. During spaceflight, a malfunctioning PHS presented no danger, though it did waste fuel that might be needed for spacecraft orientation during reentry. If he needed to, an astronaut could quickly line the horizon up with a scribe mark etched in the cockpit window. Using manual control to establish proper yaw attitude was, however, another matter. After in-flight experimentation with this important navigational task, Glenn and Carpenter both reported that manual yaw control required up to six minutes of a pilot's sustained attention to navigational clues either through the periscope or out the window. This difficulty in establishing zero-degree yaw was one reason why reentry was generally conducted on autopilot. It quickly established proper retro-attitude in both pitch (thirty-four degrees, nose down) and yaw (exactly zero degrees)—when the automatic navigational tools were functioning properly.

At the end of his third orbit, as Carpenter readied for retro-sequence, he reverted to autopilot as demanded by procedures. The spacecraft immediately lurched out of position in both pitch and yaw, the nose canting sharply to one side. To Al Shepard, California CapCom, Carpenter reported: "ASCS is bad. I'm on fly-by-wire and manual."

"Roger," Shepard replied, "we concur."

On ASCS, at those attitudes, Carpenter's safe reentry was questionable at best; a manually controlled reentry was his only option. Again, a pilot's eye on the horizon (and his hand on the control stick) can establish proper pitch attitude almost instantly. Yaw is much more difficult and time-consuming. But astronauts make their best guess, and rely on their training and guts.

At the moment of retrofire, then, *Aurora 7* was off in yaw by about twenty-five degrees. Although Carpenter had nailed pitch, the discrepancy in yaw meant he would overshoot the landing site. Working with what he later recalled was surprising detachment, however, he continued correcting for yaw, eventually bringing it to zero. Long years of training were paying off. They *had* to. Without a manually controlled reentry, *Aurora 7* could have incinerated during reentry. It could have missed reentry altogether, drifting around the planet, until even Carpenter's oxygen supplies were gone. Still, as he plummeted earthward, watching his fuel gauges dwindle to zero, Carpenter was strangely comforted by the capsule's positive stability. Aerodynamic forces were correcting the craft's attitude. He later commented:

A lot of people know about John Glenn and Al Shepard. But they don't hear a lot about Max Faget, who in the 1950s probably knew more about the physics of manned space flight than anyone else in the world. Well, Faget designed a reentry body that was aerodynamically stable. He did that because he knew, sooner or later, the reaction control system would fail, and the automatic controls, and because of these potentially fatal failures, he needed to design a craft capable of righting itself during reentry. He was right. He had proved the theory in the wind tunnels at Langley Field. And with Aurora 7, *I was able to inadvertently prove Max's theory on an actual reentry. It worked!*

Bobbing on the ocean inside his compact spacecraft, Carpenter was hot. He was also 250 miles down range of his planned splashdown zone. *Aurora 7* was rocking through a roughly sixty-degree angle in swelling seas, and he decided to get out. The well-practiced egress took four minutes—but not through the hatch, which would have sunk his ship. In a well-practiced maneuver, Carpenter clambered out through the top of his spacecraft—a tight squeeze—after which he deployed and inflated his life raft. He knew his Search and Rescue and Homing (SARAH) electronic radio beacon had begun transmitting his position automatically, and would soon bring aircraft to the site. All he could do was wait.

Meanwhile, a public relations debacle was developing back home. Although NASA appears to have known Carpenter's location, it neglected to inform the press—other than to say that he had splashed down long in an unpatrolled area of the Atlantic. As a result, forty minutes of intense speculation about a "lost" astronaut swept anxious Americans up in a space-age drama. Watching expectantly for the first surveillance aircraft, the astronaut finally saw a familiar silhouette—a navy P2V. He signaled to the crew with a hand-held mirror. Following the pilot's report that Carpenter was "resting comfortably in his life raft," triumphant news bulletins hit the nation's airwaves. The "lost astronaut" had been located. An ecstatic nation rejoiced at Carpenter's safe recovery, and in a spontaneous display of relief, New Yorkers threw a few tons of ticker tape and confetti out of Wall Street windows. "He's alive!" exultant voices shouted to each other. "Alive!"

Two hours and forty-nine minutes after the spacecraft had splashed down, and seven hours and forty-five minutes after launch, a recovery helicopter from the USS *Intrepid* arrived to pluck Carpenter from his life raft and winch him aboard. Meanwhile, the USS *John R. Pierce* moved in to retrieve his spacecraft. When a NASA doctor asked the drenched astronaut how he felt, Carpenter looked up and smiled. "Fine," he said. The understatement was soon being flashed around the world.

Postflight tests showed that Carpenter had no signs of physical and mental stress. In fact, some of his postflight scores were higher than his preflight numbers. His flight, and Glenn's, clearly demonstrated that manned Mercury flights, with hours of weightlessness and the high g's of liftoff and reentry, were possible. One could even swallow and digest solid food items in space—findings that today seem ordinary but at the time answered an elementary mystery.

Both Glenn and Carpenter flew three-orbit missions in 1962 in nearly identical spacecraft. But the similarities end there. Glenn's job was to prove it could be done, and he did. He endured a faulty indicator light, information-stingy ground controllers, and during reentry a thumping, incinerating retropack. The experience eventually led to greater wisdom about candid ground-to-air communications.

The pilot of *Aurora 7* had both less and more to do. Replicating the flight of MA-6 was the easy part. Carpenter had to make three orbits, just as Glenn had. But there was more than replication. There were experiments, cameras, and photometers to juggle. There was a temperamental control stick to manage, as well as a mysterious, intermittent malfunction, somehow fuel-related and impervious to troubleshooting. Like all his colleagues, Carpenter contended with heat and sweat. Unlike them, he faced reentry with nearly empty tanks, joking about them (in a tribute to José Jiménez), "Oh, I *hope* not," as temperatures outside approached nearly four thousand degrees Fahrenheit.

Most important, Carpenter proved that human pilots were essential to spaceflight. When the automatic systems falter, and failure is assured, humans provide a surefire backup system. Without Carpenter's guiding hand, *Aurora 7* would never have made it back to Earth. A confident NASA could now move on to longer orbital missions.

Despite the perils of his pioneering flight, or perhaps because of them, Carpenter cherished his five hours in space: "I sat for a long time just thinking about what I'd been through. I couldn't believe it had all happened. It had been a tremendous experience, and though I could never really share it with anyone, I looked forward to telling others as much about it as I could." Carpenter had found space fascinating and a flight through it so thrilling and even overwhelming that he truly wished he could get up the next morning and relive the whole amazing day. "I wanted to be weightless again," he said, "and see the sunsets and sunrises, and watch the stars drop through the luminous layer, and learn to master that machine a little better so I could stay up longer . . . space is a fabulous frontier."

Fabulous frontier or not, Carpenter would never fly there again. As the fourth American in space plummeted earthward, out of fuel and manually piloting his spacecraft to an uncertain landing site, the flight director blew his stack. Chris Kraft wrote in his memoirs that he "swore an oath that Scott Carpenter would never again fly in space." Tom Wolfe describes the scene vividly in *The Right Stuff*, while numerous other space memoirs corroborate the flight director's fit of pique. In his memoirs, Kraft suggests that he made good on his vow, pointing out that Carpenter, in fact, never did fly in space again. Others are not so sure. Slayton, who by 1962 had been superannuated to head of the astronaut office, sidesteps the issue in his autobiography, *Deke!* In a discussion of crew assignments for Gemini, he writes merely that Carpenter "was unacceptable to management." Perhaps he meant Kraft.

Reflecting on the controversy years later, Carpenter explained that the flight director stubbornly missed the point of his achievements aboard *Aurora 7*. He successfully brought home a spacecraft that had fuel, control stick, and attitude indicator problems. Was it a "textbook" flight? No. But his pitch horizon scanner was hardly a "textbook" navigational tool. It malfunctioned. The textbooks would have to be revised. This is what all astronaut-explorers are trained to do: journey to perilous places on dangerous missions—sometimes meeting the unexpected—and return, alive, with new truths, new knowledge, and new postflight wisdom on how to make future flights safer and better. That's what Carpenter did, and what Kraft to this day contests. It is one of the most persistent controversies of twentieth-century spaceflight.

Characteristically positive, even in the face of disapproval from "management," Carpenter was not one to fret. He was young and confident he could make plenty of other contributions to engineering and exploration. In 1963 he left his active-duty astronaut status for service with his parent command, the U.S. Navy. Another frontier in the deep ocean now beckoned. The navy had begun a research program, called Sealab, in underwater habitation and exploration. It relied on a fascinating new technique called saturation diving, as Carpenter explained:

My involvement in Sealab followed from my contact with Jacques Cousteau, in 1963. I had followed his work closely for years. In the long downtime after Mercury—there was some training, and a lot of design and development well into 1965—it occurred to me that the technology NASA *was building to keep men alive in space was directly applicable to keeping men alive in the ocean. I asked Bob Gilruth if I could meet with Cousteau about these ideas and proposed a leave of absence from* NASA*. So Cousteau and I talked, and he finally said, "Well look, we can't pay you very much, and you don't speak the right lan-*

guage, but you have got a good idea about sharing the technology. Why don't you do it with your own navy?"

Cousteau introduced me to Sealab and Captain George F. Bond. It is important to stress that this marvelous hypothesis involving saturation diving came from Bond and the navy. Cousteau first used it in the open ocean, but its genesis was at New London, Connecticut, under the direction of the U.S. Navy, and those techniques are in use around the world today. That's one reason for my pride in my parent service, the United States Navy.

Throwing himself into the rugged regimen required of an elite saturation diver, Carpenter hoped to participate in Sealab I, the navy's experiment in underwater habitation. It involved lowering a submersible chamber, which would hold five divers, to a depth of 192 feet in the waters off Bermuda. But on 16 July a bad motorcycle accident in Hamilton, Bermuda, left the astronaut-aquanaut with a mangled left arm and a broken left toe. The breaks would not heal in time for the extended dive. Captain Bond scratched his name from the roster for Sealab I and sent a crestfallen Carpenter back to Houston. Later, the news only got worse. Orthopedists at Houston's M.D. Anderson Hospital found that the break had healed *too* well, limiting the rotational ability in his left arm. It was a grounding injury, and Carpenter's spaceflight days were over.

Carpenter was stoic; the injury would not prevent him from diving. He was given command of *Sealab II* to be submerged off the coast of San Diego in the summer of 1965.

On 28 August 1965, the staging vessel *Berkone*, anchored half a mile off the coast of La Jolla, California, was swarming with sailors and aquanauts. On that day Carpenter and nine fellow divers swam down to the Sealab habitat on the ocean floor. The ten men were the first of three U.S. Navy teams involved in the *Sealab II* program. Linked to shore by television, they would initiate a forty-five day series of experiments conducted in and outside of a pressurized 57-foot-by-12-foot undersea laboratory perched on a ledge 210 feet below.

Scientific tests carried out on and by the three teams during their fifteen-day tenures would help to determine how well humans could adapt to living and working under a pressure equivalent to seven times that experienced on land. Nine of the divers heading into the depths that day would stay down for fifteen days, but Carpenter would voluntarily remain on board the *Sealab II* laboratory and resurface with the second team after a total of four weeks beneath the sea.

Venturing outside the laboratory for two or more hours a day, the divers had a special helper at hand—a trained porpoise named Tuffy, who carried lines from one aquanaut to another. It was welcome assistance, as visibility was limited to just a few feet, and mobility was restricted both by protocol and the cumbersome scuba gear. Underwater tools and equipment and a simulated portion of a ship's hull had also been sunk nearby for salvage experiments.

Navy spokesmen were careful to point out that the divers were there, not to set any endurance records, but to establish techniques enabling people to live and work comfortably at a great depth. Despite this, the *Sealab II* teams would actually break a nineteen-day underwater record set some time earlier by Cousteau. The navy divers would also work 110 feet deeper than Cousteau's team.

The hard work was sometimes punctuated by more lighthearted moments. Playing a ukulele and singing with helium-affected vocal chords, Carpenter serenaded his men with "Goodnight, Irene." A day after the ten divers entered their underwater laboratory, Carpenter and his Project Mercury colleague Gordon Cooper arranged a historic underwater-to-space conversation. Flying with the exuberant Pete Conrad, Cooper was on orbit 117 of the *Gemini 5* mission. The widely reported four-minute conversation was on relay, however, from spacecraft to ship to submersible and back again, making it nearly unintelligible for both men. Before the two-man spacecraft moved out of range, the explorers relayed the usual good wishes.

After fifteen days, the first Sealab team surfaced and was replaced by the second team of nine divers. Carpenter assisted the transfer by adjusting a cable on the personnel transfer capsule from within a shark cage outside the laboratory. Suddenly, a scorpion fish lurking nearby delivered an excruciating sting through Carpenter's wetsuit glove. Dr. Robert Sonnburg administered sedative drugs for the pain, and after extensive discussions it was decided that Carpenter would remain below with the second team. At the conclusion of the second team's sojourn aboard the underwater laboratory the entire group surfaced, spending thirty-three hours relaxing in a decompression chamber. While biding his time there, Carpenter took a congratulatory phone call from President Lyndon B. Johnson.

"Scott," the president said, "you have convinced me and all the nation that whether you are going up or down, you have the skill to do a fine job."

"Well, thank you very much," Carpenter replied. Recordings of the historic conversation sound as though the president was speaking not with an aquanaut but with Donald Duck. All the team members had been breathing oxygen-helium gas, both aboard Sealab and during decompression, which affected their vocal cords. Carpenter added, with his customary, if high-pitched, good manners: "There were a lot of other people who demonstrated the same kind of courage."

In his log, Captain Bond, an earwitness, later recorded that President Johnson, while gracious, hadn't understood a word Carpenter had said. Cooper, aboard *Gemini 5*, had encountered the same difficulty. But all the same it was good fun for both astronaut and president to talk with a hero of ocean and space—a hero with the Donald Duck delivery.

With Carpenter's now vast experience in underwater habitation, Bob Gilruth tapped him to design a neutral buoyancy training program for extravehicular activities, such as spacewalks. Carpenter knew, for example, that astronauts needed adequate restraints to hold them in place if they were to have any hope of doing productive work, and Gilruth agreed. As a result, Carpenter helped to design NASA's zero-g training tank, which, in a much grander form, is still in use today.

By mid-1967, with all hopes of another flight gone, and the *Apollo 1* fire still a fresh tragedy, Carpenter considered what lay ahead. He was only forty-two years old, and the thought of a desk job at a changed NASA held little appeal. He'd made valuable contributions to astronaut training and had worked on the Lunar Module cockpit. But now other challenges beckoned, so on 10 August 1967 he submitted his resignation from NASA. Two years later, he resigned his commission in the U.S. Navy and returned to private life, spending the next few decades in lower-key roles, such as consulting on engineering and environmental matters. He would become a successful novelist, writing techno-thrillers set in the unexplored ocean depths.

Following his own Mercury flight, John Glenn's fame knew no bounds. Also boundless were the demands on his time. The most famous, most recognizable man in the world was more than the first American to orbit the Earth and far more than an engineering test pilot. Glenn was the walking, talking, smiling embodiment of America's can-do spirit. NASA sent him everywhere—there were goodwill missions abroad and testimony before Congress to request funds or explain NASA's objectives. There were morale-boosting trips to space-related facilities, now employing, nationwide, nearly a million Americans. And everywhere, handshaking, posing for photographs, signing autographs, travel and more travel.

With no post-Mercury adventures in space scheduled until 1965, an adventure anywhere may have sounded appealing to Glenn. Like Carpenter, Glenn had a number of talents—and a number of opportunities. His outsized fame and obvious national appeal made a career in politics a distinct possibility—a career shift the Kennedy brothers, John and Bobby, discussed with him. Paul E. Purser, Gilruth's assistant, gave him a copy of the hugely popular political novel *Advise and Consent*. But there were also abundant business opportunities. What to do was topic number one for John and Annie for most of 1963.

As Carpenter chafed during what he describes as "the long downtime after Mercury," Glenn too was casting about for fresh challenges. In fact, according to his memoirs, he even made a bid for another mission as Project Mercury ran down. With only a fifteen-minute ballistic flight to his name, Shepard was also lobbying hard for the same Mercury flight, MA-10, and he pressed his case with a number of people. Glenn was quieter, and he asked only Gilruth. Neither astronaut got what he wanted. Project Mercury was over. The space agency would concentrate on Gemini and Apollo and on getting men to the moon and back.

For Glenn, the "postorbital remorse"—the tedium and uncertainty of life after *Friendship 7*—was resolved tragically with the assassination of President John F. Kennedy. He and Annie soon decided John would run for a seat in the U.S. Senate. He tendered his resignation from NASA in January 1964, five weeks after Kennedy was murdered. Barely a month into the campaign, however, Glenn slipped and fell in the bathroom—a humiliating injury for the iron marine who had survived aerial combat in two wars and then a pioneering spaceflight. Unable to stand, let alone campaign for national office, Glenn endured lurching, agonizing spells of vertigo. With her debilitating stutter, Annie was unable to deliver speeches in his stead.

Glenn withdrew from the Senate race at the end of March 1964 to focus on his recovery. Doctors told him that if this injury did not heal within twelve months, it never would. Glenn recovered in eleven months and three weeks, using the marine's grueling regimen so roundly hated by Carpenter: running and *more* running.

After regaining his active-duty flight status, Glenn resigned his commission in the marines and entered the business world, biding his time until the possibilities brightened for a third career—one in national politics.

In 1974 the time was right. Glenn won a seat in the U.S. Senate, carrying all of Ohio's eighty-eight counties. He was reelected in 1980 with the biggest winning margin in the state's history. In 1986 Ohio returned him to a third term in the Senate, and six years later he won a fourth consecutive term.

In 1984 Glenn ran for the office of president of the United States with high hopes. But the indefatigable marine—a decorated war veteran, celebrated Mercury astronaut, successful businessman, hardworking U.S. Senator—was not the electrifying stump speaker Americans had expected. After defeats in the Iowa and New Hampshire primaries, and with campaign debts of nearly $3 million, Glenn announced his withdrawal from the race. "Although my campaign for the presidency will end," he told a packed news conference, "my campaign for a better America will continue." It did, and it has.

On the thirty-fifth anniversary of what was then his only spaceflight, Glenn announced his retirement from the Senate, to take effect on 3 January 1999. But he still had his eyes on another mission.

In 1998, after years of persistent prodding, Glenn was invited to fly as a crew member aboard space shuttle *Discovery.* On 29 October, he began his second spaceflight and at age seventy-seven became the oldest person ever to venture into space. At the launch, adding special significance to the moment was Glenn's old colleague Scott Carpenter, on hand to reprise his famous invocation. "Good luck, have a safe flight," Carpenter told the *Discovery* crew, "and . . . *Godspeed, John Glenn.*"

The careers of the two friends had come nearly full circle. Both had solitary Mercury flights and then left NASA for new challenges. Both found new careers and opportunities after sustaining serious injuries. Capping a career in the U.S. Senate, Glenn had now left the launch pad once again, rising above the Earth one last time. Would Carpenter have liked to fly in Glenn's place? His response is quick and delivered with a smile: "Sure, but NASA wouldn't want to send me, I told John. I wasn't *old* enough!"

Glenn's flight aboard *Discovery* and Carpenter's reprised salutation were reported with some fondness in the international press. His "Godspeed" reminded many of the seemingly limitless possibilities of the 1960s, the glory days of Project Mercury, and the first steps on the moon that these pioneering spaceflights made possible. Three Mercury astronauts—Gus Grissom, Deke Slayton, and Alan Shepard—did not live to see Glenn's shuttle mission. Shepard had, in fact, passed away only three months before launch. But Scott Carpenter, Gordo Cooper, and Wally Schirra were there to carry the torch they had helped to light decades before.

For his part, Carpenter wished his flight aboard *Aurora 7* were known for more than the decades-long controversy surrounding the overshoot—and Chris Kraft's famed, invective-laced oath. With the publication of his bestselling autobiography in 2003, Carpenter's wish came true. He was seventy-seven years old when his book was launched—the same age Glenn had been when he was launched, the second time, into space. *For Spacious Skies: The Uncommon Journey of a Mercury Astronaut* was the last, and many say among the best, of the Mercury astronaut memoirs.

Carpenter is free, once again, to concentrate on the future. Having spent a lifetime satisfying his own curiosities, he is excited to see others doing the same, but retains feelings of humility and gratitude for his rare opportunities. At a book signing at the Reuben H. Fleet Science Center in San Diego, Carpenter expressed his hopes for our future in space:

I am an ordinary mortal with at least the normal quota of frailties. No one is more keenly aware of this than I. That I had been given this opportunity to add to mankind's fund of knowledge was a stroke of sheer good fortune. For months, my thoughts, plans and hopes had been only that my performance would live up to my opportunity.

After you fly in space, eventually, you come to a point where you can't do it anymore. But then you have the opportunity to tell the youngsters about it, and that is a sacred trust. That's where it's at for all of us adults, to teach the youngsters. I am so glad to see that there is this interest in the teaching of science and technology, particularly that related to manned space exploration. We

A contented Scott Carpenter in 2004. Courtesy Francis French.

will produce, with that kind of work, the first people to walk on Mars. They are maybe five years old at this time—but they are in your hands. I tell this to a lot of youngsters—when you go to Mars, which you can *do if you work hard enough, I want you to write me a letter and tell me what it looks like! It is as sure to happen as we stand here today.*

When the Mercury astronaut talks about Mars, he does so with characteristic curiosity. What will Mars look like? What will it feel like? Carpenter wants to know. Astronauts, he once explained, are more like prospectors than pioneers. It is their lot to scratch the back of the universe, just as prospectors scratch the back of mountains looking for the mother lode. Their primary mission is to return to Earth laden with as many scientific riches—including new knowledge—as can be crammed onto their spacecraft or into their heads. On the day Carpenter returned to Earth, he was ready to go back up, to keep exploring, to experience a little more, as soon as the next morning.

Perhaps it is trite to say, once more, that Carpenter is not your usual astronaut. But maybe it bears repeating. The man *is* different. Even today, he possesses the same calm he took into space back in 1962; his eyes still alight with curiosity. Carpenter was a successful engineer and pilot, one of his nation's finest, chosen for a new and dangerous undertaking. Yet he also brought something else to the program—it was not a requirement, yet it embodied the spirit of the entire endeavor. Carpenter brought curiosity. He did more than fly in space and return safely to the Earth; he saw, he experienced, what was out there.

One hopes the first person on Mars has the same gleam in his or her eye, the same thirst for knowledge that Scott Carpenter still shows and shares, more than forty years after his sole journey into space.

6. Heavenly Twins

Any sufficiently advanced technology
is indistinguishable from magic.

Sir Arthur C. Clarke

At 11:30 a.m. on 11 August 1962, the seventh person to leave the Earth in a spacecraft was launched into orbit. It was a Soviet national holiday known as Physical Culture Day, and the much-rumored launch was dramatically announced in a special Radio Moscow broadcast to the nation, interrupting regular programs. The report had actually been held back for one hour and twenty-six minutes, in the event any problems arose with the launch or orbital insertion.

News of this latest Soviet manned launch caused tremendous excitement in the streets and homes of Moscow. Radios in parked cars were turned up to full volume, attracting crowds of excited listeners. Meanwhile, Americans were just waking to start their weekend. Throughout the day, reports flowed from Moscow that cosmonaut Maj. Andrian Nikolayev, in his *Vostok 3* spacecraft, was orbiting the Earth every 88.5 minutes, with a perigee (low point) of 113 miles and an apogee (high point) of 156 miles. His call sign was "Sokol" (Falcon), and he had established solid communications with his control center, reporting "I feel well; all systems of the ship are functioning perfectly. The Earth is visible through the porthole."

In a surprise move at the time of the launch, but one it would repeat on subsequent manned flights, the Soviet government broadcast an appeal to the United States to refrain from conducting any high-altitude nuclear testing while its cosmonaut was in orbit. They were quickly assured that no such tests would take place while Nikolayev was aloft. Nine hours into Nikolayev's flight, television images of a human space traveler were broadcast for the first time ever directly to the public. It was an unprecedented move by the normally secretive Soviet space agency. To viewers in Russia and the West alike, Moscow Television Center relayed the historic if snowy images showing unearthly things such as a pencil on the end of a string floating in front of Nikolayev's smiling face. Radio Moscow meanwhile announced that the cosmonaut's duties would include a study of the effects of weightlessness. Speculation mostly centered on whether he would attempt to surpass Titov's daylong flight record aboard *Vostok 2*. After all, each Soviet flight so far had been a step up from the one before, assuring a sound propaganda victory each time. A manned mission lasting longer than a day would certainly provide this.

In space, things seemed to be going well as Nikolayev demonstrated a healthy appetite by making his way through some bite-size chicken fillets, roast veal, pies, and sandwiches. By the time he finally settled down to sleep at 10:00 p.m., Moscow time, his Vostok spacecraft had completed seven orbits, covering some 186,000 miles. However, the Soviets were aiming for something even more spectacular than record-breaking distance. The following day they stunned the world by sending up a second cosmonaut, Lt. Col. Pavel Popovich, using the call sign "Berkut" (Golden Eagle), in *Vostok 4*. Once again an official announcement came nearly an hour and a half after launch, by which time two-way contact had been established between the cosmonauts. Both flights had been launched with an orbital inclination of 64.57 degrees—exactly the same as the two previous manned Vostok missions.

News of the double triumph exploded like a bombshell over America, leaving President Kennedy, NASA administrators, astronauts, scientists, and everyday Americans awed by the sheer audacity of the Soviet accomplishment. To further add to America's woes, the dual flight came less than a week after former president Dwight Eisenhower had publicly criticized President Kennedy's crash space program, calling it "a mad effort to win a stunt race." Kennedy stoically called the dual flight "an exceptional technical feat" and saluted the courage of the two cosmonauts, while NASA officials gloomily shook their heads and pondered the miracle it might take to beat the Russians to the moon. The implications were immediately evident: the Russians had revealed a level of competence that NASA could not yet match. As astronaut Tom Stafford later remarked, "We couldn't have put two Mercury-Atlases into space one after the other like that."

Western observers were amazed that the Soviets had the capability to launch two manned spacecraft within a day of each other, an effort that would have required synchronization of Herculean proportions at the launch site. *Vostok 4*, they deduced, must have been launched at exactly the right moment to ensure the near-perfect rendezvous. More amazingly, the Soviets had accomplished this on only their fourth manned space flight. The stunned observers would have been even more astounded had they known that the same launch pad was used. How much further would the Soviets go next? With two spacecraft in near-identical orbits, credible speculation suggested *Vostok 3* and *4* might even attempt to dock. Later evidence revealed that this would not have been possible, though the Soviets did not mind at all if the West thought otherwise.

Decades later, Vasily Mishin, who was spacecraft designer Sergei Korolev's chief deputy and ultimately his successor, was still pleased with his international rival's confusion. In a candid 1990 interview for Moscow's *Ogonyok* magazine he revealed that "with all the secrecy, we didn't tell the whole truth. The Western experts, who hadn't figured it out, thought that our Vostok was already equipped with orbital approach equipment. As they say, a sleight of hand isn't any kind of fraud. It was more like our competitors deceived themselves all on their own. Of course, we didn't shatter their illusions."

Whether the spacecraft rendezvoused or not, two precise launches so close together, from the same pad, and so clearly without any launch delays, was accomplishment enough. Even Vladimir Suvorov, filming events at the launch pad, had been amazed at the quiet efficiency of the changeover. "They did not give us any time to be bored," he wrote. "Only a few hours ago did we send Andrian off to space, and now they already have another booster waiting on the launching pad, and the ritual procedure of 'passing' the rocket to the cosmonaut is under way."

Meanwhile, the two cosmonauts continued their historic tandem journey, as the world's press quickly labeled them the "Heavenly Twins." They even became the first people to sing in space, when Nikolayev suddenly burst into song about the Volga River in his native region, and a delighted Popovich responded with his favorite Ukrainian song. While the politicians and media back on Earth concerned themselves with the political aspects of the flight, the duo did their part, as trained pilots, to focus on the task assigned to them. They came from totally different backgrounds, yet both were shining stars of the Soviet Union's cosmonaut team.

Andrian Grigorievich Nikolayev, the son of a collective farmer, was born in Shorshely on 5 September 1929. The peaceful mountain village, whose name translates to "pure springs," is located some seven hundred miles east of Moscow in the middle of the Volga basin, within the Soviet Union's Chuvash Autonomous Republic. As a Chuvashi, Andrian came from a long line of Bulgar people who used the Turkish language. Both his first and last names were common to the region.

As the cosmonaut revealed to the authors years later, he actually grew up under a different family name. He was only a small boy when a tax official bearing the same last name came to live and work in Shorshely. For some unexplained reason this was such a source of immediate concern for the young family that Grigori Nikolayev took the unusual step of changing the family name. He filed the necessary papers, and soon the whole family, including Andrian and his older brother, Ivan, were officially given the name Zaitsev, for reasons even the cosmonaut did not know.

Nikolayev's mother Anna Alekseevna Zaitseva, widowed in 1944, lived all of her life in Shorshely. Following her son's first spaceflight she recalled something of his early life for the magazine *Soviet Weekly*:

When the first collective farm was formed in our region, my husband took a job as a stableman and I worked as a dairymaid there. I used to come home very late, but Andrian looked after the house. He would chop wood, bring water, and light the stove. Andrian studied very hard and was very fond of retelling the stories he read. Even as a little boy he loved to go to the forest to gather mushrooms, nuts, and berries. He was never afraid of going far into the forest and losing his way as some other boys were. Once, I remember, some of our village boys lost their way in the thick of the nut groves, and my Andrian helped them to find their way out.

At the age of seven, after seeing a fighter taxiing on a landing field, Andrian became fascinated by airplanes and flying. Not long after, he is said to have climbed a tall tree and to the astonishment of his friends on the ground called out that he was going to fly down. Fortunately, they talked him out of this foolhardy venture. "Our family was a large and friendly one," Anna Alekseevna continues. "I lost my husband very early, and my sons Ivan and Andrian (the younger of the two) did all they could to help me and the babies, Zina and Pyotr. During the war after the death of my husband it was hard to make ends meet." When Andrian's mother finally told Andrian the story of the tax agent and how his family had changed their name, he decided that with his father gone he wanted to become a Nikolayev once again, and completed the necessary forms. Many years later, after his spaceflight, his brothers also chose to revert to the original family name, although his sister did not have to—she was already married. It is believed that his mother may have carried the name Zaitsev to her grave in 1987, although the Soviet press always referred to her as Anna Nikolayeva to avoid confusion.

Following the loss of his father, Andrian told his mother that he wanted to leave school and find employment to help the family. She would not hear of him cutting short his education, despite their plight: "We had little to eat and not much to wear. Andrian finished the village seven-year school and went to Tsivilsk because he wanted to become a doctor's assistant and there was a medical school there, but he changed his mind and went to his brother Ivan who lived in Mariinsky Posad." Ivan was a lumberjack, and he soon talked his younger brother into looking at forestry as a career option. Andrian enrolled at the Mariinsko-Posad Forestry Institute in his native Chuvash, graduating in 1947 as a forest technician. He then left his Shorshely home and traveled to a forest industrial unit in Karelia, where he worked amid the birch and fir trees as a lumberjack. Later, he became a timber-procurement foreman of the Derevyanski Timber Industrial Farm, at a unit called the "Uzhkarelles Trust."

Then in 1950, military conscription and other influences caused a dramatic shift in his life. "I liked the new place," he later said of the industrial timber farm. "I thought then that I would stay for a long time, but the time for army service drew near. At the draft center the doctors examined me for a long time, almost an hour. And then I heard the verdict: 'To the Air Force.'" As an officer cadet Nikolayev first served as deputy commander of a platoon of air gunners and radio operators. Then followed several months' training as a gunner and radio operator on Tupolev Tu-2 bombers at the Kirovobad Higher Air Force School, where he joined the Young Communist League. In 1951, with a firm desire to become a fighter pilot, he entered the Chernigov Air Force Pilots School and later transferred to the Frunze Air Force Pilots School, graduating in December 1954. Over the next five years Nikolayev served as a MiG-15 pilot, senior pilot, and squadron adjutant with the 401st Air Regiment in the Moscow military area. Before joining the cosmonaut team he also served for a time as a test pilot.

Nikolayev's determination and coolness under pressure became evident during one exercise early in his fighter training. On a routine flight at thirty thousand feet, the engine of his MiG fighter jet suddenly died. As the stricken plane swept toward the ground in silence, Nikolayev managed to keep it in a glide pattern while he tried repeatedly to restart the engine. It was all to no avail. He maintained communication with the airfield, but was eventually told he would not make it back and should eject. He felt confident, however, that he could still land his airplane. "I was excited, and at the same time not excited," he later remembered. "Keep calm, I kept telling myself." He spotted a suitable field to his right, swung the MiG around, and leveled out the wings. "There were no trees, the place looked even. But coming nearer the ground, I saw that the landing strip was in fact uneven and bristling with bushes." Nikolayev believed he could still land, but "suddenly I saw a dip just beyond the field . . . beyond that, possibly a precipice. If I overran, that was that—the end." He knew he was too low to the ground to initiate an ejection, so he steeled himself and did what he could to bring the MiG down safely. It landed heavily, "plowing into the field with a terrific noise," but finally came to a shuddering rest a few feet short of what was indeed a precipice. The underside of the jet was damaged, but in rejecting the option of bailing out he had saved it from total destruction. After walking away from the airplane he calmly started jotting down notes about the malfunctioning engine.

Years later, Gherman Titov would ask him about the incident. "How did you manage it?" he inquired. "What helped you?" According to Titov, Nikolayev's answer came with an unassuming shrug. "I kept my head, that's all." A few days after the forced landing Nikolayev's regiment commander presented him with an engraved Pobeda watch, citing in his speech "the bravery and self-control shown by him in a grave situation arising during flight." The incident apparently featured prominently in Nikolayev's later appointment to the cosmonaut team.

At the time of his selection, finalized on 25 February 1960, Nikolayev was still unmarried and a senior lieutenant in the Soviet Air Force. He was one of twenty recruits chosen to train as cosmonauts in his nation's space program. Short, dark-eyed, and analytical, Nikolayev initially made less of an impression than some of his more outgoing colleagues. What brought him to the attention of his cosmonaut trainers was his outstanding physical endurance. He was easily the best of the group when undergoing centrifuge tests, and survived the second-longest number of days, four, in the claustrophobic isolation chamber—beaten only by Titov. Nikolayev spent much of his time in the chamber calmly painting winter landscapes, imagining himself in those cold environments during the discomforting heat tests carried out by the trainers. "I am like everyone else, and I get nervous like others," the cosmonaut said of himself. "But I control myself."

He was soon selected as one of Korolev's "First Six" cosmonauts—the so-called group of immediate preparedness who would receive advanced training for the Vostok missions. Nikolayev then served as backup pilot for Gherman Titov on the *Vostok 2* mission. "He speaks so seldom," Titov remarked of his quiet, reserved colleague, whom he called "the embodiment of composure. When he does talk we listen to every word. Invariably, his few sentences result from careful, long thinking." Titov further characterized Nikolayev as "the calmest man in an emergency I have ever known . . . which is necessary for the commander of a spaceship—a man of iron endurance and courageous determination." Gagarin echoed this opinion, referring to Nikolayev as "the most unflappable man in a crisis I know." The rest of Nikolayev's cosmonaut group also appreciated the taciturn Chuvash pilot and quickly became familiar with his favorite expression: "The main thing is to keep calm." It was a maxim that always seemed to serve him well. After the medical concerns of Titov's flight, these qualities made him the obvious choice for the next, longer mission.

Earlier, when Nikolayev had reported for duty as a cosmonaut, the man who greeted him had already been there for a month. He was a cheery, outgoing man named Pavel Popovich. Although they didn't know it then, their careers and lives would soon be closely linked. The two would never share a spacecraft cabin, yet they became famous as the first people to fly in space at the same time. In contrast to the introspective, focused Nikolayev, Pavel Popovich had spent his time in isolation training singing and whistling—reportedly from morning to night—and even dancing a little. One of his doctors recalled that while Popovich was in the chamber, he "gave such concerts, that one would listen with delight. For hours on end he would sing from operas, musical comedies, or folk songs." Colleagues noted that he was curious and restless, seemingly unable to sit still for more than an hour. Considering his difficult upbringing, it is amazing how relentlessly cheerful and positive he seemed. But then, he came from a family of hardworking survivors.

Pavel Romanovich Popovich was born on 5 October 1930 and began school at the age of seven. His family lived close to an air force base, and he formed friendships with many of the pilots there. He idolized these brave men, and dreamed about becoming a pilot himself one day, although he did not believe it would ever be possible. Possessed of a large bristling handlebar moustache that was the envy of many, Pavel's father, Roman Porfirevich Popovich, came from a long line of Ukrainian working people. Born in 1905, his parents and all his relatives were farmworkers, so even as a youth Roman knew the hardships of living and working in a poor peasant family. In his teens Roman met Feodosia Kasyanovna, two years his junior, and they married when he was twenty. The young couple set up a home in the Ukrainian settlement of Uzin in western Russia, near Kiev, where Roman found work in 1927 as a furnace stoker at a sugar refinery plant. He and Feodosia would eventually have five children, including the inquisitive, good-natured Pavel.

Rumors of war began to intensify as Pavel continued his schooling. When World War II eventually broke out, the fighting was far away to the west, but when he was eleven years old the dark shadows of a brutal Nazi occupation rolled like a dire wave over Uzin, and his school was closed down. One day in 1941, while Pavel and his father were out walking, they witnessed an aerial skirmish. An Ilyushin 2 Sturmovik ground-attack airplane was desperately trying to fend off several German fighters. The Luftwaffe referred to the Il-2 as the Beton Flugzeug, or "concrete plane." Neither fast nor agile, the Il-2 could absorb a tremendous amount of punishment. On this occasion, however, the pilot was hopelessly overwhelmed and was soon shot down. They heard a huge explosion as the Sturmovik hit the ground and rushed to the scene. There was nothing they could do, but Pavel's father helped some locals bury the unfortunate pilot. Pavel was shielded from the sight of the man's mangled, charred body, but what he had witnessed that day would remain a lasting and bitter memory for him.

Along with other local children, Pavel was virtually forced at gunpoint to attend a new propaganda school organized by the Nazis. He stubbornly refused to take part in any studies, instead stuffing cotton wool in his ears so he could not hear his teachers. He was soon expelled for his contrary behavior and faced the very real threat of being shipped to a slave labor camp in Germany. Because it took so many soldiers to force the unwilling children to attend compulsory classes, the school closed down two months after its start. With this extra time on their hands, Pavel and other young boys continued their defiance, stealing German bullets and grenades and passing on their spoils to local resistance fighters.

With a massive Soviet advance slowly closing in, the Nazis became even more desperate and brutal. They began carrying out mass executions of dissidents and uncooperative but innocent villagers. The Germans were highly suspicious of any grown Soviet man and warned schoolboys about the fatal consequences of aiding the guerrillas. On one occasion in 1943, after several local boys had been summarily executed, Pavel's mother took the drastic step of dressing him in old frocks and passing him off as a girl until the danger passed.

The surrender of over ninety thousand German troops in Stalingrad in January 1943 would prove a significant turning point in the war, and relentless Soviet counteroffensives then began driving Nazi forces out of Russia. Popovich wanted to resume his education, but life remained hard with four children in the house. Fortunately, his father was still working, but the burdens increased when Pavel's older sister returned from a slave labor camp. Although they were intensely relieved at her safe return, she was completely debilitated and had to be slowly nursed back to health. For a time Pavel worked as a herdsman, roaming the fields and hills on his hardened bare feet. He then became a weigher and assistant to the fireman at the sugar factory where his father worked. His meager wages went to his mother to help support the family. Meanwhile, Pavel became interested in making up his years of lost education and in August 1943 resumed part-time studies at evening school.

In 1945, finally completing his sixth-grade class studies, Pavlusha (as he was known) entered a trade school in the town of Belaya Tserkov, on the Ros River south of Kiev, then later joined the local Komsomol, or Young Communist League. In July 1947 he passed the seventh grade of his evening school and was awarded the qualification of Joiner—Fifth Class. Pavel later credited joiner-cabinetmaker Pyotr Timofeyevich Ivanov, then the evening school's teacher of production training, for helping him to understand and enjoy carpentry. For his part, Ivanov recalled a student eager to learn but even then with a strong desire to be more than a craftsman. "It seems only recently I held in my hands a carefully made stool, Pavlusha's first independent job. 'Well now,' I told him, 'you already have the experience; soon you can learn to make cupboards and chairs.' But he only replied: 'I want to study further.'"

The following year Popovich gained entry into the Magnitogorsk Industrial Polytechnic near the southern Ural Mountains, deep in the heart of the Russian Steppes. Winning admission was not an easy task, given the large number of applicants, but Pavel had his supporters, including Ivanov and the other trainees at the Belaya Tserkov trade school. Up to this time Popovich had been instructed only in his native Ukrainian; since the Polytechnic conducted classes only in Russian, he had to study even harder just to keep up.

For recreation, Popovich began singing in an amateur choir. Even five decades later he still possessed a fine tenor voice and often spontaneously broke into song at social functions. During this period, he also rediscovered another great and enduring passion—flying. He was soon training at the local Magnitogorsk Flying Club, where his days would begin at 4 a.m. with a hurried journey to the airfield. For him, "aviation was a magnet against which there was no protection." After training at the flying club he made his way to the polytechnic, where he began his practical lessons in carpentry, plastering, and stonework at 9 a.m.

In 1951 Popovich was awarded his diploma, qualifying him as a building technician. Part of his diploma work involved redesigning the dormitory in which he lived. That September he also received a certificate of course completion from the flying club. Subsequently drafted into the Soviet army, Popovich exploited his flight training to win transfer to the famed Myasnikov Air Force Flight School in Kacha, near Sevastopol. His air force career began on 3 October. He pushed himself hard in his training. "I wanted to fly like my teachers," he recalls, "and sometimes I repeated the same flight maneuver ten, twenty, thirty times." He remained at Kacha until 1954 and then served in the Soviet air force as a MiG-15, MiG-17, and MiG-19 pilot. He flew in the Far East, Siberia, Karelia, and Moscow regions, at one time serving as a squadron adjutant. In June 1957 Popovich was made a full member of the Communist Party of the Soviet Union.

While on duty in Siberia the young pilot met his first wife, a pretty green-eyed Siberian blonde named Marina Lavrentyevna Vasilyeva, born 30 July 1931. Marina was a woodcutter's daughter who also happened to be an accomplished stunt flier. They met at Marina's graduation dance, held one evening at a local aero club near Popovich's air force station. Although they didn't meet again for another four years the romance survived, and in 1955 they married. Pavel and Marina had a daughter named Natalya, born 30 April 1956, and much later another daughter named Oksana, born 8 October 1969, seven years after her father's first spaceflight.

In 1963, because of Marina's renowned flying abilities (as a test pilot and engineer she set thirteen world records and became an air force colonel), it was widely speculated in the West that she might become the first woman cosmonaut. This hopeful conjecture turned out to be false, although it was later reported that she had applied for cosmonaut training but was not selected. Popovich confirmed this to the authors. "Marina did apply in 1962," he volunteered with a twinkle in his eye, "but she was not considered, as she was told one cosmonaut in the family is enough!"

Popovich was the first candidate from the first cosmonaut detachment to enter the Star City training center outside of Moscow, where he served as unofficial greeter and quartermaster for the other cosmonauts as they arrived. A sturdy, fair-haired man with merry eyes, a fondness for song, and a hardworking, persistent nature, Popovich was the perfect ambassador to greet and familiarize the others with their new workplace. He soon became known to everyone at the center by the nickname "Romanych." But he also proved he had a mischievous side. On one occasion Popovich and some fellow trainees were literally caught red-handed—picking and eating cherries from a tree in a nearby orchard. The center's doctors were horrified that their cosseted medical specimens had indulged in any unwashed fruit. The other cosmonauts either scattered or pretended they had not eaten any.

Popovich, on the other hand, quietly made his way deeper into the orchard and continued to eat the "forbidden" fruit.

At one time Popovich was among the leading candidates for the history-making first *Vostok* flight, but like the others lost out to the undeniably popular Yuri Gagarin. One likely factor against Popovich was the possible negative propaganda fallout of his Ukrainian background. The same applied for Nikolayev, a Chuvash. Political pressure would ensure that the first man to pave the way into space was a pure-bred Russian, with Russian parents and grandparents. In any event, Popovich and the other cosmonauts had privately concurred that Gagarin was always the outstanding candidate. Nikolayev and Popovich were subsequently advised that they would definitely be used on later Vostok flights to demonstrate the "friendship of peoples" in the USSR.

Swallowing his disappointment, Popovich served with distinction as Gagarin's CapCom during the flight of the first *Vostok*. He later revealed that he found the experience nerve-wracking. The hour before Gagarin's launch was "really one of the most endlessly long hours of my life," he admitted. Yet when about to begin his own spaceflight, he characteristically laughed and joked all the way to the launch pad, even grabbing a microphone and singing with Gherman Titov during the short bus journey. Not satisfied with that, he was soon conducting the entire bus in an impromptu singalong. He would further upset his medical escorts by insisting on shaking hands with the hordes of technicians, officials, and well-wishers who had gathered by the waiting rocket. He was delighted to have his turn to fly at last.

Western observers often speculated on the mysterious time gap between Titov's flight and that of *Vostok 3* and *4*. It is now known that Popovich had been provisionally slated to fly a solo orbital mission as early as November 1961, but this flight fell victim to the race into space. The Kremlin hierarchy was well aware of NASA's declared plans to send John Glenn into orbit at the end of 1961 and began pushing for a more spectacular manned flight to upstage this accomplishment—and not just another manned orbital mission. One potentially achievable goal was to launch two cosmonauts separately into space on a dual flight. The day after Glenn's mission concluded successfully, Commission Chairman Dmitri Ustinov contacted Nikolai Kamanin, head of cosmonaut training, and began exerting considerable pressure for a tandem manned flight to take place as early as 10–12 March, just two

The Heavenly Twins, flanked by their cosmonaut predecessors:
from left, Yuri Gagarin, Andrian Nikolayev, Pavel Popovich, and Gherman Titov.
Courtesy Colin Burgess Collection.

weeks hence. However, the resources, facilities, and cosmonauts were simply not prepared, and other delays resulted when spacecraft and launch vehicles were diverted to the highly secret *Zenit-2* spy satellite program, which was fully occupying Korolev and his team. On 1 June, the second in a planned series of unmanned reconnaissance flights came to an untimely and violent end after liftoff, necessitating considerable rescheduling and repair work. Eventually, Korolev was able to refocus his attention on the Vostok program. From a pool of seven cosmonauts then in training for the planned tandem mission, Nikolayev and Popovich were selected for the task.

Astronaut Scott Carpenter, who had flown his MA-7 mission three months earlier, was holidaying in Denver, Colorado, when he heard the news that *Vostok 3* had been launched into orbit. He immediately sent a message of greetings and congratulations to the world's newest spacefarer. When asked how he felt about this new spaceflight Carpenter unabashedly said he envied Nikolayev. "Man, how I'd like to be that Russian fellow up there," he mused. "I envy him the wonderful experiences he's going through right now—and I know just how wonderful they are." Carpenter also suggested that he was not surprised that the flight's prime objective was said to be a study of the effects of weightlessness on a human being. "I'll be interested to know what they find out about blood pressure on a flight of long duration," he reflected. "There's one of the unexplained mysteries of the flights of John Glenn and myself." The following day, reporters were once again on Carpenter's doorstep. "Wow, another ship in sight," Carpenter stated on hearing the news. "That's quite a feat! I'm sure it would be an exciting thing to be able to see and hear a fellow traveler in space."

Indeed, Nikolayev had manually turned his spacecraft in hopes of observing the launch of *Vostok 4* but was unable to see his colleague racing up to meet him. *Vostok 4*, carrying Popovich, had been launched into a slightly higher orbit, placing him ahead of Nikolayev. Over the next day, *Vostok 3*'s lower, faster orbit gradually closed the gap between the craft, so that on their eighteenth orbit together they were just 3.1 miles apart. Nikolayev even reported that he could now see Popovich's craft through his viewing window. Although the duo teased each other by radio that they should share their food rations, this would be as close as *Vostok 3* and *4* came during the flight. "Western observers leapt to the conclusion that the Soviets had accomplished a rendezvous between two vehicles," astronaut Tom Stafford later remarked. "It wasn't remotely true, of course: Vostok couldn't maneuver. But we didn't know that for years."

Shortly after the two cosmonauts radioed to ground controllers the news of their proximity to each other, Radio Moscow dramatically announced that *Vostok 3* and *4* had commenced a "group flight," which sounded far more impressive than the reality. It also failed to mention that the two spacecraft were drifting apart due to differences in their orbital patterns. Despite this, the Soviets were able to claim mission objectives such as radio contact between two manned ships and comparing the effects of spaceflight on two people simultaneously—tests that the Americans would not be able to attempt or replicate for several years.

As wild euphoria exploded across the Soviet Union, speculation in the West, based on inadequate information, led to some rather bizarre forecasts. The *London Daily Herald*'s science writer naïvely suggested, for example, that the intensely complicated art of space rendezvous was actually "as easy as meeting your girl outside a cinema." He predicted (rightly, as it turned out) that future cosmonauts would be able to transfer from one spacecraft to another in what he called "a dazzling trapeze act." But he also suggested that there were more sinister implications to contemplate. "If the Russians can find each other in space," he suggested, "it shouldn't take them long to find the Americans' controversial spy-in-the-sky satellites and put them out of action."

These views reflected some of the considerable unease felt by many outside of the Soviet Union. Though there was undoubtedly a degree of admiration for their historic achievement, it came hand in hand with concern about Russia's new assertiveness, best symbolized by the construction the year before of the Berlin Wall within a week of Gherman Titov's flight aboard *Vostok 2*. The *New York Times* declared that though the United States was not irretrievably behind in space exploration, it was further back than had been imagined. It acknowledged that putting two spacecraft into close proximity, however briefly, was a scientific achievement "that represents another triumph of human genius in which all men can take pride." Yet the global celebration of Russia's dual flight was tempered by the caution and suspicion of political realities, together with a distinct feeling that these flights pushed the Soviet Union well ahead in the race to put humans on the moon.

Not everything about the tandem flight went according to plan, of course. On Nikolayev's twenty-ninth orbit he reported that the temperature in his spacecraft had dropped to around fifty-six degrees Fahrenheit. The cabin felt unexpectedly cool for the cosmonaut, threatening earlier plans to extend the flight from three days to four. Fortunately, everything else in the spacecraft was working well, and Nikolayev later told ground controllers that he was happy to fly an extra day. The senior spacecraft designers were consulted and agreed that the prolonged flight plan could continue as long as the cosmonaut remained in good health and spirits.

With every passing hour, the dual mission broke more space records. A little more than fifty-seven hours after liftoff, Nikolayev had completed his thirty-ninth orbit and had become history's first "space millionaire" in terms of miles traveled, achieving a distance equal to two round trips to the moon. Tass News Agency reported that both cosmonauts were eating normal food as well as special food squeezed from tubes. The menu also included salted fish, prompting Popovich to joke during training that he'd like to take along some beer to wash it down. Unlike the unfortunate Titov, both men reported that their appetites were good and that they felt well. His usual cheery self, Popovich exclaimed: "My spirits are wonderful!"

In a very human moment that demonstrated the camaraderie among the cosmonauts, Nikolayev found that Yuri Gagarin had stowed away a little joke for him, very much like the small gags American astronauts played on their colleagues. "I picked up the logbook," he recalled, "turned to a fresh page, and my mouth dropped open in surprise: a page covered with brightly colored road signs was lying in the book. Where had it come from? I looked more closely, began reading, and immediately guessed that it was a joke of Yuri's—it was his handwriting. He had put a sheet of road signs in the book, added comments to each, and beneath this was the recommendation 'Learn your road signs! Space is not the Earth—you have to know the rules of the road well . . . we are with you at heart.'"

There was other, more serious work to be done. Both spacecraft carried a variety of biological specimens, and much of the cosmonauts' time was spent studying their reactions and conducting experiments on them. They also filmed the Earth, its cloud formations, and stars and planets. Popovich carried out experiments on the behavior of liquids in weightlessness using a container filled with air and water. He watched fascinated as the air gradually formed one large bubble in the center of the fluid, only to break into hundreds of tiny bubbles when the container was shaken.

As the flights of both spacecraft continued far beyond the length of any previous manned missions, Western space experts pondered new concerns. They felt that the flight clearly demonstrated Russia's superior rocket and propellant power, as well as the ability to launch spacecraft equipped to maintain life for long periods, yet also roomy enough for the occupants to move around quite freely. By comparison, the Mercury astronauts were still jammed into tiny capsules and confined to their couches. It certainly took some of the shine off the upcoming MA-8 flight of Wally Schirra aboard *Sigma 7*, then scheduled for the following month.

The *New York Times* also quoted the Marshall Spaceflight Center's deputy director for scientific and technical matters, Dr. Eberhard Rees: "The Russians have done what we do not intend to do for two years—but they are not necessarily ahead of us in the race to the moon. The Russians probably do not have the rocket power to get to the moon. They still have to develop it." Rees must have said this with a confidence he did not truly feel. The simple truth is that no one in the West knew what the Russians were working on, or their timetable. But the popular view prevailed that a space rendezvous had occurred, and the world remained in awe of the feat. Back in Moscow, hopeful rumors had even begun circulating that other manned spacecraft were about to be launched to join *Vostok 3* and *4* in orbit. In a nationwide radio and television speech on 13 August, President John Kennedy praised Russia's latest achievement in space. He attempted to reassure the American people by stating that the United States had "started late in the 1950s" and would be behind for a period in the future. "But," he added, "we are making a major effort now, and this country will be heard from in space as well as in other areas in the coming months and years."

Meanwhile, in orbit, the two cosmonauts were having a great deal of fun. In contrast to their spacefaring predecessors, Popovich and Nikolayev were allowed to release their harnesses and float about their cabins. The fact that they could work comfortably while unharnessed was important news for medical observers both in the Soviet Union and the United States. After the flight, Popovich laughingly advised his cosmonaut colleagues to do this slowly—he had instinctively tried to stand up in his seat and bumped his head on the cabin wall. To the relief of the physicians, despite having moved around their spacecraft far more than the two previous cosmonauts, neither Popovich nor Nikolayev showed any of the symptoms that had plagued Titov. Nikolayev later described the novel sensation:

It is very interesting to experience this condition when one is not strapped to the chair. During every orbit in conformity with the program I unstrapped myself and got out of the chair. In free soaring a person simply hangs in space touching neither the walls nor the floor. If one does a turn then he begins to turn around his axis like a spinning top.

When it is necessary to move, one simply pushes lightly against the wall and he floats lightly and evenly. The movements of the arms and legs retain their coordination . . . I carried out communications and ate. In a condition of weightlessness one can live and work completely.

Weightlessness is an amazingly pleasant state of both body and soul, not to be compared to anything else. You don't weigh anything, you aren't supported by anything, and yet you can do everything. Your mind is clear, your thoughts precise. All your movements are coordinated. Both vision and hearing are perfect. You see everything and hear everything transmitted from the ground. The devil is not as black as Titov painted him.

Despite his euphoria, by the fourth day of his marathon flight it seemed Nikolayev was showing signs of strain, openly quarreling, for example, with a Soviet ground station that provided incorrect times during their report. Popovich, meanwhile, was thoroughly enjoying the experience, and four decades later he remained in awe of the sights he glimpsed through his spacecraft's porthole: "Looking outward to the blackness of space, sprinkled with the glory of a universe of lights, I saw majesty—but no welcome. Below was the welcoming planet. There, contained in the thin, moving, incredibly fragile shell of the biosphere is everything that is dear to you; all the human drama and comedy. That's where life is; that's where all the good stuff is."

Interior cooling problems similar to those Nikolayev had reported earlier also dogged *Vostok 4*, much to Popovich's chagrin. His temperature readings were much lower than those Nikolayev had reported, indicating conditions barely above the safety limit. Popovich, too, had been happy to extend his mission to four days, but as the temperature dropped to just fifty degrees Fahrenheit, with 35 percent humidity, safety became a concern. "I've tried all necessary measures but the decline is continuing," he reported. It was decided to bring Popovich back on the following orbit, as originally planned.

Back on Earth, the propaganda machine was in full swing. Western observers had noticed something of a shift in reporting from Soviet news sources in the latter part of the dual flight. The earlier informative narrative had slowly given way to intimidating political ideology and a colorful criticism of Western bureaucracy with its "denizens of the Washington capital." Prompted in part by American testing of nuclear weapons in space, the Soviet newspaper *Pravda* criticized what it described as Western "military strategists" plotting "battles in space." *Pravda* suggested that the Soviet Union would not need to conduct its space launches in secrecy if the United States simply laid down its arms, and that "businessmen who worship the golden calf are beset by dreams of the immeasurable riches they could grab by exploring, for instance, the treasures of the moon. The atomic maniacs, for their part, are ready to transform the moon into a military base. What monstrous plans!" Such was the rhetoric of the Cold War.

Meanwhile, Popovich blissfully spent part of his last orbit watching thunderstorms through his viewing porthole. "It looked as if the clouds contained anvils, and some giants were pounding on them," he would later remark, "throwing off sparks that made the clouds flicker." He reported what he was seeing to controllers, and his words sent them into a sudden panic. Before his flight all had agreed that "observing a thunderstorm" would be his coded message for feeling unwell. A veiled query came from the ground, and Popovich suddenly realized what he had done. He added in haste, "I am feeling excellent. I observed a *meteorological* thunderstorm and lightning." His blunder could have meant a mistaken, early return had he not already been preparing for reentry. If an aborted flight had occurred Popovich would likely never have flown again; the Soviet space hierarchy was very unforgiving of such things.

At 9:24 a.m. Moscow time on 15 August, *Vostok 3*'s retrorockets fired. After being only 3.1 miles apart in the initial stages of the dual mission, the two craft were now separated by over 1,780 miles. Thirty-one minutes later Nikolayev landed beneath a parachute, after an automatic ejection from his spacecraft. His touchdown took place in the desert steppe country south of Karaganda in the Central Asian republic of Kazakhstan, about 1,500 miles southeast of Moscow. He had completed sixty-four orbits and traveled 1,639,190 miles around the Earth in ninety-four hours and twenty-seven minutes. "The weather was foul," he later reported, "but I smelled Earth, unspeakably sweet and intoxicating. And wind. How utterly delightful; wind after long days in space."

Nikolayev would later confess that he had endured some misgivings during his plunge through the atmosphere, as he watched the intense buildup of heat around his spacecraft: "It's a very interesting sensation when the craft begins to burn on reentry. Flames rage and crackle outside the window. You think you must be losing the heat-protective coating on the craft. But I knew better than that. I said to myself, 'Take it easy, let it burn, the descent is going according to plan.'"

At the same time, Popovich was reporting similar sensations: "The g-forces were really tough! Outside the portholes there is a sea of flame, of different colors: blue, orange, yellow, dark blue, red . . . very, very big g-forces." Both cosmonauts had chosen not to cover their portholes and were able to observe the intense fiery shroud of reentry. Popovich had also been unnerved by the loud crackling sound, thinking his heat shield might be failing. But unlike earlier Vostok missions, the reentries went according to plan.

Six minutes after Nikolayev touched down, and 124 miles away, Popovich also landed by parachute. He had completed forty-eight orbits in seventy-one hours and traveled 1,230,230 miles. Popovich was equally delighted to be back on Earth. "It was very pleasant," he chuckled, "a feeling that takes one's breath away. If you remember that we spent several days and nights on the craft it will be easy to understand our feelings." Both men had comprehensively smashed all existing spaceflight endurance records. In fact, Nikolayev landed with the beginnings of a dark, bristly beard. He may have been reminded of a time when, not long after he had met Yuri Gagarin, his colleague had jokingly pretended to be a gypsy fortune-teller. "I see a long journey ahead of you, dearie," Gagarin had told the bemused Nikolayev. Now it had come true—he had just completed the longest flight in human history.

The first indications that the outside world had of the two cosmonauts' landings came when Radio Moscow abruptly cancelled its regular programs and began playing patriotic music. Broadcasters then alerted listeners to "stand by for an important announcement in two minutes' time." For eleven minutes the Kremlin chimes could be heard chiming dramatically as anticipation continued to build. Finally, at 10:48 a.m., an announcer gave Muscovites the good news: "In conformity with the program of the flight for August 15, 1962, the spaceship *Vostok 3* with cosmonaut Andrian Nikolayev on board and *Vostok 4* with cosmonaut Pavel Popovich on board have been landed in the predetermined area. The spaceships have landed normally. After the spaceflight and the landing both cosmonauts feel well. The program of the flight of the spaceships has been carried out fully." Crowds of excited Muscovites began to converge on Red Square, where photographers handed out huge pictures of the two spacemen to wave at the world's press, and loudspeakers continued to blare out the news. Groups of people eagerly clustered around car radios to hear updated reports and stood by the open windows of blocks of flats where radios were playing at full volume. The successful dual flight was seen by many observers as tangible confirmation of the Soviet Union's lead in the space race, and the Soviet government crowed that "Communism is scoring one victory after another in its peaceful competition with capitalism."

Following their safe recovery, the two cosmonauts were flown to a secret reception center on the Volga River, where a medical panel gave them a thorough examination. The panel finally declared that the two men had come through their flights unscathed. Nikolayev and Popovich later gave a detailed report on their flights to a committee of doctors and scientists before flying to Moscow for an official welcome home and celebrations to mark their feats. Both men appeared alongside Premier Nikita Khrushchev on top of the Lenin Mausoleum, where they gave enthusiastic speeches to a massive, adoring crowd. They were awarded the title of Hero of the Soviet Union and received the Order of Lenin.

A few days after their safe return, on 21 August, the two cosmonauts gave a press conference to five hundred Russian and Western journalists. It was hardly a spontaneous affair, as all the questions had to be submitted beforehand in writing. Nikolayev began the conference by reading carefully from notes, pausing every few sentences while his words were translated into English. He revealed that the two spacecraft had never been closer than three miles to each other, that both cosmonauts had landed by parachute, and that their spacecraft weighed about the same as the first two Vostok missions, although more advanced and comfortable. Other than that, neither cosmonaut would give any technical details of their spacecraft.

Eventually, Nikolayev set his notes aside and began to spontaneously describe his sensations and emotions while in orbit. He reported seeing storms over the Earth and described how at night he could even make out the lights of major roads. "When I first saw the moon, I rejoiced," he said with a wide smile. He mentioned that the moon illuminated his cabin so well he could make out all of his switches. Normally the more open and fun-loving of the two men, Popovich introduced a note of serious caution into the discussion. "One must not imagine that our flights were a pleasure trip," he said. "Some gentlemen across the ocean try to deny our flight and want proof of our having made it. What can one say to such gentlemen? Follow us into space, then you will see how such flights are made." Nikolayev was also circumspect when responding to a question about potential military threats posed by the dual flight. "*Vostok 3* carried no nuclear weapons," he assured reporters, "and there is no need for such things. But if need be, we have missiles, which are splendidly capable of delivering nuclear warheads to any point." Pencils flew over notepads following this stern admonition. On a more productive note, Popovich mentioned the glowing particles seen in orbit by Gagarin, Titov, and Glenn. "We feel we understand these," he stated. "They are merely the exhaust of the rocket motors." Although it still remains unclear whether he actually saw Nikolayev's spacecraft, Popovich told the press: "Knowing where Vostok 3 would be in relation to me, once I was in orbit, I looked for it immediately and saw it at once. It was something like a very small moon in the distance."

Although both cosmonauts would undertake international tours, they largely escaped the grueling itineraries of their predecessors. They also differed from Gagarin and Titov in that both would journey into space again. In 1970 Andrian Nikolayev, after serving as a backup commander to a number of early Soyuz missions, became the first of the Heavenly Twins to fly again. He had been named to command the historic linkup flight of *Soyuz 6, 7,* and *8*, but he did so badly on the preflight examinations that he was replaced by Vladimir Shatalov. He was chosen instead to command the *Soyuz 9* research mission, although this required a postponement of the mission to allow him extra training time. The mission, pairing him with flight engineer Vitaly Sevastyanov, became known as a "solo" Soyuz flight because it was unrelated to the rendezvous or docking activities carried out on other missions.

History records that Nikolayev also came perilously close to losing this mission. One week before the launch date, General Kamanin caught him in the strictly forbidden act of smoking a cigarette. This was the second occasion on which he had been caught smoking, and a furious Kamanin gave serious thought to replacing him on the crew. Nikolayev was only saved from this embarrassment by the fact that the Central Committee had already approved the crew selection. To change this so close to the launch for a simple, disciplinary matter would have been a complex, difficult task that Kamanin was reluctant to undertake. After a severe dressing-down Nikolayev was allowed to resume training and eventually took his place alongside Sevastyanov aboard *Soyuz 9*.

On their launch day, 1 June 1970, Neil Armstrong was a special guest at a party in the Star City apartment of Soviet cosmonaut Georgi Beregovoi. Just before 10:15 p.m. Beregovoi unexpectedly switched on his television and invited a surprised Armstrong to sit and watch a manned launch. Armstrong had been unaware this was going to happen, and Beregovoi told him, "the flight is in honor of your trip to the Soviet Union." The Apollo astronaut was delighted with the gesture and responded: "It is one of the best gifts you can make to me." There was a certain irony in this, as Sergei Belotserkovsky, in charge of the cosmonauts' academic training, had once said that *Soyuz 9*'s Nikolayev was "the leading candidate" for the first Soviet flight to the moon. Armstrong had flown a mission that Nikolayev would have readily undertaken.

The flight of *Soyuz 9* underlined the differing aims of the two space nations at that time. While America continued to send men to the moon, Russia's priority was shifting to sending cosmonauts on increasingly longer-duration missions and to building orbiting space stations. Tass News Agency reported that the mission would conduct an extensive scientific research program on conditions of solitary orbital flight.

Soyuz 9 remained in orbit for nearly eighteen days, smashing the previous endurance record held by the crew of *Gemini 7* in 1965. The crew returned in very poor shape because of a lack of physical exercise during the flight. They both experienced difficulty readapting to terrestrial conditions, falling ill and having difficulty walking. The condition would later become known in medical literature as the "Nikolayev Effect," and the cosmonaut spoke about it years after in a newspaper interview:

Cramped conditions aboard our Soyuz craft ruled out appropriate exercise, spelling quick physical degeneration in every respect. My heart lost 12 percent of its initial volume over our eighteen days in flight. The first day back home was quite an ordeal.

A few days later, however, Sevastyanov offered to share a cigarette with me, in a washroom where he had hidden it during medical checkups before going up. . . . I smoked first, as mission commander, and then passed the half-smoked cigarette to Vitaly. At this moment our doctor came in, and instead of scolding us, which we certainly well deserved, he burst into jubilation. He told us he felt assured of our quick recovery seeing us smoking.

Following the mission, Nikolayev received his second Hero of the Soviet Union award and Order of Lenin and was promoted to the rank of major general.

As the Soviet Union's third cosmonaut, Nikolayev led an interesting post-Vostok life. Following his epic flight he became his nation's most eligible bachelor, especially when reports suggested that he had sacrificed an earlier love affair in order to undertake his secret cosmonaut training. Press photographs of him frying eggs in his small apartment resulted in thousands of marriage proposals from farm and factory girls. They were all to be disappointed; in 1963 he married the first female cosmonaut, Valentina Tereshkova. This union, however, would end in divorce in 1982.

From 1966 to 1968 Nikolayev served as commander of the air force cosmonaut team, and also graduated from the Zhukovsky Air Force Engineering Academy. Following the untimely death of Yuri Gagarin, Nikolayev became deputy director of the cosmonaut training center, TsPK, in July 1968.

In July 1971, Nikolayev inadvertently caused what might have become a major diplomatic incident. Always fond of a drink, he had quietly slipped out of a cocktail party at the Soviet Embassy in Paris for the less stuffy ambience of a nearby sidewalk café. When Soviet officials realized that he had disappeared they panicked, convinced that one of their senior cosmonauts had defected, valuable secrets and all, to the West. When they finally located Nikolayev he was whisked straight back to the embassy and sent home to Moscow on the first available flight the next day.

Nikolayev's active cosmonaut career effectively ended with the *Soyuz* 9 flight, although he was technically eligible to fly until January 1982, when he and a large number of veteran cosmonauts were officially removed from the unit. He remained deputy of TsPK until he went onto reserve status (effectively military retirement) in August 1992. During his twenty-two years as an administrator he had completed a candidate of technical sciences degree (1975) and earned the rating of senior scientific staffer in 1977, while publishing seventy-five technical and scientific papers. In 1993 he served as a supreme consultant to the Mandate Commission of the Russian Federation's duma (the lower chamber of parliament), and in January 2000 he was elected chairman of the duma's Defense Committee.

The world's third cosmonaut often visited his beloved home village of Shorshely, and in fact was responsible for the construction there of a large museum of cosmonautics, built in the shape of a large white spaceship. Here, young people from all over the Chuvash Republic can take courses in elementary space science and study astronomy in the museum's planetarium.

After his first spaceflight aboard *Vostok 4*, Popovich remained in training for another mission, but over a decade would pass before he once again traveled into the cosmos. From 1966 until 1969 he trained as a crew commander for the ill-fated Program L—the Soviet lunar program. Together with flight engineer Vitaly Sevastyanov, he formed the prime crew for a possible lunar orbital mission. Sevastyanov was replaced later in the training by Georgi Grechko. In 1968, during his lunar mission training, Popovich also graduated from the Zhukovsky Air Force Engineering Academy.

When Program L began to fall into oblivion in the spring of 1969, Popovich received a new assignment as the leader of a group of cosmonauts training for service on the highly secret Almaz space station—a military version of the civilian Salyut station. These stations carried powerful thirty-three-foot focal length telescopic cameras and film pods that could be ejected in orbit and returned to Earth for recovery and analysis.

Popovich was set to command the first such mission to *Salyut 2* in April 1973, but the vehicle fell victim to a propulsion failure in orbit and was unable to receive crews. In July 1974 he finally flew into space again as commander of *Soyuz 14*, together with rookie military engineer Yuri Artyukhin. They docked with the *Salyut 3* orbital space station, launched the previous month, which was effectively the first Almaz military space station. Popovich and Artyukhin would remain in residence for the next fifteen days, performing medical experiments and conducting extensive observations of intelligence targets back on Earth using fourteen different cameras.

Salyut 3 was also equipped with a secret "self-defense" cannon gun, installed in the forward section of the station. The crew could only operate the fixed weapon by changing the attitude of the entire station. However, it is believed that the gun was never actually test-fired in orbit, as any such blasts would have caused considerable reactive shaking of the station.

After their return to Earth an attempt was made to launch another crew to *Salyut 3*, but the automatic docking equipment aboard *Soyuz 15* failed, and the two-man crew was quickly brought back in an emergency night landing near Tselinograd. Following this, *Salyut 3* was abandoned. It was later deliberately deorbited and burned up on reentry in January 1975. When recently questioned by Dutch researcher Bert Vis about military applications and covert activities on this flight, Popovich simply shook a finger in warning, crossed his arms to form an X, and said "*Nyet!*" The subject was still off-limits: end of discussion.

In 1976, Popovich was promoted to major general. By this time he had become deputy head of the cosmonaut training center, and was also working on designs for the Soviet space shuttle. During the 1980s, Popovich headed a Soviet commission to investigate all reports in the Soviet Union of unidentified flying objects (UFOs). In 1987, still possessing his well-known boisterous charm, Popovich became director of what is now called the State Enterprise Russian Institute for Land Ecosystem Monitoring, which develops remote sensing techniques for satellites to survey agricultural sites and forests, land reclamation projects, and bodies of water.

In the 1980s, Popovich remarried, this time to Alevtina Oshegova. His former wife, Marina, a highly decorated test pilot with more than 180 world and Soviet aviation records to her credit, eventually became a colonel in the Soviet air force.

As one of Sergei Korolev's "First Six," a man who was struck by the incredible beauty of the Earth from orbit, Pavel Popovich is still involved with spaceflight. Now in his seventies, he continues to tour the world as an ambassador for the Russian space program, and sees his work at the Institute for Land Ecosystem Monitoring as part of a long and ongoing commitment to his home planet. When asked to look back on his first spaceflight for this book, the normally ebullient Popovich became serious and reflective. "When I look back now, I feel very elated and proud of what Andrian Grigoryevich and I achieved," he declared. "Our planet—the Earth—is very beautiful, and we must preserve it. That is my message. It is alive, and a living organism. We must not abuse it. I was able to look out of my porthole a lot, and like anyone who looks at the planet from space I realized—I am convinced—that the Earth is not the property of any one person or country. It belongs to us all."

Nikolayev had similar feelings about the view that he and Popovich shared in orbit. "Flying above our lovely planet, I was carried away by its beauty," he reminisced more than three decades after his flight. "I watched the bright colors with ecstasy. . . . Earth, in all the astonishing diversity of her beauty, might be compared to an elegant bride behind a thin, tender blue veil." In fact, Nikolayev's comments on peace and fellowship in the years before his death contrasted profoundly with the Cold War rhetoric of the infamous post-Vostok flight press conference. "Our main task," he said in 1994, "is not to allow under any circumstances a thermonuclear catastrophe in our cosmic home. It is the only one we have, at the moment, in the whole universe."

Andrian Nikolayev died of a heart attack suffered on 3 July 2004 while judging the All-Russian rural sports games at Cheboksary, the capital of his native Chuvash Autonomous Republic. Three days later, with friends, relatives, and many fellow cosmonauts including Pavel Popovich in attendance, he was buried in his home village of Shorshely. He was 74 years old.

Today, we may live in a world that is mostly blasé about spaceflight, but in 1962 the tandem flight of Nikolayev and Popovich not only created history, but immense excitement. Though both men accomplished much in their later lives, the incredible journey of the Heavenly Twins remains a significant legacy of the Space Age.

7. The Two Wallys

The secret of health for both mind and body
is not to mourn for the past,
not to worry about the future,
not to anticipate troubles,
but to live the present moment wisely and earnestly.

Buddha

The bus carrying NASA administrator Dan Goldin's group of specially invited guests cautiously made its way back from the shuttle launch viewing area through a nighttime lightning storm. Like the weather, the mood at Cape Canaveral that evening in 1999 was somber; the launch of Eileen Collins, the first American woman spacecraft commander, had just been scrubbed for the second time. On this occasion, only five minutes had remained on the countdown clock. Then, with little warning, another NASA guest bus swerved around Goldin's bus on the narrow road and sped ahead, as if determined to beat them back to the parking lot before the inevitable post-launch-attempt traffic jam. As the driver roared by, he leaned out of the window and gleefully shouted back at Goldin's vehicle, "This is Wally's bus!"

Inside the passing bus, a white-haired woman named Wally Funk laughed with merriment. Her long-held dream of seeing a woman command an American space mission would have to be postponed another night or two, but nothing was going to get her spirits down. After all, it had been almost forty years since she had aspired to become one of America's first women astronaut candidates. Another couple of days were not going to make a big difference.

When most space enthusiasts think about the early days of spaceflight and someone named Wally, Funk is not the first name that comes to mind.

Wally Schirra, who unlike Wally Funk actually did become a Mercury astronaut, may be the better known of the two, but they are surprisingly similar in temperament. Both are fun-loving and fond of a joke, and both share a lifelong love of flying. And when work must be done, both aviators are deadly serious and dedicated to precision in their work. Neither Wally Schirra nor Wally Funk has ever been afraid to be utterly frank with their bosses, whatever the ramifications.

The parallels in the careers of the two Wallys are many; for example, a place in New Mexico where their stories intersect—namely, the Lovelace Clinic, where both underwent a number of astronaut selection tests. There is one ironic difference, however. Though Wally Funk dearly wanted to become a NASA astronaut candidate and did not make it, Wally Schirra did make it but almost turned down the opportunity.

Though Funk's involvement with Lovelace's tests put her in the middle of a bitter battle over who would ride in NASA's early spacecraft, it was never Wally's battle to fight. Outside of her control, a group of dynamic and forceful personalities wrestled to determine whether NASA or politicians would be in charge of astronaut selection and whether candidates' medical or piloting profiles should be the deciding factor. Although Funk never joined NASA, she was involved in a maelstrom of events that threatened to overturn some of NASA's deepest-held beliefs.

One major reason why one Wally made it and the other did not is gender. Though it would be easy to assume that this was a case of NASA discriminating against women in favor of men, as often happens in history, the truth is far more complicated. Funk's chances of becoming an astronaut candidate were over before they even started because of institutional discrimination—but not on the part of NASA. The inequity originated in a decision made at the end of 1944 that effectively determined that no woman would pilot an American spacecraft until the very end of the century.

It could all have been so different. During World War II, women flew military aircraft, including a number of American women who assisted the armed forces by ferrying airplanes for the war effort, so male pilots could concentrate on combat duties. These women aviators risked their lives, and in thirty-eight cases lost them, in the service of their country. In addition to ferrying aircraft, they also flew as engineering test pilots evaluating aircraft that had undergone major overhauls or testing the risky, untried strategy of towing gliders filled with military personnel and their vehicles. They also flew combat-bruised aircraft in almost unflyable condition to the scrap yards, risking engine failures and disastrous tire blowouts. Though given the privileges of military officers, however, they were not formally adopted into the military and remained civil service employees.

If these skilled aviators had been allowed to fully join the services, the history of women astronauts might have been very different. By 1944, however, the number of American men trained as pilots had exceeded combat requirements, and the women flyers were considered surplus. On 20 December of that year the group of women aviators known as WASPs (Women Airforce Service Pilots) was officially disbanded by the War Department. Just thirteen days before, the last class of women pilots had graduated. At the bittersweet celebration, Gen. Henry "Hap" Arnold, who commanded the army air forces, stated that as "more than two years have passed since the WASP first started flying with the air forces, we can come to only one conclusion. The entire operation has been a success. It is on the record that women can fly as well as men." Yet women were now barred from flying military aircraft and would be for decades to come.

The decision to disband the WASPs was discrimination, pure and simple. Although it was not illegal in that very different era, by the more enlightened standards of today it was wrong. A large number of highly capable pilots were being thanklessly discarded based solely on their gender. The WASPs were hardly the only victims. As the war wound down, many of the social advances women had made were rolled back. Much of America began retreating into a cozy dream of domesticity, in which women would not be expected to work at all, let alone fly military aircraft. Some of the WASPs were offered jobs with civilian airlines after disbandment—but as stewardesses, not pilots. It would take thirty years and congressional intervention before they were allowed veterans' benefits.

This reversal of social progress in America would have a knock-on effect fifteen years later. Since women had been barred from flying the best aircraft, none would be eligible when it came time to select America's first astronauts. In a further twist of fate, at the time the WASPs were being disbanded a new aviation technology, known as the jet, was emerging. Because this technology was used almost exclusively by the military during its first few decades, women's exclusion from the advancing path of aviation and aerospace was magnified, and they found themselves confined instead to the technology of the past: propeller aircraft. Although an American woman flew a jet in 1944, it would not be until 1953 that another would get the opportunity. Her name was Jackie Cochran.

Cochran had become the first woman to fly in the Bendix transcontinental air race of 1935, and had subsequently set an impressive list of aviation records. The International League of Aviators named Cochran the world's outstanding woman flier every year from 1937 through 1950 and chose her again in 1953. She had been a powerful driving force behind political efforts to allow women to fly in the war, which were supported by her millionaire husband, Floyd Odlum. Among his many acquisitions, Odlum was the owner of RCA and General Dynamics, which would one day build the Atlas rockets that put Mercury astronauts into orbit. Odlum's wealth helped Jackie to pursue her dreams. Entering air races was expensive, but Floyd could fund her.

Jackie Cochran was a woman who had learned to navigate her way in what was still very much a man's world and had forged many powerful contacts. Through them, she generally managed to get what she wanted. Floyd Odlum also owned part of the company that tested the F-86 Sabre, and with his support she was able to gain access to one. By this time she had befriended Chuck Yeager, the first man to break the sound barrier, and convinced him to teach her how to fly the aircraft. In May 1953, Cochran maneuvered the Sabre jet into a steep dive over Edwards Air Force Base in California, reaching speeds of nearly 720 miles an hour, thus becoming the first woman ever to fly faster than sound.

Cochran was an inspiring pioneer, but was also an aberration. She was able to create firsts because she had more money, better-connected friends, and more persistence than most. Other women were forced to find alternate means to fly airplanes. Flying schools were growing in number, and civilian flying was becoming cheaper in general. If a woman wanted to work as a pilot she might, if she was lucky, find a job flying executives. Otherwise, the only realistic options were teaching in a flying school, flying freight, or crop-dusting. At the end of the 1950s, there were barely more than three thousand women in America with private pilot licenses. It was in this environment that NASA began looking for its first astronauts.

Originally, NASA had considered screening civilian test pilots in its search for astronauts—after all, President Eisenhower was insisting that NASA be a civilian organization. However, in December 1958 Eisenhower agreed with NASA Administrator T. Keith Glennan that active-duty military test pilots would be ideal astronaut candidates. It was a wise choice, especially given the hurried pace of the space race. Military test pilots were the country's best pilots, cleared for classified projects, and easy to reassign. Moreover, they were used to the discipline required of military service, had extensive engineering experience, and routinely flew the world's cutting-edge aerospace vehicles. Their test-pilot skills included the ability to identify problems in flight, clearly communicate those problems to the ground, and work out the cause.

As NASA stated in 1958, astronauts would have to show "a willingness to accept hazards comparable to those encountered in modern research airplane flight," and test pilots had already proved themselves in this very field. Choosing them would also simplify the selection process, as qualified military test pilots were so few in number they all could have fit into a medium-sized lecture hall. In a more prudent vein, military pilots could also be returned to their respective services if America decided not to pursue a long-term space program.

NASA did not state gender in its selection requirements, but more than two decades of discrimination by the military didn't give the agency any qualified choices other than men. Not only did the military still bar women from flying high-performance aircraft except in extremely rare circumstances, but civilian companies rarely hired women pilots either, let alone trained them as test pilots.

Being a military test pilot wasn't enough to become an astronaut, however; it merely meant that your name would be considered. If you were too tall, not healthy enough, or even socially awkward, your name would be taken off the list. A barrage of tests awaited those who wished to apply. One of those who chose to undergo them was a navy test pilot named Wally Schirra.

Born in Hackensack, New Jersey, on 12 March 1923, Walter Schirra came from a flying background—on both sides of his family. His father, Walter senior, was a World War I flying ace who in the 1920s traveled across the United States with his wife Florence performing aerial stunts at county fairs. "My interest in flying was apparently bred into me," Schirra later remembered. "Before my sister and I were born, my parents had a fine time barnstorming in a Curtiss Jenny. Mother was a wing walker. With Dad at the controls, she would dance on the lower wing of the biplane, using the struts for support. It looked hair-raising, and no doubt was." Florence also worked on the aircraft's maintenance and fueling, and even wing-walked during the first few weeks of pregnancy. "She only stopped wing-walking when I was in the hangar," Schirra said in tribute to his pioneering mother. "I was flying before I was born."

Schirra would cycle the long distance to Teterboro's airfield just to watch aircraft take off and land and by age fifteen was flying the family airplane. Being around airplanes also meant he got to meet many of the famous early aviation pioneers. "Dad went to Canada to learn how to fly with the Royal Canadian Air Force," he recalled, "then flew with the precursor to the RAF, before the United States joined World War One. So he had quite an interesting background. He took me on my first airplane ride, where I could have a hand on the stick. I recall going down to Teterboro, New Jersey, where aviation pioneer Clarence Chamberlin took me up on a monster airplane. I also remember seeing war hero Jimmy Doolittle fly a Gee Bee racer there. He was my childhood hero."

Schirra graduated from Dwight Morrow High School in Englewood, New Jersey, in 1940 and then attended the Newark College of Engineering. He entered the U.S. Naval Academy on 1 July 1942, later earning a bachelor of science degree. In the 1946 academy yearbook, *Lucky Bag*, Schirra's classmates wrote the following about the future astronaut: "Here is a guy who could make anyone laugh. His never-ending sense of humor, descriptions and ability to execute new pranks have kept us either amused or holding the bag . . . meanwhile his big brown eyes still have that fields-to-conquer look." At about the time the WASPs were disbanded Schirra entered active navy duty, becoming the first in his academy class to commence flight training. As a naval aviator, he made ninety combat flights in Korea, shooting down two MiGs. Afterward, he hoped to be sent to the navy's Test Pilot School, but was instead assigned to China Lake, California, to test the Sidewinder air-to-air missile. It turned out to be very interesting work that included some challenging flying, but he still hoped for more.

Eventually, Schirra got his wish, and was assigned to the navy's Test Pilot School at Patuxent River. There, he studied aerodynamic theory and worked closely with engineers, experience that no doubt contributed to the next stage in his career: he was given orders to report to the Pentagon, with no reason given. On arrival, he discovered that he was being considered as a candidate for Project Mercury.

Schirra was very skeptical at first, viewing it as a career interruption that he could not afford. He was at the age when a navy career really kicks into high gear—or not. Many a promising aviator had been sidetracked with interesting-sounding projects that ended up going nowhere, and Schirra thought Mercury sounded just like one of those:

We were in a meeting in Washington DC. On the stage were two engineers and a psychologist, telling us how neat it would be to get in a capsule on top of a rocket. I said, I want out of here. I was looking for the "no" desk. As a child I went to a circus in New York City. They had a man shot out of a cannon into a net. At this briefing I said, that's who you want. You don't want one of these engineering test pilots. You just want a dummy to get into a capsule on top of a rocket—I'm not going to do that! I was a test pilot, a truly committed flyboy.

Schirra didn't say no that day, however—although some of his peers did. Jim Lovell recalled Schirra saying, "I'm not so worried about putting my ass on the line. I'm worried about putting my career on the line." But as Schirra weighed the project, it began to sound more intriguing to him:

As I talked to my peer group, they kept saying, "Do you want to go higher, farther, and faster—to push the frontiers out? This is the way to go about it." We talked about how space was a totally different environment. I felt that I had the experience and qualifications for the job. I was certain that my years of combat flying and test piloting had prepared me to handle any kind of emergency. So I went along with it for a while, and finally became intrigued with what was going on. I accepted the challenge, but with great reluctance.

Schirra was also influenced by something that had happened to him not long before: "While I was in Test Pilot School at Patuxent, I was writing a report on the F-4D, an airplane that just barely exceeded the speed of sound. One night, I looked up and saw the booster of *Sputnik* flying by, and I said, that's doing Mach 25, what am I doing working on this slow airplane? Mach

25, that's a pretty good deal. I forgot about it, until I was ordered to Washington. Then I began thinking, maybe this is the way we should be going, not just sitting back waiting for something to happen—become part of it." Schirra allowed his name to go forward for the next series of tests. To this day, he is not sure if that was the best decision for his career:

It was a tough decision to make, because I realized I was going to lose a lot of opportunities. I aspired to having command, and was about ready to go to a squadron to be exec or skipper—my first command. As a naval officer, I was trained, essentially bred, to be a military aviator. I was a naval officer on assignment, not an employee of NASA—I had made a commitment to be a naval officer. But I made a decision that was apparently not retraceable, or it turned out that way anyway. By the time the second astronaut group arrived, they started sliding into the NASA family. Our first group, we didn't know what would happen: no one did. At the end of our NASA careers, it turned out that no one had a place for us in the military. I never had a command, and I regret that. A number of times, I have regretted that.

By saying yes, Schirra had joined a group that would steadily be whittled down to an elite few. He had begun as one of the 508 eligible test pilots on the list supplied by the Department of Defense. As the number was reduced by careful evaluation and the 69 remaining pilots were invited to Washington for the briefing, 8 pilots declined to be considered for the program, most probably for the same reasons that concerned Schirra. During the written and psychiatric examinations and medical history reviews that followed, a further 8 pilots chose to remove themselves from the running. Only after this very extensive selection process was Schirra allowed to progress to the grueling medical examinations conducted at the Lovelace Foundation for Medical Education and Research in New Mexico. He was one of only 32 candidates left.

NASA was still very much a young agency, with few specialized departments of its own yet. For this reason, it looked to Dr. Randolph (Randy) Lovelace and his clinic to take care of the medical-testing aspects of astronaut selection. It was a fateful decision that would push the question of women astronauts into the political arena years before there were qualified candidates. Lovelace, a prominent space scientist who was also the chairman of NASA's life sciences committee, had helped create medical test procedures for the astronaut candidates. He'd been a pioneer in aerospace medicine for decades, helping to develop pilot g-suits in World War II and studying the effects of rapid loss of cabin pressure. Throughout the 1950s, Lovelace had helped organize conferences that, for the first time, looked into issues that might face pilots making space flights. His primary strength was not as a researcher but as a leader who could motivate and change the direction that medicine was taking.

Lovelace's clinic was not part of NASA—rather he operated as a NASA contractor, free to take on other contracts. One of the reasons that NASA chose Lovelace to perform the candidate evaluations was that he was good at keeping secrets. Since aviators were concerned that applying to the astronaut program might turn up medical conditions that could terminate their flying careers, the clinic provided an independent venue. Lovelace was under no obligation to report medical problems back to the pilots' respective service. Applicants felt they could participate without risking their careers.

Not knowing what effects space travel might have on the human body, Lovelace put the candidates through every conceivable trial at his New Mexico clinic. He did not, however, want to limit himself to the job NASA had tasked him with. He was also interested in testing women for spaceflight, and as he owned an independent clinic he was free to do so. Lovelace was not a social crusader, simply a curious scientist who was more concerned with the physical than the social limits of what was possible. From his previous work with NASA, he knew that every ounce of weight saved in the spacecraft was vital. He also knew that women consumed less oxygen, water, and food and on average were lighter than men. What he did not know, as very little medical data existed, was whether women could stand up to the physical and mental stresses anticipated in spaceflight. Lovelace decided that the best way to find out was to test women in the same way he was testing the men sent by NASA, then compare the results. He would do this for himself, not for NASA, and if the results were encouraging, he would approach NASA with his findings. As no one had yet flown into space, it seemed possible that NASA might be open to new ideas.

In the meantime, Wally Schirra had begun the tests. "I allowed myself to be subjected to the innumerable indignities perpetrated on prospective astronauts," he later recalled with a grimace. "I was a tailhook aviator, not a lab rat! I let the doctors play around with me to their heart's content. I am afraid, however, that they got too much data on us to ever figure out. I still feel that the physical exams were an embarrassment, a degrading experience. It was a rare, almost unheard-of situation in which so many healthy individuals submitted to an array of tortures. I have said this many times—and meant it—that it was a case of sick doctors working on well patients." To Schirra, the whole ordeal was a painful, humiliating waste of time. It certainly did not seem to advance the selection process much—the candidates had already been so thoroughly prescreened medically that the Lovelace tests only disqualified one of the thirty-two. The culling of candidates to the final seven had to be done in a more traditional and less painful way—looking over their records once again.

According to Schirra, Lovelace's records were of no use to NASA even after the examinations, since they were not maintained in a way that NASA could use. "The doctors at Lovelace," Schirra recalled, "were trying to establish a physiological and psychological baseline to be used in tests during a spaceflight. That was a valid excuse, I will admit. It might even have amounted to justification for our agony except for one thing—it did not work." Dr. Charles (Chuck) Berry, who was part of the Mercury selection process and went on to become NASA's director for medical research, agrees that the tests did not help. As he later told a NASA historian, the data did not help them narrow down the numbers any more than had background checks and interviews. In fact, during the process Berry claimed that he learned these traditional methods could be far more valuable than Lovelace's medical tests. "This wasn't a medical program," he recalled. "This was an engineering-oriented program." It is a surprising admission for a medical doctor but an honest one.

The tests helped Schirra in one important respect, however. He gained a clue that he was under serious consideration when the Lovelace doctors seemed very keen to have him undergo a minor throat operation as soon as possible, so he could proceed without any medical reservations:

This doctor had been playing around with my ears and nose, and now he was looking further down my throat than anyone had ever looked in my life. I had a node on my vocal cord. It was a tip-off, going for the operation, though I already knew I was being considered. They said, if you are willing to do this visit to Bethesda, which is an unusual place for a young naval officer to go, we'd consider you for a candidate. The doctor who operated said something to the effect of, I have never worked on somebody this junior; you must be going to the moon or something! By then my tongue was being held by forceps, I couldn't say a word!

In the fall of 1959, not long after the Mercury astronauts were selected, Lovelace was at a conference in Miami Beach when he met a young woman pilot named Jerrie Cobb. He was still considering putting women through his astronaut medical tests, and Cobb seemed an ideal candidate to him. Before asking her to undergo any testing, he checked into her flying credentials and found them to be impeccable. Cobb was, in fact, one of the country's top female aviators. She had over seven thousand hours in her flight log, held three speed and altitude records, and had been planning a solo flight around the world via the North and South poles. She did, however, have a total of only one flight in a jet. Though the aircraft was an impressive one, an F-102 Delta Dagger, she had not piloted it solo, having shared the controls with an experienced air force pilot. The Mercury astronauts had between three hundred and two thousand hours in jets, many of them in challenging test piloting conditions. Cobb had handled the controls for forty minutes.

Lovelace's focus on female pilots was, at first glance, an odd one. If he was interested only in seeing whether women could pass the same medical tests given to male astronaut candidates, why did he care if they were pilots? Would the results not be the same if they were healthy women of any profession? The answer appears to be that Lovelace was thinking far beyond the needs of his immediate tests. He had in mind evaluating a group of women who were not only physically capable but who were also the best women pilots in the country—just as the Mercury astronauts were for their gender. It seems likely that he hoped these women test subjects would be selected by NASA as astronauts.

That Lovelace kept his real reasons secret was perhaps due to the experiences of another woman aviator who had just undergone tests with the air force. In August 1959, pioneering aviator Ruth Nichols had undertaken many of the astronaut medical tests at the Wright-Patterson Aeromedical Laboratory, driven largely by her own interest and personal contacts, before the air force put a stop to it. It had no interest in women astronauts, and in fact was largely horrified at the idea. Another very accomplished aviator, Betty Skelton, had also been able to try some aspects of astronaut tests in 1959, this time with the help of NASA and even the Mercury 7 astronauts. Skelton's test was primarily conducted for a magazine article, however, and was never intended to lead to the selection of women astronauts. Wally Schirra assisted Skelton in the spacecraft simulator, the F-104 cockpit, and other parts of the testing. In an interview with a NASA historian, Skelton remembered Schirra as very friendly, smiling, and happy to talk and work with her. In fact, because Schirra's father had performed aerobatics similar to her own piloting specialties, she felt that they even developed something of a kinship.

Lovelace probably had other reasons for keeping quiet, at first, about his testing plans. Keeping his program informal and unofficial meant that he could bypass the hurdles that barred women from access to aerospace technology. But whatever reasons Lovelace had for keeping his project and his motivations quiet, they were ultimately largely the reason why it ended. Not only did Lovelace's secrecy and the way he finally chose to disclose his project's existence get him into trouble with NASA, but the lack of involvement by anyone outside the clinic led most of the women Lovelace selected to believe that they were part of a real training program for NASA astronauts. After all, the head of NASA's life sciences committee was inviting them to his testing facility, where he had evaluated NASA's male candidates. Lovelace, apparently, never corrected this misleading impression.

In February 1960, Jerrie Cobb undertook Lovelace's medical evaluations, which were identical to the ones the NASA Special Committee on Life Sciences had created for the men, but with the addition of a gynecological examination. However, the tests at Lovelace's clinic comprised only one very small part of the testing the men had undergone in their selection process. Cobb did not have to pass through the barrage of engineering, personality, psychology, and other medical screenings that had excluded most male candidates before they even attended the Lovelace clinic. This process had also removed anyone with an IQ of less than 130. Cobb passed Lovelace's tests, with only two minor problems detected—a slight hearing loss in her left ear and the common circulatory disorder known as "cold feet." Though her tests were not enough to qualify a person as an astronaut, they provided useful medical data nonetheless.

When Lovelace announced Cobb's results at the Space and Naval Medicine Conference in August 1960, he highlighted what he saw as the medical advantages of using women astronauts. He supported this by referring to the tests in which Cobb had rated better than many male candidates. She had consumed less oxygen, had a higher tolerance for pain, and was able to endure radiation exposure with less risk. After Lovelace's announcement the story took on a life of its own, and quickly began to enter the realm of fantasy. The press instantly decided that Jerrie was America's first woman astronaut and besieged her with calls and interview requests. They did not care that the tests were only for Lovelace and not for NASA—they liked the sound of a "lady astronaut" story, and if it didn't exist they'd invent it. Media reports appeared that used inventive terms such as *Astronette* and *Moon Maid*. When Cobb returned to Oklahoma, the governor gave her a formal greeting ceremony at the airport and commissioned her Oklahoma's "Ambassador to the Moon." Offers to endorse commercial products poured in.

Lovelace had helped create this mass misimpression by telling the press that Cobb was now qualified to "live, observe and do optimal work in the environment of space, and return safely to Earth." To take one small portion of astronaut physical testing and say it proved a person could pilot a spacecraft and conduct a successful space mission was, at best, an extrapolation. But Lovelace was an expert, and Cobb had no reason not to believe him. She took Lovelace's assessment as an incentive and grew determined to become a Mercury astronaut.

It was at this point that Jackie Cochran began to take an interest in the program. A close friend of Lovelace's, having first met him back in 1937, she had become increasingly curious about the ways in which he sought to explore and advance the frontiers of high-altitude flying. She was particularly interested if they could help her set more aviation records. As chairman of the Lovelace Foundation's board of trustees, Cochran's husband, Floyd Odlum, was funding much of Lovelace's aerospace research to the tune of thousands of dollars. This gave Cochran some particularly powerful leverage to become involved in Lovelace's work. Finally, Lovelace had also won aviation's Collier Trophy primarily because of Cochran's active campaigning. In short, he owed her a lot.

Offering to use her husband's money to underwrite the female "astronaut" testing, Cochran asked that she be involved in the selection process for the other women who would take the tests. It was not long before she and Cobb were wrestling behind the scenes to decide who was actually leading the women's testing program. Cochran even put herself through some of the tests and was not at all happy when told that she was no longer in good enough shape to pass them. For all her personal ambitions, however, Cochran was very realistic about what the women's testing program could achieve. From the outset, she believed that it needed to be a long-term project that drew on a large group of women. She also knew that it might be quite some time before the first American woman flew into space.

With Cobb's story appearing in countless newspapers, NASA called a news conference to say that it had no plans then or "in the foreseeable future" to choose women astronauts. The press did not seem to care and continued to refer to Lovelace's program as a NASA initiative and to Cobb as an astronaut. Cobb did not go out of her way to set the media straight, and in fact used the publicity to try and persuade NASA to allow her into their astronaut program. However, the stories had in fact annoyed many at NASA, who preferred to see media coverage on the actual programs they were undertaking. NASA's only impression of Cobb at that time was as a minor irritant who seemed to be generating stories that they continually had to refute. Cobb's relationship with NASA was not off to a good start.

At the same time, Cobb was helping Lovelace select other women to test—a necessary next step, Lovelace felt, to prove that Cobb's results were not just a "one-off." Once again, those chosen were among the country's top women pilots, but none had ever piloted a jet solo. One woman, Jean Hixson, was a former WASP who had made two flights in 1957 and 1958, but as a backseat passenger. Lovelace never gave his project a formal title, but Jerrie Cobb, deciding the participants deserved some sort of name, settled on the acronym FLATs, which stood for Fellow Lady Astronaut Trainees. Though she liked it, the program's other women appear to have shunned it. Decades later the media invented its own name for them—the "Mercury 13." But despite all the forced group identity, the women aviators went through the testing at different times, and many never even met, either then or in the decades that followed.

The letter that Lovelace sent to many women inviting them to apply referred to the tests as "The Woman In Space Program" and as "the initial examinations for female astronaut candidates." Any woman pilot receiving one of these letters who wasn't privy to NASA's policy could naturally have formed the impression that she was being selected for a NASA astronaut program. As they arrived for the tests, many dreamed of a bright future exploring a new frontier. The youngest selected was barely twenty-two years old: Wally Funk was about to face the same doctors Wally Schirra had endured not long before.

Wally Funk was born sixteen years after Schirra and grew up in Taos, New Mexico. Like many families in the region, her parents had moved to the warmer climate to combat poor health—in Funk's case, that of her father. The family ran a store in town, and Wally sold carvings, bows, and arrows at her own stall outside. Mary Wallace Funk II was her full given name, after her great grandmother. But when her family celebrated her first Christmas the name was too long to write on her stocking, so her grandfather wrote "Wally" instead, and the nickname stuck.

Wally's mother may have expected her to be ladylike and demure, but she was never going to be that way. At age three, she wandered out of the house one day and befriended some workers laying a pipe in the road. Wally's mother was horrified to find her mud-encrusted daughter in the company of laborers, but to Wally it was just the first of many adventures in life. She grew up fishing, riding, and hunting with the local Taos Indians, very much the tomboy. "The Indians taught me a lot about my spirit, how to live life," she remembered later. "They taught me how to survive the wilderness. A kid today in a city hasn't experienced skiing, snow, or air like I did. I played with snakes and rode horses bareback from the age of five. I was very adventurous, a free spirit."

Though Taos had only one airplane at that time, Wally dreamed of flying as a child. "I was five years old when I put on my Superman cape before climbing out onto the roof of our barn at home," she recalled. "I jumped off the top of the roof into a haystack and imagined I was really flying. For me, flying is really about freedom." Her father encouraged her interest by bringing home model airplanes for her to build, and soon many were hanging from her bedroom ceiling. She frequently persuaded her mother to drive her to the local airport just so she could see the lone aircraft there. Wally studied it intently, gathering ideas for her aircraft models. By the time she was a teenager, she was also putting her energies into sports, including representing the southwest United States in downhill skiing at Olympic team trials. "I was skiing as soon as I was learning to walk," Funk recalls with a smile. By the age of fourteen she had also earned a Distinguished Rifleman shooting award from President Eisenhower.

Wally's parents encouraged her to see more of the world than Taos could offer, so she headed for Stephens College, a two-year girls' school in Missouri. It was hard for the tomboy to fit in at first—she was expected to wear hats, hose, heels, and gloves—but she soon discovered a new passion quite by accident. It came as the result of a skiing accident: "I fell twenty feet and broke my back," Funk explains. The injury forced her to find a less physical activity during her recuperation. A college advisor suggested an aeronautical sciences class, and she gave it a try. At first, Wally, still confined to a half-body cast, could only take part in the course's classroom component. Eventually, she was fit enough to try the real thing—flying—and did so with her parents' permission. She was hooked before her first flight was over. Her parents agreed to pay for lessons, despite the huge financial strain this put on the family. By the time she was seventeen, Wally had her flying license. "I knew then," she said later, "that I wanted to fly for life. I am married to airplanes—my whole life has been aviation. I eat it, live it, and breathe it. I get withdrawal symptoms if I don't fly every three days, at least."

It was only in the 1990s that Funk discovered why her mother agreed to let her pursue so unladylike a hobby. In 1919, Virginia Funk had taken a brief barnstorming plane ride herself and had also asked her parents if she could learn to fly. Her father had told her no; she was to learn instead to be a "good wife and mother." So she buried her ambitions for all of those years—until her daughter asked her the same question. "She passed along those flying genes to me," Funk remembers, "and she supported me all the way."

After graduating from Stephens College, Funk transferred to Oklahoma State University, which she says "was the best flight school in the United States at that time." By her senior year she was winning national piloting trophies, and for two years running was the college's top female pilot. Though she left college in 1960 qualified to be a schoolteacher and certified as a flight instructor, she had already decided she would rather be a pilot. There was disappointment ahead, however. She applied to two commercial airlines, but was turned down. In a reflection of the times, the reason

Wally Funk as a flight instructor at a flying club attached to Fort Sill military base. Courtesy Wally Funk.

she was given was that "there were no ladies' bathrooms in the training facilities." Wally's childhood and education had been very similar to Wally Schirra's, with supportive parents encouraging an interest in aviation. Yet while Schirra could join the military to advance his career, Funk found those opportunities closed to her.

Despite the limited options for women pilots, Funk found a great aviation job right out of college. Though Wally wasn't allowed to become a military pilot, the flying club attached to Fort Sill military base in Oklahoma offered her a job as a flight instructor. She became the first female instructor in their history. Right from the outset, she was inundated by servicemen who wanted to learn to fly, and she subsequently trained hundreds of them to solo. She enjoyed the military life around her, and though she was not an official part of it she enjoyed the place she had found in it.

Funk's name was not on the list of pilots that Cobb began submitting to Lovelace in late 1959. As Cobb later wrote, "Wally was only twenty-two; an age we hadn't dreamed would produce a woman pilot with enough flying experience to qualify." Instead, Funk read about the tests taking place in her home state while reading an October 1960 *Life* magazine at the base quarters. "I read about Jerrie Cobb," she recalled, "and I was really taken by the heart—I really wanted to be an astronaut." By then, despite her youth, Funk had accumulated a staggering three thousand hours of flying time, and she could already claim an extremely impressive résumé. She'd never given much thought to space flight before then. "I didn't even know about it," she admits. "It was just as new as new. I didn't know what 'astronaut' meant, but I knew it was for me. I've always been an explorer, and I wanted to explore the new frontier of space."

Once she had discovered who was behind the tests Wally wrote to Lovelace directly, expressing a wish to be tested. "If I see an opportunity," Funk explains, "I go and seek it out. I wrote a letter and said I wanted to be a pioneer in space—could I please apply?" The *Life* article's wording suggested that the tests were a prelude to a spaceflight for Cobb, who was described as a "prospective space pilot." So in her letter to Lovelace Funk stated that "I am most interested in these tests to become an astronaut." Her follow-up letter to Lovelace said that "my interest now lies in the direction of the astronaut and space program." She had no idea that the media had distorted the story, and when Lovelace replied in January 1961, inviting her to be tested, he did not correct her. In fact, Lovelace wrote that "examination of potential women astronauts is continuing," and subsequent letters to her referred to the "Woman-in-Space program."

Lovelace was particularly interested in testing Funk. Not only was she a competitive skier, but because she had been raised in Taos she was also physiologically adapted to higher altitudes. These factors intrigued him from a medical standpoint. Funk began a self-imposed regimen of exercise, believing that she was on the first rung of consideration as a NASA astronaut. "I wasn't aware of the test pilot requirement at the time," she remembers. "I was so out of the loop. Too bad I wasn't told that to begin with."

In February 1961, NASA was making early preparations to send the first American into space. That same month, Funk's mother drove her to the Lovelace clinic, where she became the third woman to undergo the testing program. "I was the youngest too," she adds. "They just called me The Kid." The tests kept her on a strict timetable, specifying when and what she could eat—and long stretches of time when she could not eat at all. During that grueling week, Funk underwent dozens of x-rays and fifty other physical examinations, which took eight hours each day. "Every part of our body was x-rayed. Everything that you saw in *The Right Stuff*, that's what we did—pretty barbaric," she recalls. "I had no preconceived ideas of what was going to happen in any of these tests; I had no idea of the things that could be done to my body and mind. Being a grownup now, I might have had some reservations going in!"

Her adaptability, stamina, and general physical health were closely scrutinized. The tests included pelvic examinations, drinking radioactive water, and having needles stuck into her head to record brain waves. To test her nerve response, a needle was also slid into her hand and then an electric current was passed through it. In the evenings, she was required to give herself enemas:

A lot of the tests were very challenging. I took a lot of pain. The things I was tested on aren't tested anymore—they went to the extreme. Our hands and feet were submerged in icy water, we drank radioactive water, and had barium enemas. The only thing that really hurt was when they injected supercooled water into my ear, to create disorientation. That was really painful. I really wondered if it was worth it when I had to swallow three feet of rubber tubing, castor oil, and barium. But I took it all in my stride. I love challenges, and would have endured anything. It was going to get me one step closer to space, I thought, and that was where I wanted to go.

Funk remembers Lovelace being very supportive and helpful throughout the week—he wanted her to pass. She didn't ask what all the tests were for, emulating the attitude instilled at the military base: "Don't ask if you don't need to know. I didn't question the doctors," she recalled. "I just wanted to pass, with very high marks."

Donald Kilgore, one of the doctors who administered the tests to the women, remembers Funk well. "I was impressed with her motivation and vitality," he later reflected. "She was the youngest candidate and an Olympic-caliber skier." He also remembers Funk and Cobb as being among the women who were most dedicated to passing the tests.

At the end of the week, Funk's scores were evaluated, and she was told how she had done. Kilgore remembers that "she passed the tests with flying colors." Jackie Cochran also wrote to Funk soon afterward, telling her she was "delighted that your tests were so satisfactory." The tests had indeed proved that the women pilots were as physically resilient and capable of enduring the tests as the male astronaut candidates who had been tested. Lovelace told each of the women who passed that there was no reason for NASA to exclude them from becoming an astronaut based on their performance on the tests. In her autobiography, Bernice Steadman, one of the female pilots tested, recalled that Lovelace told her that "after our jet training was finished, he was positive that we would be included by NASA in what he described as a Woman In Space program. His enthusiasm was contagious; . . . we had a reasonable expectation that he knew what he was talking about." Lovelace did not tell them that in reality NASA had no plans to include them. Instead, he asked each of the women who passed if they were prepared to take a second phase of tests. Funk and the others all said yes. If Funk had seen an article by Jackie Cochran in April 1961's *Parade* magazine, she might have decided differently. In it, Cochran stated that the testing was private, unofficial, and "just a gleam in the eyes of the doctors." She also predicted that it would be six or seven years before women would fly in space.

Lovelace and Cobb had arranged for the examinations to continue with a psychiatric evaluation at a college of medicine in Oklahoma. The Mercury men had gone through stress tests at Wright-Patterson Air Force Base; Lovelace, improvising as he went, believed that the Oklahoma assessments would be a close-enough equivalent. Psychologists were concerned that long-duration spaceflights might lead to feelings of isolation, depression, and hallucinations, and the evaluations were designed to test candidates on these potential unknowns. Cobb had been through three days of evaluation there in September 1960, which included general intelligence examinations, Rorschach tests, electroencephalograms (EEGs), and personality analyses. Now, almost a year later, Lovelace was ready for the other women to follow.

The Oklahoma evaluations included testing sensory deprivation with an isolation tank rather than isolation in a silent, dark room as the Mercury men had experienced. The Oklahoma trials used a lightproof water tank, eight and a half feet deep. Floating pillows supported the body, while the water was carefully heated to the subject's body temperature, creating a feeling of weightlessness. It became difficult to tell where the body ended and the water began. Cobb had set a record by floating in the tank for nine hours. Funk was again the third woman to be tested. "When they put me in the tank," Funk remembers, "the water and air temperatures were exactly the same as my body temperature. I couldn't see, hear, or feel a thing. Most people start to hallucinate after a while, but I didn't; I took a couple of naps." Funk stayed in the tank for over ten and a half hours. She didn't say a word the entire time.

Lovelace was able to organize—unofficially—a third phase of testing at the U.S. Naval School of Aviation Medicine in Pensacola, Florida. By this stage, he had performed all the trials he was able to carry out at his clinic and needed access to advanced military equipment to proceed further. Funk was asked to be ready to do this in June 1961, then July, before it was postponed into September. When Lovelace told the women about the Pensacola tests, he also advised them to begin thinking about how they wanted to handle the publicity that he believed would be coming their way. He said that they could expect it to be similar to what the Mercury 7 astronauts had endured and that signing a deal with *Life* magazine might be a good option. Once again, Lovelace was intimating that his unofficial testing would become a true astronaut training program by making direct comparisons with "the male astronauts" and discussing "group policy." As a result, Funk recalls that "I absolutely believed that if I passed I would be part of the NASA corps."

Jerrie Cobb went through the Pensacola trials in May 1961. These tests examined the physical stresses associated with space, such as g-forces and motion sickness. They also involved simulating conditions up to sixty thousand feet in an altitude chamber while wearing a pressure suit. Among many tests, she was fired along a rail track in an ejection seat, asked to perform tasks in a rotating room, and propelled into a pool of water while inside a simulated aircraft cockpit. The climax of the evaluation was a backseat jet ride in a Douglas Skyraider being taken through a variety of dizzying loops and rolls. An EEG monitored Cobb's brain waves while a camera recorded her facial movements. Once again, she passed the tests. Cobb dreamed of what she hoped would come next. "We had reached the top of the ladder," she later explained. Despite the harsh reality, she believed that "the next step would be—would have to be—an official astronaut training program for women."

The world of spaceflight that Randy Lovelace thought he knew was changing fast. Nineteen sixty-one was turning into a very busy year: Yuri Gagarin made mankind's first spaceflight in April, closely followed by America's Alan Shepard. Cobb was hoping to capitalize on the momentum of the year, and shortly after Shepard's flight in May she wrote to NASA's newly elected administrator, James Webb, suggesting that a woman be sent up on a Redstone that summer before the Soviet Union beat them to it. She told Webb she was sure the Mercury 7 astronauts would not mind waiting out one flight so a woman could achieve this. She could not have been more wrong.

Shepard's flight had in fact given NASA a clear new mandate that would keep it occupied for the next eight years. In May, President Kennedy had directed NASA to land on the moon by the end of the decade. Any form of human spaceflight not related to that particular goal was not going to occur. There was no political will to send a woman into space just for the sake of doing it, but there was now tremendous political pressure to beat the Russians to the moon. NASA, a federal agency whose funding was decided at the whim of politicians, knew what it had to do; for now, it wouldn't have time for women astronauts.

Moreover, another argument for using women—that they generally weigh less than men—was now proving to be irrelevant. NASA had clearly demonstrated that it had spacecraft and rockets capable of launching its male astronauts. Shepard's flight, though short, had also removed many of the aerospace medical community's concerns. Anxiety over what the unexplored environment could do to a person, physically and psychologically, fell sharply as the first astronauts returned fit and happy. Medical tests that had seemed so important in the original astronaut selection process now seemed far less important than other attributes—such as piloting skills.

The Mercury astronauts had, in fact, been persistently pushing the Mercury spacecraft designers from the beginning to make the pilot more central to the program. They insisted that, rather than being mere passengers and medical subjects, they should be able to use to the fullest their test piloting skills. This meant, in effect, having as much control over the spacecraft as could be given to them. Allowing pilots who had never even flown a jet aircraft—male or female—into the astronaut corps in the middle of such a debate, based only on the results of medical tests, would have been, for them, a step backward. Donald Gregory, who became NASA's executive officer of flight crew operations, explained their thinking in an interview with a NASA historian: "The early guys went through testing that was unbelievable, that was ridiculous. You look back and you say, what on Earth? What the hell were those guys thinking? We've got to put them through this rigorous testing? Yeah, right. What did it show? Nothing. There were a lot of things that you could look back on and see we should never have got involved in. But maybe it was best that they did, because it was shown how ridiculous some of that was, to get the medics off our backs and make it easier for the later guys."

Yet the same month as Shepard's flight, James Webb suddenly and unexpectedly gave Cobb a glimmer of hope. At a conference in Tulsa, Oklahoma, that both attended, Webb led Cobb to believe that she would be appointed a special consultant to NASA. Though he had not asked her before the conference, nor worked out any of the details of what this role would mean, Cobb took it as a sign that her campaign for a flight was working.

However, it is not clear that Cobb was actually asked to be a consultant that day. A close reading of Webb's speech shows that he said: "I expect to ask her to serve as a consultant on the role of women in the space program." Expecting to ask is, of course, not the same as asking. Ultimately, Cobb was never sent any formal paperwork from NASA regarding a consulting position, never asked to perform any official duties, never told what she was supposed to consult about, and was never paid the standard NASA consulting fee (nor was a salary ever discussed). Assuming that her task would be to work on the women-in-space issue, Cobb began preparing a report on that subject. But her services were never requested. It seems likely that Webb only made his surprising announcement in the hope that Cobb and her pronouncements would stay more in step with NASA's plans. If this was his intention—and his true intent can never now be known—it did not work.

Cobb believes that she *was* a consultant. Indeed, in her autobiography she describes taking the oath of allegiance to serve NASA at a "swearing-in ceremony in Washington." However, no record of this ceremony can be found. Martha Ackmann, author of *The Mercury 13* and the foremost researcher studying the women tested by Lovelace, doubts the ceremony ever took place. "I don't believe Cobb was sworn in," she told the authors. "I compared Cobb's letters to Webb and Cochran's letters about meeting Cobb in Paris and ascertained that right after the Tulsa talk, Cobb departed for France. I researched Cobb's file in the NASA history office in DC, and the only mention they have of her as a consultant is a slip of paper that said she was paid nothing."

Meanwhile, Cochran, feeling slighted by the possibility of Cobb's appointment and wanting to be more involved, persuaded Lovelace to delay the Pensacola examinations until September so she could be there. She offered to pay for the tests, thus bringing them more under her control. She also started to look at the whole project a little more closely, wondering if they would interfere with the existing astronaut program. When she began questioning her contacts in the navy about this, the navy began to pay more attention to the tests they had casually agreed to. They made inquiries to NASA—would they have a need for women astronauts in the near future? For the first time, NASA was officially asked to comment on the usefulness of Lovelace's private testing program.

In the meantime, Wally Funk stepped up her physical training program. In July, a letter from Jackie Cochran gave her her first indication that Lovelace had been too ambitious about the results of his testing. Though Cochran's letter "strongly urged" Funk to take the examinations, it also told her plainly that no astronaut program for women existed at that time and that the tests were "purely experimental and in the nature of research, fostered by some of the doctors and their associates interested in aerospace medicine." She stressed that no program for women had been officially adopted by any government entity. Furthermore, Cochran explained that the current testing did not include enough women to be meaningful in her opinion, but that she hoped a "properly organized" program would be developed in the future. The same letter was sent to all of the other women participating in the evaluations.

The letter was unexpected news for Funk. "I was heart and soul into the whole program," she remembers. "I thought we would go to Pensacola, do more tests, and become part of the Mercury team. I'd never heard the word 'promise,' but all of Lovelace's letters until then read as if we would become women astronauts." On September 12, six days before she was due to travel to Pensacola, Funk received a telegram from Lovelace: the tests had been cancelled. NASA had responded to the navy's inquiry and made it quite clear that it was not interested at that time in selecting a group of women astronauts. The navy saw no reason to use national resources to test them. Lovelace had managed to get one person—Cobb—to undertake the Pensacola tests. But with the "women astronauts" story now in the national media and Lovelace asking for twelve more women to be tested, he could no longer expect to quietly gain approval from the navy. In fact, the whole process had been so unofficial that the tests were never officially "cancelled" by the navy—they were simply "not required."

Funk took it in her stride. "I was disappointed," she remembers, "but you go on with your life, on to your next assignment—and that's what I did. I have no regrets." Jerrie Cobb, however, saw the test cancellation as an affront and a personal challenge. She believed that women deserved an immediate place in space, and she was determined to prove it. Despite her public pronouncements that her campaign was for women in space in general, Cobb's letters to Webb that year make it clear that she had undergone the testing "in the eventuality that you might need a qualified woman in the space program." She wanted to be the one to fly and was frank about this in her autobiography: "I can't deny it—I want to be the first woman in space. I am not fooling about wanting to go all the way—all the way into space as an active participant." She was surprised that the consulting possibility had never materialized. Nothing seemed to have come of it except a letter from Webb in September 1961, stating, "I know we have been slow in working out arrangements for utilizing your services."

When the tests were cancelled Cobb flew to Washington DC. In her autobiography she described how she talked to every NASA and military official she could "corner long enough to listen to me. To say that I went about it with zeal and enthusiasm is an understatement," she recalled. "I built fires wherever I went." She met with the Chief of Naval Operations in Washington—after turning up at the Pentagon and talking to anyone who might have influence with him—until he finally agreed to a meeting. He could only repeat what she had already been told by everyone else: NASA had no requirement for the tests to be carried out.

To keep the pressure on Webb, Cobb began making a series of public speeches across the country. Rather than persuade Webb, this fresh campaign had the opposite effect on the NASA administrator; he withdrew even more from the idea of using her as a consultant. In December 1961, he told a conference organizer point-blank that Cobb was not a consultant for NASA. Within days he had reiterated this in an official letter to Cobb, stressing that he had not found any assignment for the possible consulting role "that you and I had in mind."

On 20 February 1962, John Glenn became the first American to orbit the Earth. The same day, NASA instituted an equal opportunity employment policy. In it, Webb stated his intention "to take positive steps to ensure equal opportunity for employment and advancement for all qualified persons on the sole basis of merit and fitness without discrimination on the basis of sex." Following this declaration, NASA actively increased the number of women engineers and technicians it employed. The agency also publicized its most notable women workers so as to encourage other women and girls to enter scientific professions. Yet because women were still not being admitted to test pilot schools, the policy specifically listing "qualified persons" could have no effect on astronaut selection.

NASA was indeed employing the best-qualified candidates for any position, whoever they were, and this often meant that they were breaking down racial and gender barriers ahead of other federal agencies. Gene Kranz, in an interview with a NASA historian, recalled that Mission Control contained a very diverse group of highly qualified people, of all races, who were focused solely on getting the job done. Kranz did not care where they came from as long as they were the best asset available. "We added women into this very critical equation," he remembered, "and they had no problem not only in measuring up but taking the lead. In Mission Control today, we are at forty percent women. We were probably one of the first truly equal opportunity employers within the federal government."

Astronauts' nurse Dee O'Hara agrees, emphasizing that discrimination, particularly against women, had no place in the agency when she first arrived at NASA:

Back then it was completely a man's world, and at the Cape it was totally a man's world. It was the early 1960s, and there were not many women involved in di-

verse careers as engineers, physicians, and jet jockeys. As we know, that came later. The people at NASA *flying airplanes, designing spacecraft, and working on booster rockets were all men. I believe there were a couple of women secretaries working in Hangar S at the Cape in addition to myself during the Mercury program, but that was it—I was almost the only female there. But never once did I sense or feel discrimination in any form. I was never made to feel uncomfortable. It was in fact a wonderful working environment; I can say with all honesty that* NASA *did not discriminate against women. I was treated with respect; we joked and kidded around, but we were also very professional. Everyone had a job to do, and we did it. More importantly, we worked as a team—it was always a team effort. It wouldn't have worked otherwise.*

It all boiled down to the best-suited person for a position, regardless of what the position was. For the Mercury program, the astronaut candidates had to have a minimum number of flying hours in a variety of jet aircraft, and have been through test pilot school, along with many other requirements. It really was not NASA*'s fault that women were not qualified back then.*

Dr. Lovelace, in the meantime, was doing his best to mend fences with Jim Webb. To NASA, it seemed as if the researcher had been trying to create his own semiofficial group of astronaut candidates, which he would then try to thrust upon them. That wasn't the arrangement NASA wanted with him, and the situation threatened Lovelace's position as chair of NASA's life sciences committee. The loose conglomeration of talents and organizations that comprised NASA at its inception was rapidly changing. With time, experience, and Kennedy's strong directive to go to the moon, the agency was streamlining, strengthening, and focusing. Individual interests such as Lovelace's were not going to be considered if they were not directly related to landing on the moon. Realizing how politically precarious his situation had become, Lovelace made his apologies to Webb and carefully backed away from the whole issue of women astronauts.

Cobb did the exact opposite. She kept pushing for meetings with NASA executives, White House officials, and even Wernher von Braun, who also told her that "jet-test experience" remained a compulsory requirement. She had another ally in her efforts to apply political pressure, however. Jane Hart was one of the women who had undertaken the tests, but she was also the wife of Senator Philip Hart and very politically active in her own right.

"We talked to everyone but the president," Cobb recalled, "and that wasn't because we didn't try." The two women also pushed hard for an appointment with Vice President Johnson, and finally got one. The same month that Johnson agreed to meet with them, Jackie Cochran wrote to Cobb and said of women astronauts: "their time will come, and pushing too hard just now could possibly retard rather than speed that date." Yet Cobb persevered. "Our strategy was to surround NASA," she wrote shortly afterward, "to force a decision one way or the other, though one way—our way—was what we had in mind. Relentlessly, we applied pressure."

Johnson agreed to meet with Hart and Cobb on 15 March 1962. At the same time, he started making inquiries at NASA, asking why women were not part of the program. After weighing all the factors, he decided that the selection process had to remain a NASA decision rather than being pulled into the political arena. At their meeting, Johnson was polite to Hart and Cobb but firm—he was not willing to help. Following their meeting, Cobb continued to write and press Johnson for answers. Eventually, he had to spell it out to her—it was up to NASA. The vice president patiently told Cobb that "there are no standards for selection that exclude women, but the requirements are so exacting that only trained and experienced high-performance jet aircraft test pilots can be expected to succeed." Johnson's staff had drafted a letter for him, in case he chose to ask Jim Webb to look into the question of women astronauts. Instead of signing it, the exasperated vice president simply wrote four words across it—"Let's stop this now!" Cobb and Hart weren't to know it yet, but the effort to apply political pressure had not only failed—it had in fact backfired.

Jackie Cochran was also trying to tell Cobb the evident truth in her letters at this point. In one, she put it as bluntly as she could: "Two tests or checks do not constitute a program." She even took Cobb out to lunch and told her she wasn't going to support her campaign. But Cobb wasn't prepared to accept this—in fact, she and Hart were still trying to build up political pressure for what would be the final showdown. In March 1962, Hart was busy writing to both the Senate and the House of Representatives, asking their space committees to look into the issue of women astronauts. The Senate was not interested, but the House of Representatives agreed to hold hearings in July on the issue of whether women were being discriminated against in the space program. Cobb and Hart would present their viewpoints in person, as would Cochran, while John Glenn and Scott Carpenter would represent NASA's viewpoint.

Lovelace did not attend the hearing, but he did share his thoughts with Jackie Cochran beforehand. He now told her that he did not believe a woman should be immediately placed within NASA's astronaut program. Rather, medical testing should continue, and gaining meaningful data might take as long as five years. His aim, he stressed, was more to create the right data and conditions for the future, not to initiate a crash program. Cochran fully agreed with him. Lovelace then read and approved a statement that Cochran had prepared for the hearing.

Just before the hearings Cochran wrote to James Webb, suggesting that this might be the time to draw the line, for NASA to state definitively that the women were not eligible. "I agree with your decision that women should not be included in the Mercury program at this time," she wrote. "My idea is not to keep women out of the space program but to get them in at the proper time and in the proper way." She added that the idea of a hurried program to beat the Soviets into sending a woman into space did not make sense to her "from the standpoint of overall national interests."

On the first day of congressional testimony, Jerrie Cobb began by laying out the facts clearly and correctly: "I could not answer the minimum qualification for an astronaut because I am not qualified. . . . The qualifications that the authorities of NASA have set down have made it impossible for women to qualify as astronauts or ever demonstrate their capabilities for space flight; . . . some of us have worked as test pilots, but it is impossible for a woman in this country to be a jet test pilot because there are no women pilots in the military services and the test pilot schools are operated solely by the military services." Cobb had hit precisely on the real issue—one that was not NASA's responsibility. Her proposed solution was not to correct discrimination in the military, however, but to bypass the whole pilot training process altogether. She suggested that the selected women be allowed to finish the medical tests and that one of them should then, bypassing NASA's existing astronauts, be given the next spaceflight.

In answer to the objection that these women totally lacked jet test piloting experience, Cobb argued that "equivalent experience" was good enough—flying hours in other kinds of aircraft gave them enough piloting experience to qualify. It was a highly contentious point, and one that nobody within

NASA or the Kennedy administration had yet accepted, despite Cobb and Hart's determined efforts. To say that women should be allowed to pilot jets was a valid point. To say that they did not need to do so to be considered for spaceflight was a huge stretch. This was the point that infuriated those who were jet test pilots and turned some potentially supportive voices against Cobb's campaign.

Schirra, for one, was furious. Years later, he was still angry about it. He had never been impressed by total pilot hours as a measurement of ability; he knew that a fighter pilot might log one-fifth the time per flight as a transport pilot, but those hours included not only a far riskier takeoff and landing but far more advanced flying in between. Schirra looked at his own impressive total flying hours, looked at those of the women, and saw a huge difference. He fumed to historian Stephanie Nolen, "yet some of these women have the gall to say they were not given a chance to be part of the Mercury program!"

As expected, when Cochran testified she said that any forced effort to introduce women into the astronaut program would only delay NASA in its immediate goal of voyaging to the moon. Instead, she proposed that a new, larger group of women test subjects be selected, both pilots and nonpilots. A slow and careful medical study could then be conducted. She further stated that she did not believe women had been discriminated against and that it was quite correct that NASA was choosing from the pilots who had the most relevant experience, whoever they were. Women should be included, she said, not just because they are women, but because they have finally gained the skills to advance the space exploration process. At the present moment, she concluded, they would only slow it down.

John Glenn and Scott Carpenter claimed that they did not know much about Lovelace's tests of would-be women astronauts. Glenn also stated that he was not qualified to judge whether putting a woman in space should be a national goal. What the two aviators could do, they said, was to explain very clearly what was currently required to become a NASA astronaut. They explained that candidates who were not test pilots would have to spend years achieving the required level of piloting skill before it could be fully proved that they had the necessary capabilities. NASA was already able to choose from the best of the test pilots; so it had no interest in selecting unqualified people and then training them up to that level. On the question of multiple hours in propeller aircraft being equivalent to jet test piloting, Carpenter gave a simple but powerful analogy: "A person can't enter a backstroke swimming race and by swimming twice the distance in a crawl qualify as a backstroker," he explained. "A preponderance of hours received in normal civilian flying does not compensate for a lack of military jet test flying."Glenn agreed: "To say that a person can float around in light planes . . . and run into the same type of emergencies that he is asked to cope with in just a normal six-month or one-year tour in test flying is just not realistic."

In 1996, Glenn responded to a question on the gender issue in a National Space Society question-and-answer session. He clearly felt that the same factors applied as in 1962. "I have always supported equal opportunities for all Americans," he declared, and summarized the many contributions that NASA's women astronauts had made to space exploration and space science. Reflecting on the Lovelace tests, he explained that he was busy doing his job as an astronaut at the time, and had little interaction with the issue. "Over the years," he added, "those of us in the Mercury program have come under criticism—unfairly I might add—because these women were not selected for a flight. However, none of us in the original seven had any say over who was selected or what criteria were used for selection. NASA set the criteria, which included substantial test pilot experience in supersonic aircraft, combat experience, and experience working under hazardous conditions."

Glenn has since been even more pointed in his comments. "They've made me the hit guy," he said, "for why women were not in space." The veteran astronaut did not believe that NASA deliberately sought to exclude women. Though agreeing that Jerrie Cobb was an experienced pilot, Glenn pointed out that she was not experienced in supersonic jet aircraft. NASA wanted the best, he concluded, and he and the other test pilots of the Mercury 7 were far more experienced. Glenn is generally known to be a very humble man who will only make such strongly defensive statements when he feels he is 100 percent correct on the issue.

Carpenter also clarified his opinion on the gender issue in an interview with the authors. Though his words reflect a certain ambivalence toward the issue of women's equality, he is far more open-minded than most men of his generation. As he explains, it was not up to him to say whether the women were qualified or not—he was before the congressional committee to explain NASA's position:

The women did great work, and I don't really like the thought that I didn't believe they were qualified. I remember saying in those early days, and I still think it is true, when weight is so critical, we ought to fly women, because they are at least as smart as men, and they come in lighter packages. But they didn't meet the arbitrary requirements set by Eisenhower and NASA at that time. They were neither test pilots—there may have been some graduate engineers—but they didn't have the flying requirements that were mandated by NASA and Eisenhower. Eisenhower decided at that juncture that we had to get men in space—that was part and parcel of supremacy. He said that we should take the candidates for this feat from the ranks of military, jet-qualified test pilots with a degree in aeronautical engineering or a related science.

Learning to fly at that level takes a long time. So it was too late that time to accept these ladies, who clearly did not meet the qualifications. Why should we have let the ladies in who did not meet the qualifications, and not let a lot of men in who didn't meet the qualifications? Namely, Chuck Yeager. You all know Chuck Yeager, he is the dean of all test pilots. But he didn't have a degree, he didn't meet the arbitrary qualifications, but he didn't try to beat the system—he went with it. These two ladies did not.

I understand the gender problem that was faced by the ladies. But sometimes I thought their protests tended to be for the purpose of protest. Women faced that problem everywhere. But women finally broke down the barriers to spaceflight, and that's a good thing—that's great. Those qualifications have been changed—as they should have been.

The committee was wrapped up after only two days; the expected third day was cancelled. In October, the committee's official report was released. It concluded that NASA's selection criteria were sound and recommended that research into the possibility of women astronauts should take place "sometime in the future." Once again, it was over.

At least six of the thirteen women Lovelace tested agreed with Cochran's thoughts on the matter, and told her so. They believed that women would go into space one day, but only after the wider discrimination issues in aviation were addressed. Betty Skelton knew it too. "It was just not to be, and I realized that," she recalled. "I didn't feel that it would ever go anywhere." Jerrie Cobb, however, was not going to let the issue go. She repeatedly tried to arrange a meeting with President Kennedy, and was repeatedly turned down. She bombarded Webb with pleading letters, and asked friends to send telegrams to the president. She tried to rally the other twelve women to appear in public and make statements to keep some momentum going for her cause.

Publicly criticizing a federal agency while at the same time seeking to alter its rules and qualifications so she could participate at its most prestigious level was never going to work. Enough was enough. Webb summoned Cobb to a meeting and laid it out in the plainest of terms. She was *not* a NASA consultant and was to immediately stop using the title. The issue, Webb told her, was settled, and he would appreciate it if she desisted in criticizing his agency. It was his opinion, he added, that her public statements were causing more harm than help to the issue of women astronauts. On that day, Cobb was the last person—male or female—that an irate and disgruntled Webb would have allowed to join NASA as an astronaut.

Why was Cobb pushing for something that was quite clearly never going to happen? The answer can be found in her own words. In the many letters she sent to politicians and other people of influence, she referred to what she was doing as a very personal "cause." In her letters to Webb, she stated that flying in space "means more to me than life itself" and "I would willingly give my life for it." She even wrote in some that she'd be willing to fly into space even if it meant not coming back. What she was expecting to find in space, what personal revelation she was hoping to experience, is unclear. But her words consistently demonstrated that flying into space was a hugely personal crusade, but also that she had no clear concept of what she would actually do when she got there. In her autobiography, Cobb says she was "restlessly anxious to escape from, not conform to, a ground-level existence." She was not someone who appeared to want the publicity or glory—on the contrary, she was often painfully shy. Instead, she seemed to have a deep personal belief that the overriding purpose of her life was to journey into space. Her last letters to Webb on the subject included pleas such as "I beg of you."

Wally Funk wasn't aware of the political decision-making that was taking place. "I was not made aware of the congressional hearings," she later recalled. "I didn't know what Jerrie and Janie Hart were preparing themselves to do. I was out of the country, so buried in my work that I didn't realize that the hearings had gone on until after it was over with." She had, in fact, been so divorced from the political debates that she only discovered that jet test pilot experience was an astronaut requirement after the hearings had concluded.

Setting that disappointing experience behind her, Funk thought that it might be worthwhile to find ways to continue a little unofficial aerospace testing. If nothing else, she figured it would be interesting and good experience. In 1963, at the University of Southern California, she therefore trained for and took a series of centrifuge tests. One of her flying students was also able to make arrangements for her to try the ejection seat test and high-altitude chamber at the El Toro Marine Base in California. They were great experiences, and certainly far more comfortable than the Lovelace tests. But they were not going to lead to a career as an astronaut. As Wally put it: "No regrets from this kid. I was young and I was happy; I realized I still had a lot to achieve and do. I was disappointed, but I was never bitter. I was brought up that when things don't work out, you go to your alternative. Let it pass, keep moving forward. As a professional pilot, I was always taught to have an alternative plan." Instead of being resentful, she lived up to a phrase she learned in childhood from the Taos Indians—she "threw it a fish" and moved on. Some of the other women responded similarly. One of the other medical test subjects, Gene Nora Jessen, characterizes her own Lovelace experience as "that miniscule touch with the astronaut program . . . simply a tiny, forgotten footnote in history."

A year after the congressional hearings, Valentina Tereshkova became the first woman to fly in space. One of Cobb's arguments—that the Soviets could be beaten to this propaganda victory—had now vanished. Despite treading a cautious path, NASA was nevertheless interested in achieving some firsts and had been disappointed to miss out on the first manned orbital flight. It would soon find it had also lost the race for the first multiperson flight and first extravehicular activity (EVA). These, however, were operational firsts, and were not dependent on who flew them. NASA never had any interest in putting the first woman up in space just for the sake of it.

Though Cobb regarded putting the first woman in orbit as a "scientific spaceflight feat," NASA disagreed. In response to one of Jerrie's many letters, Robert Gilruth, director of NASA's Space Task Group, told her in 1962: "We cannot select women just because they are women." He also told her that the program was designed not to create "propaganda stunts" but instead "sound technical and scientific information." His view was echoed by feminist Betty Friedan, who wrote after Tereshkova's flight that "We are so far from really seeing women in terms of their full potential that to send up an American woman would seem to be a publicity stunt. I am not for using women as publicity stunts."

What Cobb also did not realize is that America would never have won a propaganda race to put a woman into space. General Kamanin, who oversaw the cosmonaut group, wrote in his journal in mid-1961 that "under no circumstances should an American become the first woman in space." Any introduction of a woman into NASA's astronaut program would have been front-page news. The Russians, who did believe in propaganda "firsts" just for the sake of them, would have had few problems in secretly selecting and flying a Soviet woman long before the Americans could have trained a newcomer.

Cobb did not take Tereshkova's flight very well. "If I had lived in Russia," she mused, "I would have been the first woman in space!" Trying to keep interest alive, Cobb wrote an autobiography, optimistically titled *Woman into Space*. This push, once again, did nothing to further her goal, and with reluctance she finally withdrew from her campaigning—for a few decades, at least. "I had wasted three years of my life preparing to be an astronaut," she would later reflect. "I felt lost."

As the Mercury program progressed and his colleagues made the initial flights, Wally Schirra had stayed true to his philosophy of test piloting excellence. He'd also developed a social reputation within NASA as a prankster and punster of renown, which sometimes made him appear less serious than his fellow astronauts. When it came to work, however, he was a serious, focused aviator. "Humor is the lubricant of crises," he explains. "My rambunctious approach to the off-duty aspect of life may have fooled some people, but this was not a game, I often said to myself. This is for real. I was not interested in the glamour of being a space hero. Instead, I was interested in getting up and getting back." This attitude colored everything Schirra did. When asked about his religious faith at a press conference, he answered, "I think my faith relates to the machine age, to our technical accomplishments, and to the people in the project." Gordon Cooper confirmed in his

Flight Director Chris Kraft ensures that Mercury astronaut Wally Schirra goes by the book on his first flight. Courtesy NASA.

autobiography that Schirra "was outgoing and congenial and loved to make people laugh, but he was formidable when the time came and could focus with engineer-like tenacity on the technical intricacies of space flight. . . . He believed in training very hard, and was very precise."

Schirra found that his degree in engineering allowed him to exchange ideas with the spacecraft engineers so they would constantly improve and refine the design while ensuring that the pilot stayed central to its operation. "We definitely had to make the spacecraft pilot-oriented, as opposed to chimpanzee-configured," he says with a laugh. "We had all been working test pilots long enough to know that the engineering fraternity was capable of designing an aircraft which was perfect as far as the theories involved, but which no pilot could possibly fly."

On 3 October 1962, Schirra was given the opportunity to apply his training, by making America's third orbital spaceflight. Having been Slayton's backup, then Carpenter's, he was extremely well prepared. After the system failures of the previous mission, redesign work had been performed to overcome these problems. It was now time to make a pure test flight, building on the valuable experience and feedback of Glenn and Carpenter. Schirra did not want an overburdening science program or any other distractions during his six-orbit flight. Ensuring that the spacecraft worked and flew well would be his major goal, together with resolving the propellant issue. Keeping this in mind, he named his spacecraft accordingly: "I called my spacecraft *Sigma 7*. Sigma, a Greek symbol for the sum of the elements of an equation, stands for engineering excellence. Not a fancy name like *Freedom* or *Faith*. Not that I didn't appreciate those names, but I wanted to prove that it was a team of people working together to make this vehicle go. That was my goal—engineering excellence. I would not settle for less. I thought that it was a very well-made machine, and very, very carefully designed."

The flight got off to a less than perfect start. "Ten seconds into the flight," Schirra recalls, "the clockwise roll rate of the Atlas was greater than planned, and it startled people in Mercury control." The booster's incorrect roll rate took it close to an abort, but luckily this was not necessary. "I was fortunate—the Atlas stayed within safe limits." The rest of the ride to orbit was, according to Schirra, "beautiful." In fact, the booster kept firing beyond its predicted shutdown point, putting *Sigma 7* into the highest Mercury orbit yet. Now in this strange new flight environment for the first time, Schirra's test work began. The control thrusters had been modified, replacing the system that had severely malfunctioned on *Aurora 7*. On the first orbit Schirra tried them out and was delighted by how crisply the thrusters responded as they turned the spacecraft. He found that a light, delicate touch, like he had used in aerial combat gunnery, was best.

A similarly light touch also resolved the issue of rising spacesuit temperatures, which had caused problems on prior flights. By very carefully adjusting the spacesuit's control knob, Schirra ensured that the suit system maintained its equilibrium. If he had not solved this, Flight Director Chris Kraft had been prepared to bring the spacecraft back after one orbit. By bringing the suit temperature down, Schirra effectively saved the mission. Fellow astronaut Buzz Aldrin later praised his colleague's work, saying that Schirra analyzed and solved the problem like a precise, serious test engineer.

At one point, Schirra deliberately took the spacecraft far out of alignment, then deliberately ignoring his instruments and using only visual cues from features on the Earth far below carefully and precisely brought it back to the exact position needed. On the shadowed side of the planet he did the same, this time using star fields as his only reference. As all good test pilots are trained to do, he kept up a running commentary on the results of his testing. Throughout the tests Schirra's new control thrusters performed perfectly, allowing him to make precise use of his fuel in a way not previously possible. Halfway through the flight, he still had 92 percent of his manual fuel remaining and 82 percent of his automatic propellant. Shepard and Slayton were impressed. "He conserved fuel in a way that amazed Mission Control," they later recounted in their coauthored memoir. "In the process he went through his checklist with an efficiency that would have turned a robot green with envy." Schirra was delighted with the smooth progress of the mission. "It was a honey of a machine," he explains. "Test pilots describe their reaction to the performance of an aircraft in terms of the harmony that develops between a person and a machine. If you ever achieve exquisite harmony, you have reached a level of absolute confidence. Man and machine have become one. There is no limit to what they can do together."

Next came a real test of trust in the hardware. Schirra allowed his spacecraft to drift randomly while he powered down the electrical systems. In order to safely return to Earth, the spacecraft would need to work perfectly when he powered it up again and return to the exact reentry alignment. *Sigma 7* passed these crucial tests with flying colors.

With the spacecraft once again at the correct angle, Schirra made a precise retrofire. As he plummeted back through the atmosphere, he continued to coolly describe the conditions he was experiencing. Just before splashdown, he still had 50 percent of his manual fuel and 60 percent of the automatic supply. It had taken a spacecraft redesign, but the technical issues of propellant had been comprehensively resolved and overcome.

"I punched off the small drogue chute manually when we reached 40,000 feet," Schirra stated after the flight. "It is programmed to open automatically at 21,000 feet, but we wanted to test it out at this higher altitude and see how it worked. What was the most beautiful view of the flight? That parachute." Sailors onboard the recovery carrier, USS *Kearsage*, heard *Sigma 7*'s sonic boom and watched its contrail until the parachutes deployed. The spacecraft dropped into the ocean a mere four and a half miles away. "I'm still convinced the ship was four miles out of position," Schirra says with a laugh. In true navy style, the astronaut stayed in his spacecraft until it was lifted to the deck of the recovery ship and only after he'd radioed the captain for permission to come aboard. He was intent on making a carrier landing as an aviator. Gordon Cooper, Schirra's backup, was delighted with his colleague's performance: "His mission was considered a textbook flight and went so well that NASA decided to advance the flight schedule, to jump to a one-day mission—and it was to be *mine*."

Reflecting on his Mercury spaceflight, the first of three missions that he would command, Schirra believes his pre-NASA test pilot experience was vital. It was, he says, "a prescreening experience. If I hadn't been through it as a fighter pilot and test pilot, I might not have made it as an astronaut. I had a couple of hairy moments in the space program—but I was ready. I had to face up to it without hesitation, without panic, coolly, using my brain. We'd trained out fear."

The flight controllers were also very pleased. "Schirra's six orbits would be the bridge needed to go to a full twenty-four-hour mission," Gene Kranz exulted. But if the flight controllers thought they had found an astronaut willing to agree with every pronouncement they made, they were wrong. Though Schirra was perfectly happy to work closely with controllers on the ground to ensure mission success, once in orbit he considered himself the commander. Schirra was proud of his successful mission, but he was just as proud that he hadn't needed to talk to the ground much to complete it. "I wasn't up there flying for Mission Control," he explains. "I'd flicked a switch that turned off all the automatic sequences, including any capability of the ground to control my retrofire and bring me down. The spacecraft was all mine. I could not have been brought back by signals from the Earth until I turned them on again. No astronaut before me had been permitted such freedom."

Chris Kraft delighted in pointing out what he believed to be the differences between Carpenter and Schirra and in downplaying the major differences in their missions and spacecraft design. But he'd misunderstood how similar the two naval aviators actually were. He was in for quite a surprise, but it would be five years before Schirra would spring it on him.

NASA had made a giant leap forward in engineering success, but other advances were also taking place within the agency. Now as president, the same Lyndon Johnson who had dismissed Cobb and Hart signed the 1964 Civil Rights Act into law, meaning that NASA, along with all other federal institutions, would be examined very closely for any discriminatory hiring practices. In areas where NASA had been deficient in hiring anyone other than white males it was criticized. After all, the space agency may have had a valid excuse for "discrimination" when it came to selecting astronauts, but the vast majority of NASA employees had duties that did not require such specific and exclusive qualifications.

More importantly for the future of women in space, new laws and attitudes demonstrated that the entire country was changing. Cobb's single-minded focus on one small aspect of one job for one federal agency may have failed, but the general issues of inequality were being skillfully argued and campaigned for by others. American society as a whole was changing, and as it did so would hiring practices—and so would NASA.

When NASA made its first astronaut selection of the 1970s, it actively encouraged applications from women and nonwhites. Although no women were selected as pilots—the navy and air force were still not allowing women to be test pilots—NASA's first astronaut intake of the new decade included six women.

Wally Funk continued to follow the space program—for both men and women—with interest and excitement. However, rather than fixating on what had not been possible, she was creating more firsts in the wider world of aviation. In 1971, she was appointed the first female general aviation operations inspector ever hired by the Federal Aviation Administration (FAA) and in 1975 became one of the first female accident investigators for the National Transportation Safety Board (NTSB). "It happened quite by accident," she recalls. "Being in the right place at the right time. I was a chief pilot for a company out in California. I was applying for a university professorship, and I'd used an FAA chap's name that I had known. He said, 'Wally, what is this? You be in my office at nine o'clock Monday morning. You're set up for an interview next Monday because I want you to be an inspector for the FAA'. I had a great time with them for four years: then the NTSB stole me over."

At the NTSB she investigated air accidents and determined causes, which enabled the board to make any needed national policy changes and avoid a repetition of such disasters:

I became an investigator for the rest of my tenure until I retired. I've done over 450 accidents, whether large or small. It was a great challenge, but very interesting; every accident investigation was different. I got to many accidents by hiking, by horse, by mule, by helicopter, by rappelling, by boat—you name it, I got there. We weren't in the office much. I was on the road every third or fourth day; we had three states and the entire South Pacific as our territory. I was schooled at every aircraft manufacturer, engine manufacturer, propeller manufacturer, anywhere they made the parts that go with any engine. One of the reasons I retired early was because I wanted to take my safety presentation around the world, educating pilots about why they have aircraft accidents and how to avoid them.

Though women were being accepted into a space program that would ultimately put them in the space shuttle—the world's most advanced winged vehicle at the time—the rest of the aerospace world took longer to catch up. For a while at least, NASA led the way in gender opportunities. It was not until 1983 that the U.S. Navy accepted its first woman test pilot. And although a woman had been accepted into the U.S. Air Force Test Pilot School in 1974, her role was flight test engineer; it was not until 1988 that air force training became available for women pilots.

The first woman pilot selected, Jackie Parker, had worked for NASA beforehand—as its youngest-ever flight controller. "I worked with astronauts, mission controllers, and some of the brightest people I have ever met," she recalled for a magazine article. "They treated me with tremendous respect even though I was very young." The next two women pilots to graduate from Edwards' Test Pilot School were almost immediately picked up by NASA as pilot astronauts. "Jackie Parker was the first person to go through the Air Force Test Pilot School as a pilot," NASA pilot astronaut Pamela Melroy explains. "I was the third—Eileen Collins was the second." As soon as qualified women were available, NASA wasted no time in recruiting them; Collins was selected within a year of graduation.

Eileen Collins was very aware of the women who had tried to become pilot astronauts before her. Many of the questions the media asked her centered on her place in this history, and she was proud to be the fulfillment of so many people's hopes. She even invited the surviving women from the Lovelace tests to each of her launches, along with Betty Skelton and others.

Small mementos were taken up for some of them, including for Jerrie Cobb and Wally Funk. Funk also enjoyed several visits to Collins's home and accepted VIP invitations to all of her launches. They became firm friends: "I first met Eileen in the very early nineties, although we already knew of each other through magazine articles. We became instant friends, and have always stayed in touch—she calls and writes several times a year."

Collins's life was not dissimilar to Funk's. She had always been fascinated by flying as a child and had saved up enough money to start taking flying lessons when she was nineteen. Like Cobb, Schirra, and Funk, Collins says she "wanted to fly more airplanes, go farther, faster, higher, and do more." Unlike Cobb and Funk, however, she was able to do this by entering the air force. "I went to test pilot school in 1989–1990," she describes, "and that was the year I applied to be an astronaut, and was fortunate enough to be accepted here. I have since been convinced that this is the job I was born to do." In 1995, Collins flew as the first woman shuttle pilot on mission STS-63, reprising that role in 1997 on STS-84. With this experience under her belt, she was then assigned to command shuttle mission STS-93, which flew in 1999. The penultimate manned mission of the twentieth century would be NASA's first commanded by a woman. "I think it was a big step to have a woman shuttle commander," she said after the flight. "It was going to happen eventually." Following February 2003's *Columbia* disaster, Collins found that she had been tasked with returning the shuttle to space by commanding the next mission—perhaps the ultimate responsibility any shuttle commander could have.

When plans were announced for former astronaut John Glenn to fly in space again on a 1998 shuttle mission, Jerrie Cobb sensed a new opportunity to realize her long-held dream. Once again she asked the other surviving Lovelace women to petition for her, but this time it would be for her alone. She began asking NASA to send her on a shuttle flight as a subject for tests on the effects of weightlessness on aging, just as they were doing with Glenn. Almost forty years after her original petitioning efforts Cobb was still pushing for a flight into space.

Her campaign drew the support of the National Organization of Women and even First Lady Hillary Clinton. Cobb was quoted widely in the press at the time as saying, "It is my destiny—I've been waiting thirty-eight years. I've thought about it all my life. I'll do whatever it takes. I'd give my life to fly in space. I would do it then and I will now. It would mean everything in the world to me." Her quest to fly in space was as personal a crusade as ever.

Just as it had three decades before, her campaigning secured Jerrie a meeting with the incumbent NASA administrator—who was now Dan Goldin. The meeting, however, was only arranged for public relations purposes; Goldin told Cobb point-blank that he was only meeting with her because the First Lady had requested it. There was as little reason to fly Jerrie Cobb into space as ever, and once again a good deal of political and bureaucratic energy had been expended for nothing. As a NASA spokesperson said at the time: "It would be nice to be able to fly Jerrie Cobb as a consolation, but that's not going to happen. These women . . . were misled into believing that they would become astronaut candidates. We weren't the ones who misled them, and we can't make up for it."

In July 1999 Wally Funk stood on the doorstep of her home in Texas and watched history unfold. In the skies above, minutes away from a landing at Cape Canaveral, Eileen Collins piloted the shuttle *Columbia*. As Wally watched the glowing point of light cross the sky, she shouted "Go, girl, go!" Funk recalled later: "It was so very real, as she streaked across the sky, like I could reach out and touch her contrail. I felt she took me up in spirit—I am so proud of her. I knew she was heading for home, and I was willing her to kiss the runway with that shuttle nose. It was a great finale!"

In retrospect, Dr. Lovelace's tests on Cobb, Funk, and the other women had some positive results. Female pilots had been given access to the world's most cutting-edge aerospace medicine testing, and a large number of medical assumptions about women's physical inferiority had been proved false. "These women showed they could take the stresses of spaceflight better than the men. They generally outdid the men," Dr. Kilgore recalled in summarizing the tests.

In official circles, the whole testing episode had certainly raised the issue of the lack of women astronauts. As a result, the exclusion of women was never again going to be an assumption but a debate with a history. On the negative side, the way in which the complex interplay between Cochran, Cobb, Lovelace, and Webb had unfolded threw the question into the political rather than the operational arena. The resulting antagonism and ill feeling meant that although, as Paul Haney puts it, NASA "never said never" to women astronauts, the space agency did not actively involve itself in the question until well into the next decade. Because of this, opportunities were missed.

In June 1963, less than a year after the congressional hearings on spaceflight and gender, NASA selected its third group of astronauts. The agency was evolving beyond the initial pioneering flights, so this time although the successful candidates all had extensive jet experience the test pilot requirement had been dropped. The following year, NASA sought applications for a group of scientist-astronauts who were not required to have any piloting experience whatsoever. The shortlist of suitable applicants that NASA submitted to the National Academy of Sciences for further evaluation included the names of four women. If there could have been an instance in which political will could persuade NASA to include a woman in the final selection, this would have been it. For in this selection the women were competing on the same qualification levels as the men. However, the entire issue of women in space had such a bitter recent experience for both NASA and the White House, that when none of the women scientists made it to the next round, there was no outcry.

If the women who went through Lovelace's tests had been given the opportunity to obtain jet test piloting experience, we cannot know if any would have become NASA astronauts. They would not only have had to pass the exacting piloting and physical tests of a military test pilot, but they would also have had to keep themselves at the top of the pack so they could build the required experience and level of excellence to be selected. Certainly, most had the same drive, determination, and love of flying as the Mercury men, and all had proved themselves to be among the best woman aviators in the world. If any women in the country could have passed, it would have been them.

They also provided inspiration for those who came later. "I didn't get here alone," Eileen Collins noted when she was named the first woman astronaut commander. "There are so many women . . . that have gone before me . . . women military air force service pilots from World War II, the Mercury women from back in the early 1960s that went through all the tough medical testing . . . all these women have been my role models and my inspiration." Having said that, Collins also later added: "After the mission, I'll think about the fact that I am a woman commander. Before the mission, I have to think of myself as a shuttle commander." Wally Funk was "overwhelmed that Eileen could think of us as a support group who loves aviation and space exploration—she is so sensitive to us. Being a commander is a responsibility—and an inspiration." Wally Schirra was also pleased to see Collins succeed in her chosen career. "She has done very well," he agrees. "She made the mark."

What many of the Lovelace women also proved is that it is important to move forward in life despite setbacks and make the best of what is possible. Some remain bitter about their experiences, feeling that NASA cheated them out of what should have been theirs. When John Glenn made his first flight, one of them even said that she hoped he'd fail. Funk wasn't against Glenn's second flight, but she was bothered that his flight probably took a seat from another, newer NASA astronaut who had trained for years for such an opportunity. Glenn's flight created much the same feeling about "jumping ahead in line" that had made many at NASA dismiss Cobb's plans to get a flight in the early 1960s—and disregard them again in the 1990s.

Wally Funk never gave up enjoying an active and interesting life. These days she competes in nationwide shooting competitions dressed in cowboy clothing of the 1860s. She also lectures on air safety around the country and continues to work as a flight instructor. Many of the young students she teaches, both women and men, end up at the Air Force Academy and Naval Academy. "It's great to make a living doing something I love," she says with a broad smile. On occasion she competes in air races and enjoys bungee jumping, and parachuting, hang gliding, and jet skiing—and even drives at a local NASCAR track. "Girls can do anything they want to do!" she laughs.

In June 2000, she was offered a trip to Star City. It would be her second visit, following a 1988 meeting at the facility with Valentina Tereshkova. This time, she undertook a week of the cosmonaut training that's offered to tourists. It was paid for by the Discovery Channel, who in return filmed Funk training for a television documentary. This time around, she knew that the tests she was taking were just for her own enjoyment—she wasn't being promised anything that could not be granted. Her training week, supervised by cosmonaut Gennady Manakov, allowed her to climb into a Sokol spacesuit, tour the space station simulators, endure a ride on the centrifuge, and manually dock the Soyuz spacecraft simulator. "The docking control thrusters were incredibly sensitive," Funk found, "and it was difficult to make adjustments with our bare hands. I can't imagine how the cosmonauts do it while wearing pressurized spacesuit gloves."

During the ten-hour training days Funk studied celestial navigation and even the intricacies of the space toilet. The highlight for her was a flight in an Ilyushin 76 aircraft, which flew ten parabolas carefully designed to create weightlessness. In zero gravity at last, she had the time of her life, tumbling and spinning with a huge grin on her face. "I wish I could have gone for ten more," she remembers with delight. She pushed her arms out and flew like Superman, just as she had imagined doing as a little girl in Taos.

Funk learned another thing from her Star City experiences. "My attitude is much different now," she wrote not long afterward. "Now that I have had a taste of the real stuff in Star City, I can see much more clearly how NASA looked at things back in the 1960s. Women hadn't been given the chance to prove themselves yet, which is too bad—but it's okay with me. We were never promised the moon—I just wanted to pass all the tests and have a chance to be with NASA. Maybe I was naive!"

Forty years after the Lovelace tests, a substantial amount of private investment funding is now riding on Funk successfully piloting a spacecraft. She has been involved in several private space ventures over the years, but Interorbital Systems is the first to have named her as a rocket pilot. The company plans to have Funk fly their proposed Solaris X rocket plane on a spaceflight. Interorbital is just one of many space tourism developers that are operating in the Mojave Desert close to Edwards Air Force Base, and only time will tell which companies will succeed. But, as ever, Funk is enjoying the experience. "It's really just another kind of space race," she says.

Eileen Collins has told the press that she thought about Funk while in space and knows that if she gets there she'll really enjoy the experience. "She's an amazing person," she adds. "A real sharp lady." Interorbital, meanwhile, has stated that having Wally pilot their spacecraft would provide "the perfect ending" to her story. In the meantime, she continues to teach flying.

Three Lovelace "graduates" at a social function in January 2003: *from left*, Scott Carpenter, Wally Funk, and Gordon Cooper. Courtesy Colin Burgess Collection.

"I'm giving back the safety that I've learned over my years of instruction to the kids just starting out in flying," she says. "So it has all come around full circle."

Funk has been invited to social events hosted by Scott Carpenter and has shared a stage with Gordon Cooper at speaking events, but the two Wallys have only ever met once, as far as they recall. In 1997, the pair met at the Forest of Friendship in Atchison, Kansas, where granite plaques are placed in the winding pathways to honor individuals who have contributed to the advancement of aerospace achievement. Wally Funk had been inducted in 1983 and happened to be there when Schirra's turn came fourteen years later. "I went up and introduced myself," Funk recalls, "but I think he was in a hurry, or didn't want to talk about the past. He made our conversation very short."

What Funk didn't know was that other women from Lovelace's program, including Jerrie Cobb, had cornered Schirra in the past and criticized him for keeping them, they believed, out of the space program. Wally Schirra turned the issue around on them by revealing that he had been instrumental in having women test pilots inducted into the Society for Experimental Test Pilots. But the confrontations had left him with a bad taste in his mouth when it came to meeting any of that group of women again:

I was quite annoyed that they felt that they were comparable to us. They weren't test pilots, very few had college education, and they were not engineers. They had no credentials other than they could fly an airplane. It's the same thing as Chuck Yeager—he didn't have an engineering degree. Chuck wasn't eligible! Neither were these women. With all the hundreds of people that were looked at, if any one of these women had met the standards of the hundreds of people they looked at—but they didn't and I think it was a damn shame. But they tried to put themselves in the same format—I don't accept it. Jerrie Cobb wanted to be another Valentina Tereshkova, but we weren't on that track at all—that wasn't our concept.

In recent years, however, Schirra has come to realize that not all of the women are as combative, and in fact that some were fully accepting of NASA's criteria: "It turned out that it's not fair to dismiss them all, because one of the women wrote me a letter and said, it was not NASA's fault, that's all wrong, we were brought for physicals at Lovelace clinic and that was all. It was Dr. Lovelace who had the idea, not NASA." Indeed, Schirra is delighted to hear that Funk is still flying, and wishes her the best of luck with her Interorbital plans. "She hasn't given up yet," he says with a congenial smile. "Oh wow—that is good!"

Funk has been an inspiration to many young women looking for a challenging career, but she was never very interested in the women's movement:

Although I was used to kicking down doors, I wasn't a feminist or a political person. We were just born twenty years ahead of Sally Ride, and thirty years ahead of Eileen Collins. The time wasn't right. I've never been discriminated against in my life. I was very serious about becoming an astronaut, but I didn't have the jet time or the engineering degree. When I found out the parameters were jet test pilot experience to be an astronaut, that let us out of the league. We were just guinea pigs for the space program medics. Having since read about some of NASA's internal politics, maybe I am glad I was not involved! I have never understood all the amazement people have about me going through the tests—it was just part of life. People send me Mercury program envelopes to sign and I think, shouldn't they be asking John Glenn instead? Failure is not part of my makeup—my philosophy has always been to get over it, and move on. I love life, I've loved every bit of it, wouldn't change it for the world. I've got a lot to do—fifty more years of stuff to do. I plan on flying until I'm ninety!

Schirra, when looking back, sounds remarkably like Funk. "I love life," he says. "What I felt about my first ride into space so long ago just about sums up the way I feel about life; I've really had one hell of a good time."

Others have tried to read a story of government discrimination into Funk's fleeting brush with the early space program. Gene Nora Jessen, who'd also endured the Lovelace trials, dismisses such conclusions. "It was a testing program," she remembered in a magazine article, "which is quite different from a training program. Some folks presumed we were actually part of the Mercury program, and our tenuous connection to the world of astronauts grew into a plethora of inaccurate and downright fictional accounts of our long-ago adventure."

In truth, the stories of both Wallys are stories about making the best of what life offers you—and making sure you enjoy it all the way.

8. A Change of Attitude

Circumstances do not make the man, they reveal him.

James Allen

It could almost have been an action scene straight from *The Right Stuff*, a movie that has attracted a cult following since it was first released in 1983. An overlaid caption tells us that we are at Cape Canaveral the day before the planned Mercury flight of acclaimed astronaut Gordon Cooper—the self-professed "greatest pilot you ever saw," according to the movie's scriptwriters. We can see that it is a warm, languid Florida spring day as we take in a sweeping panorama of the launch facilities and military air base at the Cape circa 1963. Occasional clouds billow in a brilliant azure sky, while closer to the ground a mild heat haze lends a soft shimmer to a near-barren landscape, dotted with slash pines and palmetto shrubs.

Suddenly, on the horizon, a dancing, silvery dot approaches in a wide, sweeping turn. The glimmering object quickly comes into focus; it is a sleek, low-flying jet fighter, an F-102 out of Patrick Air Force Base, twenty miles away. The airplane's wings, now flying level with the ground, seem to brush the fringes of black mangroves growing along the estuaries as it rapidly approaches, hurtling toward the camera, tearing a relentless, eerily silent path over the sparse Cape wilderness. Finally reaching his target area, the pilot grins with impish delight and kicks in full afterburners. The powerful jet's needle nose rises into the sky, and then a massive, deafening blanket of sound catches up to us, consuming everything in its path. In an instant the audience is flayed with an apocalyptic scream, as the pilot rips his beast into a near-vertical climb.

We moviegoers would be transfixed in silent awe as the banshee howl of jet engines slams through the windows and doors of an administration building like a physical thing. In a heartbeat the drama's characters have all stopped what they are doing. One aggrieved clerk curses loudly as hot coffee slops onto the day's paperwork and down the front of his neat white nylon shirt. Someone else is shown ducking for cover beneath his desk. On the rust-red launch tower, the team preparing Cooper's Atlas rocket and Mercury spacecraft quickly hugs the nearest steel strut as the launch complex shakes and shudders, and we hear bolts and rivets rattling in protest.

Meanwhile, in his second-floor office, Mercury operations director Walt Williams is seen conferring with flight director Chris Kraft when they are both nearly blown out of their chairs. The two men rush to the window and see the offending jet quickly receding into the sky, leaving chaos in its wake. For the next few moments there is speechless confusion; then a phone on Williams's desk begins to ring. We are left in no doubt that it is someone demanding to know just who the hell that goddamned lunatic was, creating goddamned havoc, breaking every goddamned rule in the book, and doing supercharged cowboy stunts in a totally restricted area. The caller wants blood—he wants that son-of-a-bitch pilot hauled over the coals and grounded immediately.

The Walt Williams character knows exactly who that "son-of-a-bitch pilot" is, and he gnashes his teeth in fury. Gordon goddamn Cooper. And Gordon Cooper is just about to become the sorriest goddamned pilot, goddamned astronaut, on the face of the Earth. Forget that he is due to fly into orbit the next day on America's longest space mission or that he is on the verge of becoming the last NASA astronaut to launch into orbit alone. Right now, Maj. Leroy Gordon Cooper is not only in deep, deep shit—he's submerged in it.

According to the real-life Gordon Cooper, whose true character was very different from the swaggering, superconfident fighter jock portrayed in Hollywood's *The Right Stuff*, that Hollywood-like incident actually happened, but it was, he says, "nothing really." Far from flaunting his flying prowess, Cooper explained that the event was just a harmless way to let off a little prelaunch steam. What better place for a fully trained-up, super-ready astronaut to be on the day before flying into space than sitting in the familiar confines of a sleek jet, hitting the Florida skies, and enjoying the exhilaration of putting a powerful bird through its paces?

Cooper had another reason, too. He was angry, and rightly so. Earlier, on the morning of Monday, 13 May 1963, the thirty-six-year-old U.S. Air Force pilot had arrived at Cape Canaveral's Hangar S, ready for his first space mission, which was set for the next day. It was then that he received a nasty surprise. He was told that a last-minute modification had been made to the custom-made pressure suit he would wear on his orbital flight. As he later told the authors, he was absolutely furious. This, in his opinion, violated a very firm rule:

By good common-sense agreement, we had decided that nobody *would do anything to a suit if you didn't have a chance to run it in an altitude chamber and test things out. That suit was all ready to go for the flight, and they came along at the last minute and cut this big damn hole in it, which really teed me off no end. Still makes me mad to think about it. The doctors decided they wanted to add a different kind of blood pressure cuff. They cut a hole in my spacesuit, put a new fitting in, which rubbed right on one rib. My way to let off steam was with an airplane!*

And fly an airplane he did. Cooper clambered into a NASA F-102 and took to the skies. Airspace over the Cape was completely restricted, which obviously meant that low-level flying was forbidden, but this minor irritant meant little to him as he pushed the fighter hard through some satisfying loops and dives offshore. He then pointed the nose down and headed for the area near his launch pad, leveled out, threw on full afterburner, and went into a howling vertical climb. It was an episode that probably should have made the film.

Though he was never quite the gregarious superjock portrayed in the film version of *The Right Stuff*, Cooper went along with it to a certain extent, albeit grudgingly at times. Whenever someone threw that "greatest pilot you ever saw" line at him, he quietly answered as expected, in his soft Oklahoma accent, "You're lookin' at him!" However, it was understood that he much preferred his characterization in the original Tom Wolfe book.

Scott Carpenter also openly admires Wolfe's book and regards it as accurate, but he is always careful to distinguish between it and the movie, which he refers to as a "docudrama." On the subject of Gordo's character, played by Dennis Quaid, Carpenter knows what director Philip Kaufman was trying to achieve by injecting this outspoken, brash, hotshot fighter jock into the movie. "Somebody," he muses, "*had* to play that role." It turned out to be Gordo, and thus Gordon Cooper was reinvented for the big screen.

The real Gordon Cooper was known for using very few superlatives to describe anything. In fact, space historian William Shelton, in his *Soviet Space Exploration: The First Decade*, compared Cooper with a soft-spoken movie star bearing the same last name: "Among American astronauts, Gordon Cooper, who is every bit as laconic as the screen characterizations of Gary Cooper, epitomizes the modest virtue of understatement . . . from the outsider's point of view." Dee O'Hara agreed, describing Cooper as "a really sweet guy. 'Swaggering' is not a word I would ever use to describe Gordon; quite the opposite."

"Laconic" and "sweet" he may have been, but Gordo managed to rub a few people the wrong way, and for many reasons. In another book about early space flight, *The Race*, space reporter James Schefter characterized Cooper as "an independent cuss" given to racing fast boats, flying his own airplane, and "turning up in newspaper photos doing something that might break a bone." Schefter wrote that Cooper succeeded in attracting the ire of the bureaucrats at NASA in many ways. His "dry Oklahoma twang" proved especially irritating to some of the NASA headquarters people, "which was increasingly being populated by bureaucrats far removed from the operational end of things." According to Schefter, they saw Cooper as "a yokel, ignoring his engineering and flying skills while concentrating on an accent they considered anything but cosmopolitan."

Yet those who knew Gordon Cooper best are the ones who *truly* knew him. As a staff member at the Museum of Flight in Seattle, Washington, Jake Shultz had the opportunity to work many times with the former astronaut. He referred to him as "a class act, a true gentleman, a very accomplished pilot and astronaut, a friend and one of the finest men I have ever met":

When asked to do so, Gordo could relate adventures ranging from flying the Mercury and Gemini spacecraft to piloting the Goodyear blimp, yet the seasoning of years brings a certain humility to the telling.

Gordon Cooper, the youngest Mercury astronaut. Courtesy NASA.

Gordo was clearly a pioneer, and flowed with the first-hand knowledge that came with that experience, but there was a lot more to this gentle man. With the respect due that of an elder, his stories captivated those who listened. The lucky ones were keenly aware that he was there for spaceflight's beginning. There was also a certain smirk as he talked, a grin that came from his knowledge of an inside joke or a story that he seemed to know more about than he was telling. There are things he would not have shared when the antics and rivalries of early spaceflight were new, stories that may have leaked out over the years. Yet those little nuggets are part of what makes up the human experience, the reason that people in space are infinitely more interesting than machines.

Leroy Gordon Cooper Jr. was born in Tecumseh, five miles south of Shawnee, Oklahoma, on 6 March 1927. Today the highway between the two towns is named Gordon Cooper Drive in his honor, and Shawnee is also home to the Gordon Cooper Technology Center. His father, who eventually retired from the United States Air Force with the rank of colonel, was also one of the first Oklahoma-born judges to serve on the Pottawatomie County bench and had served as assistant county attorney before beginning an eight-year tenure as judge of the superior and district courts. Leroy Gordon Cooper Sr. had earlier married Hattie Lee Herd, a teacher who also shared his passion for flying. In 1930, when Cooper's father was ready to begin his term as a judge, the family moved to Shawnee.

Neighboring families soon grew accustomed to the sight of the little Cooper boy happily playing on Jefferson Circle with his trusted German shepherd dog, Max. When he was just five years old, the future astronaut experienced his first flight. On that day his father strapped him into a Curtis Robin monoplane and took to the skies from the grass runway at Shawnee's airfield. By the time he was eight, the adventurous youth, known to everyone as "Gordo," was joyfully piloting airplanes with his father. He flew solo for the first time, albeit unofficially, at the age of twelve, with the help of some seat cushions and custom pedal blocks. "I grew up flying," he reflected years later. "It was something I loved doing. My father became a pilot right after World War I—he had served in World War I, but in the navy. I started at a very early age. I just thought everybody flew airplanes."

Frank Leslie, a former *Tulsa World* reporter, knew Cooper when the two were young boys. In a 1963 article written for the *Shawnee News-Star* by Jack Reese, Leslie recalled a friend with a determination to succeed and incredible concentration. "Gordon could sit in a room filled with visiting, talking youngsters or adults and become completely absorbed in a project," he reflected. "His competitive spirit was great, although never manifested in any angry, vocal outburst of action or emotion. But you knew when it had risen to the surface. Gordon simply set the jut-jaw—his most marked facial feature—even more firmly and pushed to the goal." During one of their summer breaks the two friends became junior "campaign managers" for Cooper's father, giving up a lot of their free time to distribute "Vote for Cooper" leaflets around Pottawatomie County communities.

Academically, Cooper was a good student. He first attended Jefferson School, where he spent all seven of his elementary education years. His principal, Mrs. Risher, once told the *Shawnee News-Star* that she remembered him as a "serious young man, intent on doing what was expected of him and what he expected of himself. He was a happy child, and as so many of us look back to those years we remember the quick smile, the bright, alert eyes, the sunny disposition." At Shawnee High, Cooper was always in the upper echelon of his class. He also played tailback on the school's Wolves football team and lettered in the sport, participated in track events, and was a member of the Hi-Y and Honor Society. All of his elementary and secondary education was taken in Shawnee, with the exception of his sophomore year. His father had reentered military service during World War II and was stationed in Kentucky that year, so Cooper attended Murray High School in that state until he could return to Shawnee High, where he graduated on 24 May 1945.

Through his father's military career and subsequent friendships, the impressionable young boy had been introduced to many prominent aviation figures, and he thrilled at their visits and stories. He met flamboyant aviator Roscoe Turner, later admitted to having a boyhood crush on family friend Amelia Earhart, and sat in awed silence during one of Wiley Post's visits as the heroic pilot recounted some of his pioneering adventures. Desperately wanting to emulate their feats, Cooper would take any odd jobs on offer around Shawnee's modest airport to earn extra money for flying lessons. A local engineer named M. C. "Davey" Davenport taught the eager young man to fly in a J-3 Piper Cub, a fifty-horsepower airplane. Cooper learned very quickly, and it was a momentous day in his life when, at the age of sixteen (and officially this time), he flew solo and was awarded his pilot's license.

Though he spent most of his youth in Oklahoma, Cooper always referred to Carbondale, Colorado, as his true home. It was here that his father purchased a mountain ranch for the family in the mid-1940s. The young boy soon developed a lifelong passion for the great outdoors and the invigorating mountain air. His favorite activities were riding, hunting, and fishing, and he would often head off for trout fishing in the Roaring Fork, Colorado, and Frying Pan Rivers, from whose sparkling clear waters he would regularly pull plump rainbows and browns.

Cooper's innate love of flying led him to apply at age eighteen to become an air force pilot. But with the end of the European war just two months away and the war in the Pacific seemingly headed the same way, there would be an overabundance of trained pilots by year's end. As a result, neither the army nor the navy flying schools were taking on any candidates. Disappointed, but with his usual determination to overcome setbacks, Cooper enlisted in the Marine Corps and was assigned to the Naval Academy preparation school for several months as a backup for an Annapolis appointee.

Two months after Cooper's eighteenth birthday Germany capitulated, and in August the two atomic bombs dropped on the Japanese cities of Hiroshima and Nagasaki brought World War II to an end. Cooper had enjoyed his time at the Naval Academy; however, as he said in his memoirs: "the guy who was the primary appointee made the grade . . . so I was reassigned and wound up in Washington DC serving with the Presidential Honor Guard." Cooper was subsequently discharged from the marines in August 1946. Meanwhile, his father had received an assignment to Hickham Field on the Hawaiian island of Oahu, so Cooper decided to join his parents in this sun-drenched paradise, renowned for its laid-back attitude. He took up his college education at the University of Hawaii, and while studying engineering there he met pretty Trudy Olson from Seattle, Washington. Trudy was an instructor at an aero club Cooper joined, and there was a mutual attraction from the outset. Their courtship blossomed during many romantic Piper Cub flights over and around the spectacular island. They were married in Honolulu on 29 August 1947 and lived on Oahu for another two years while Cooper continued his studies.

While attending college, Cooper received a commission in the U.S. Army ROTC, and then later transferred to the U.S. Air Force. He was recalled to active duty for extended flight training in 1949, first pilot training at Perrin AFB, Texas, and then advanced training at Williams AFB in Arizona. After receiving his wings in 1950, Cooper was assigned to the 86th Fighter Bomber Squadron at Landstuhl, near Frankfurt in West Germany, where for the next four years he flew F-84 and F-86 fighter planes. He later became flight commander of the 525th Fighter Bomber Squadron.

When the young couple returned to the United States, Cooper began attending the Air Force Institute of Technology at Wright-Patterson AFB in Dayton, Ohio. Two years later he graduated with a bachelor's degree in aeronautical engineering, and in the fall of 1956, to his delight, became a student at the Experimental Test Pilot School at Edwards AFB. Edwards, located on Rogers Dry Lake in California, is known as the birthplace of supersonic flight. Back then it was home to many of the greatest research test pilots of that era, all elite flyers of the air force. These men not only had to possess a solid knowledge of engineering but had to bring to Edwards a flying background full of variety and experience as well as a dedication to improving the performance of new high-speed and high-flying airplanes. Although their work has been glamorized over the years, they were generally serious-minded, highly cautious individuals willing to take calculated risks for the advancement of aviation.

Upon graduating from test pilot school in 1957, Cooper was assigned to the performance-engineering branch of the Flight Test Division at Edwards. Within two years he was doing the type of work he loved most: experimental flight-test engineering. It was a job that other pilots did not especially covet, but Cooper saw something in it that others did not.

"It really was a dream job," he told the authors:

It was a different kind of era, when everything was new. We had project management offices—the fighter section, the cargo section, the bomber section, and the helicopter section. When you got a new airplane coming along in the very early design phase, these offices would pick up this airplane, and start planning how to flight test it. You had the first prototype made; you put a flight or two on it, and then started your flight test program. So I had the opportunity to be both project manager of several new airplanes, as well as getting to fly them.

I was working out the bugs, ensuring that the airplane had the latest technology that was available for it. We were also starting this utilization program . . . bringing in the agencies who were going to be using these airplanes . . . bringing in key people. We would start getting them checked out in an airplane, and get their input into what they really needed for their particular use. That was particularly true in the U2. The biggest part of the flight testing was being done out in the units who would be using the airplane.

Cooper's eyes lit up when he recalled this particularly exciting period of his life:

I had a big advantage . . . I was put in charge of a group where I had all of these airplanes that would fly with the test airplanes in order to observe them, take pictures of them, and calibrate their airspeed system. These airplanes were really loaded with instrumentation. I had a T-28, *a* T-37, *an* F-86, *an* F-100, *and an* F-104. *They had to be flown about once every couple of weeks. Flying time was kind of scarce at times, so being in charge of that section was a real plum to have. It was good flying, and it also allowed you to be checked out in all of these airplanes.*

He would not know it at the time, but as Cooper test-flew at Edwards his military personnel records were being screened. He was under serious consideration for a new program, to be known as Project Mercury, and he met all the requirements that NASA and President Eisenhower had agreed upon as mandatory for the first astronauts. When he was invited to attend a classified interview in Washington DC, Cooper's commanding officer at Edwards strongly advised him not to get involved in what he referred to as "some idiotic program." "Nobody knew what this was," Cooper recalled. "There hadn't been an official announcement that we were going to have a space program. At the time we were called to report to Washington, there was no official commitment to a space program. He just wanted us to keep our eyes open, and not get into something that would turn out to be some kind of fly-by-night program."

Like others at the interview sessions, the confident young test pilot had a tough decision to ponder and very little time to make up his mind. "When I was given the presentation on what the program would be, I had to think: 'I've got the best job in the world already, which I love. I'm giving this up to go into this new program. I hope that this new program is all that I think it will be.' And of course, as it turned out, it certainly was. It was a big decision to make, and we had to make the decision that day." For Cooper, the new project offered other enticements that outweighed his immediate instinct to maintain a career path in the air force. One was the undeniable attraction of the whole concept of spaceflight. As a young boy he had always been interested in rockets and space travel, eagerly following the fictional adventures of Buck Rogers in comic strips and radio programs. "My dad always said, even when I was a kid, someday we'll be traveling in space, traveling in rocket ships. That was many years before, of course. I really didn't think I would be one of the first, though. I suppose it really didn't dawn on me that we'd even get into space in my lifetime."

On that fateful day, after a lot of consideration, Cooper said yes. He realized, however, that he would be competing for a very limited number of places against a good number of other, very experienced and qualified test pilots. "I felt that this was the way that I was going to get higher and faster and further—to get into space. The only way I was going to do it was just decide, 'I'm going to make it, and whatever happens I'm going to be one of the ones selected!'" Before the decision had even been made, Cooper was so confident of his selection that he told his commanding officer at Edwards to begin looking for someone to replace him. His self-confidence was fully justified for, on 2 April 1959, he was selected as one of the seven Mercury astronauts.

Others in that exclusive group later admitted that they were a little surprised to see Cooper among them. Deke Slayton was one of them. He considered the brash young Oklahoman to be an engineer, not a test pilot. Wally Schirra told the authors that "a couple of times he was looked down on by Deke, because he wasn't a test pilot in flight test as Deke was; Gordo was involved with engineering tests." As Cooper explained, Slayton's feelings were conditioned by the way things were organized at Edwards, where Slayton had been involved in test flight operations:

There was kind of a wall there; although we were good friends, there still was that wall. There had been somewhat of a prejudice at Edwards between flight operations and engineering. Airplane by airplane, I had been made to get approval to go ahead and be allowed to fly each one. I had moved on up to a progression where I was flying the top airplanes. Flight operations were very jealous of allowing anybody in engineering to get into flight operations. It was partly competition for funding, and from a flight operations point of view I could see that they didn't want just anybody to be flying in flight operations. They were very jealous—there were times when the programs were a little scarce. As Deke and I worked as a team, we got over that feeling.

A substantial amount of prejudice was also directed against the entire Mercury program from several senior test pilots, including Chuck Yeager. As Cooper later recalled, these men did not consider riding inside an unwinged vehicle mounted on top of a missile to be any sort of piloting at all. "Of course, part of that was jealousy over funding," he stated. "It was quite a fight to get funds for any new program. Suddenly here was this new program, Project Mercury, that had top priority in the United States on funding. It was only natural that these guys who were involved in the Dyna-Soar, the X-15, and these other military programs were somewhat jealous of the funds we got in Mercury." Cooper added: "Dyna-Soar was ahead of Mercury when I entered the program. When I left my office at Edwards, my cohead in the office and I had a little bet on whether Dyna-Soar would hit orbit first or whether Mercury would do it. I won a bottle of Scotch from him. It would have been nice if we could have done Dyna-Soar also, at that time. Now we're back to developing those space planes. We could have been a lot of years ahead if we had continued on where we were then."

Before his selection, there was one little piece of personal business that Cooper had to remedy. Although it was never explicitly stated, he had quickly realized that NASA only wanted astronauts who were happily married. This didn't augur well for any applicants whose marriages were in any way shaky, as NASA publicity was gearing up to present them as God-fearing, home-loving, all-American family men. Anything less than that would be unacceptable. This presented a huge problem for Cooper because his marriage was faltering. He was not only separated from his wife, Trudy, but she was living hundreds of miles away in San Diego. Trudy, however, was still an air force wife—if in name only—and understood the duties that came with this designation. After discussing the opportunity her husband had to become an astronaut and just how much it meant to him, they decided to give NASA what it wanted. Without too much fuss they quietly got back together, although things were never the same as before their relationship faltered. Their so-called ideal air force marriage was nothing more than a public relations façade.

"I had a feeling that was the way it was going to be with NASA," Cooper later revealed. "Anybody who was in a separation or divorce process would not even be considered. It was the way America was at that time. Certainly, with a conservative organization like NASA—I think it still is probably very dominant within NASA—if it came down to a competition between someone who was currently involved in a divorce versus somebody who was supposedly happily married, then the divorce guy would lose out." Cooper quickly realized that the public relations side of being an astronaut was of immense importance to NASA and the press and that the desired wholesome public image also included the subject of religion. "I got the clue," he revealed, *when we were introduced in Washington, and John Glenn broke out talking about he and his family going to church regularly. We realized . . . by the reaction of the news media . . . that really went over big. I think John was the only one who was rather familiar with the fame, having been on that transcontinental flight, and the* TV *program he had been on. He was pretty public relations oriented. But it was kind of a jolt to the rest of us, as we had never been exposed to public relations at that level before. It took a while to get used to. I avoided all the public relations wherever I could, staying out of the limelight.*

Like the other six Mercury astronauts, Cooper introduced himself in the opening section of their collective book, *We Seven*. There, he discussed how far he hoped to go as an astronaut. "Since I am the youngest," he wrote, "I think maybe I have a better chance to have some real good fun in this program than some of the others. I've known pilots who could keep going until they were over fifty. That would give me a good twenty years to go yet. . . . I'm planning on getting to the moon. I think I'll get to Mars."

Though Cooper would certainly have fun in the program, his prescience about his career wasn't acute enough to inform him that he would never make it to the moon. Whether he could have and should have, he would nevertheless come to know both the highs and lows of the spaceflight business in a decade-long career as one of his nation's first astronauts.

Despite his quiet character, Cooper could be very articulate when it came to his feelings about the space program, as he demonstrated to science writer Joseph Bell soon after his selection: "I have faith in the people I'm working with in this program," he said, "and I know it will be a success. I think I'm motivated by the fact I'm a career officer, career pilot—and this is something new and very interesting." The word *faith* obviously meant a great deal to the gangly young test pilot, and this would become even more evident when he was later asked to name the craft that would carry him on his first journey into space.

A great deal of work was assigned to the astronauts in those early days of the Mercury program. There were big jobs, and there were smaller jobs. One of Cooper's first assignments was to help integrate the Redstone booster with the Mercury spacecraft; that was certainly a big job. But he and his fellow astronauts were also deeply involved in all engineering aspects of the program, and not just the piloting. No flight-associated detail was too small, as he told the authors—not even designing a knife: "Having an astronaut assigned to some specific portion of the program would certainly lend a little enthusiasm to it," he recalled.

I felt each assignment was very important. It was a team effort, and every guy's assignments were very important to the overall program. I got together with Bo Randall, who was considered to be the best knife maker in the world and worked with him on designing a knife. We needed a knife; every fighter pilot always carries a survival knife of some type with him. We figured we needed a knife that was strong enough to cut through the side of the spacecraft if you had to. If you landed on land, and you got jammed in, you might need to cut your way out. So we had to have a knife that would serve for that, also be able to whittle and cut wood, kill game, whatever you needed to do to survive in some remote area.

As well as working hard, Cooper and the other astronauts played hard, and this was where another side of the young adventurer from Oklahoma came to the fore. These men were test pilots and military aviators who were accustomed to staring death and danger in the face, overcoming obstacles, and testing limits—not only those of superfast, powerful machines but of themselves. They handled some of the hottest aircraft in the skies, so in most cases it was only natural that they also wanted to drive the most powerful, fuel-guzzling road beasts they could get their hands on. As a consequence, it did not take too long for Chevrolet's sleek and speedy Corvette to become synonymous with America's astronauts.

When Alan Shepard became an astronaut in 1959 he was already driving his second Corvette, having bought the first in 1953. Shortly after Shepard's suborbital flight, the president of General Motors, Ed Cole, decided to present the history-making astronaut with a 1962 Corvette. He knew that the astronauts were not permitted to make commercial endorsements, but like many other Americans he was swept up in the glory of the moment. And of course it also did not hurt business to see Shepard's photograph in newspapers and magazines driving or posing next to the sleek Corvette. This was meant to be purely a one-off gesture by General Motors, but it quickly started a trend.

Around this time a Melbourne-based Chevrolet and Cadillac dealer and professional driver named Jim Rathmann had an inspiration. The winner of the 1960 Indianapolis 500, Rathmann had met and befriended Shepard and his Mercury colleague Gus Grissom. Given the sublime success of the Shepard-Corvette association, he felt that he might be able to swing a special deal for the rest of America's newest heroes. He approached Ed Cole, and they set up a special executive lease deal for the astronauts. This "perk" gave them access to brand-new, powerful Corvettes, creating an unofficial alliance that lasted a long time.

Shepard and Cooper now began a fierce but friendly rivalry to see who could squeeze the most power out of their automobiles. Cooper, who also raced speedboats, was especially thrilled with his new car; he even held Sports Car Club of America (SCCA) and National Association for Stock Car Auto Racing (NASCAR) licenses. Glenn, predictably, was the odd one out—he politely refused Rathmann's offer and faithfully stuck with his underpowered station wagon.

Led by Shepard, Grissom, and Cooper, the astronauts were soon chewing up the roads, conspicuous in their Ray-Ban sunglasses and Ban-Lon shirts. The police along Cocoa Beach knew them and their cars well, and most simply turned a blind eye to their antics. Astronauts deserved a little leeway, they reasoned. There were also drag races up and down Cocoa Beach when beach racing was still a legal activity, and sometimes they even towed each other along the shallow water on skis. After a few hours of fun they would head off for a little relaxation at Ramon's, the Starlight Lounge, or Bernard's Surf, or race over to Henri Landwirth's Holiday Inn. Everyone loved the astronauts—they were welcome everywhere, and they always seemed to attract a bevy of beautiful women wherever they went.

When Alan Shepard passed away in 1998, Cooper recalled the competitive fun they had enjoyed in their souped-up Corvettes. "We raced many miles in identical Corvettes," Cooper said in presenting his eulogy at the Johnson Space Center, and then gave a small smile. "I'm sorry Al, but I never told you that I changed the ratio in the differential. You weren't any less a driver, it's just that I cheated a little."

To the consternation of many NASA managers, Cooper soon began entering official racing competitions. It was a risky practice that NASA frowned on but did not try to prevent—at least at that moment. These and other distractions allowed Cooper to live life to the fullest. Somewhere along the way, however, Operations Director Walt Williams began to have some serious reservations about these extracurricular antics and began to regard Cooper as too much of a maverick. For his part, the aspiring astronaut knew that he was being watched, but was reluctant to change his ways. "We had a lot of fun," he confessed, "played a lot of practical jokes on one another. A lot of humor, lots of good pranks; like a bunch of brothers. We might fight amongst ourselves on some things, but we loved one another dearly. All and all, we got along very well together. Walt just had a feeling that I was not taking things seriously enough, I guess. Somehow Wally Schirra and, later, Pete Conrad did the same and got away with it. But I got rapped on it."

Cece Bibby was one of those who fell victim to the astronauts' practical jokes. Hired by Chrysler in 1959 as a contract artist at the office of NASA Publications, she painted the "nose art" on John Glenn's spacecraft when he wanted something more stirring than the stencil-and-paint names on Shepard's and Grissom's spacecraft. The astronauts loved her work so much she later did the artwork on *Aurora 7* and *Sigma 7*. She later recalled:

One day I was at a stoplight after grocery shopping in Cocoa Beach when I heard a horn blowing behind me. It was Wally Schirra in his Austin Healy, so I waved at him and he waved back. When I got home, he pulled in the drive right behind me and helped carry my grocery bags into the kitchen. Wally saw a covered cake dish and lifted the lid to see what was in it. He thought it was a spice cake, but I told him it was a carrot cake, a new recipe that was making the rounds. I asked Wally if he'd like to try it. He said he thought it might be interesting, so he cut himself a piece and went into the living room while I finished putting away the groceries. He really liked the cake but couldn't believe there were carrots in it.

The next afternoon, when I got home from work, there was a knock at the door. It was Gordo Cooper. He said Wally told him I had baked a really good "spice cake," and he wanted to know if there was any left. I thought it was interesting that Wally had told him it was spice cake instead of carrot cake, but I figured Wally was setting Gordo up for some reason, and I would just go along with it.

Gordo finished off that slice of cake and said he'd really like some more, and was halfway through the second slice when he asked what was in it. I told him the recipe called for cinnamon and pecans, but that I also added a little nutmeg and some cloves. Then I said, "Oh, and carrots . . . grated carrots."

The fork in Gordo's hand stopped in midair. He got the strangest look on his face and he said, "Carrots? You really put carrots in this cake?" By this time the fork was back on the plate. He pushed the dish away and said he thought he'd better get on back. When I asked if he would like me to wrap up the rest of his cake to take home he said that he thought he'd had enough. I could tell by the look on his face that this was true.

The next day I saw Wally at the astronaut office. He said that Gordo had come in the previous evening looking almost ill. While Wally struggled to keep a straight face, Gordo told him, "That wasn't spice cake Cece baked. She put carrots in it! You don't put vegetables in a damn cake!" Wally, trying to look shocked, said "Really?" I was amazed at Gordo. Here was a guy who was unafraid of going into space but who couldn't bring himself to eat carrot cake.

Eventually, Cooper found out he'd been the victim of one of Schirra's famous "gotchas," and a few weeks later he had a chance to square the joke ledger with Cece, as she later recalled:

My pride and joy was my sports car; a candy-apple red AC Ace roadster with huge white racing stripes that I called my "aiming stripes." It was the only one like it in the area. I was in my garage trying to replace a burned-out taillight when Gordo came over to see what I was doing. I told him I couldn't get the new light to work, and there must be a short in the connection. He said he could fix it for me, so I left him my tool kit and some electrical tape and went into the house. A little while later he came in, said he'd found the short and the taillight was fixed.

The next day I found that he'd fixed it all right . . . he fixed it so that I had permanent brake lights. I hadn't noticed it until I got into the parking lot at work, and some guy let me know. "That rat!" I said. I called over to the astronaut office and was told that Gordo was on his way to the Mexican tracking site for a launch. It's a good thing he was out of reach because I would gladly have wrung his neck. I had to call one of my friends, a U2 pilot, to come and fix the taillight. Of course, I had to tell him why they were in such condition. He just chuckled and said, "Gordo did that, huh?"

On his return to the Cape, Gordo came over to the art department and chatted with the other artists, but kept looking over at me. He finally came over to my drawing board and asked how things were. I told him everything was fine and asked how his tracking site duty had gone. I never mentioned anything about the lights. When he left, I noticed that he stopped by my car and just stood there for a second or two. Then he turned the lights on and went to the rear of the car, kind of scratching his head. I watched him turn the lights off and go across the street to Hangar S, where the astronaut office was located at that time.

I never, ever said a word to him about the job he did "fixing" the lights. I know he must have wondered why I hadn't read him the riot act. I don't know for sure, but I think I finally put one over on Gordo. At least, he never asked.

For whatever reason, Cooper was not one of the three Mercury astronauts selected to train for the first spaceshot, the one everyone felt confident would be the first manned spaceflight ever. He was disappointed, but he got over it and pressed on. "I guess I was so overjoyed at being in the program I really picked a team approach to things. I felt that the team was the important thing. Whatever I needed to do, and whatever flight I got on, I felt it would work out okay for me. . . . I would wind up with a good flight, whichever one. Besides, I considered the last flights the best assignments of the Mercury flights." Of course, none of the Mercury astronauts became the first person in space. That honor went to cosmonaut Yuri Gagarin. "It was a blow, but it certainly wasn't unexpected," Cooper later mused:

We knew the Russians were really working hard to get somebody up into space. We knew they were a long way ahead of us, because the politicians would not let us get started on the space program for about five or six years, when we could have. A lot of people like Wernher von Braun and others were working hard to try and get Congress to authorize, fund, start up, and get things going. They wouldn't do it.

So it really shouldn't have been any surprise that the Russians were about five or six years ahead of us. Plus they had a big advantage—their nuclear warheads were neither as small as ours nor as efficient as ours. They were much larger, so they had to have bigger boosters. We didn't have to develop these big booster engines; they did. So they were a long way ahead of us. That forced us to not only miniaturize, but microminiaturize everything. We were terribly weight-limited in what we could hoist up. That's where we really got ahead of them because we went into microminiaturization and advanced electronics.

The Mercury program was soon making strides to catch up, and Cooper was heavily involved. He served as blockhouse CapCom when Alan Shepard became the first American to fly into space. Then, on the ensuing Mercury mission, he flew an F-106 "chase" plane as Gus Grissom's Redstone headed skyward. When John Glenn made the first American orbital flight and a faulty signal suggested that *Friendship 7*'s heat shield might be loose, Cooper was placed in an awkward situation. As the Muchea CapCom for the flight, he had to ask Glenn about the landing bag switch without being able to tell him why. It was an uncomfortable feeling. "Everything was new—we were all learning," he pointed out. "I personally think that they probably should have discussed it with him openly."

When Deke Slayton was grounded, Cooper was very close to resigning in support of his colleague. He maintained that when Slayton's heart problem became evident, the seven astronauts were involved in research and development on the centrifuge, running higher-level g-force loadings just to find out what might happen. They had a firm agreement in place that, in Cooper's words, "if anything adverse happened it was privileged information. It would not be used nor held against the person. So this directly violated all of our agreements. [It] really teed us all off no end. It was close—we really were all considering just resigning."

In August 1962, Cooper made a return visit to his birthplace in Shawnee to perform a pleasant but commemorative duty. He was there to dedicate an Armed Forces Reserve Center named in honor of his father, who had died in March 1960. Sadly, Leroy Gordon Cooper Sr. did not live to see his son launch into space, but he had been very proud of his selection as a Mercury astronaut. During his visit to Shawnee, and as part of the homecoming celebrations, the Otoe Indian tribe presented Cooper with a traditional war bonnet and proclaimed him to be Chief Oklahoma Space Man.

Though he was playing the space agency's PR game, Cooper was still having problems with NASA management. In one unguarded interview he had told a reporter that the astronauts were not getting enough piloting time to keep up their flying skills and the lack of available flying time meant they were losing their flight pay. While the Mercury astronauts were being heralded in the press as the nation's finest pilots, he said, they were having to fly as mere passengers on commercial airlines. "I didn't go out of my way to tell this reporter about it," Cooper explained:

He was there at Langley, looking over our training schedules, and started talking about our training. He said, "I see that flying airplanes is your number one training device, keeping up constant flying in fighter airplanes. What are you flying?" I said, "Personally, I'm going over to McGee Tyson on weekends and standing alert with them, flying F-104s. Which I have to go over there to do, because we don't have any airplanes." He said, "You mean you don't have any airplanes in NASA at all?" I said, "That's right, we don't." That started the whole thing. He was quite an investigative-type reporter and really did a lot of investigating. Of course, I'm sure that irritated some NASA people. But on the other hand, they were sitting around dragging their feet on getting our number one training device. They had all the funding; they had everything they needed to get them for us. They just wouldn't do it.

The resultant publicity, albeit unwanted by NASA, certainly played a part in the astronauts getting the type of airplanes they needed to hone their flying skills. They would be allocated T-33s, then F-102s, and then F-106s. Though the other astronauts were delighted with the results of Cooper's outspokenness, NASA managers were unhappy because it appeared that Cooper had gone outside of the agency, manipulating the press to get certain problems fixed. As far as Cooper was concerned, however, the results of this little bit of "astropower" had justified the means. Yet despite the lone maverick image Cooper seemed to have acquired and the credibility problems he was having with management, Wally Schirra really liked his fellow astronaut. "Gordo portrayed the role of the hotshot flyboy," he reflected. "He was a sharp guy."

Things finally began to happen for Cooper when he was assigned as backup for Schirra's six-orbit flight, planned for late 1962. Apart from the grounded Slayton, Cooper would be the only Mercury astronaut who hadn't flown, so the expectation was that he would fly the next, eighteen-orbit MA-9 mission. It wasn't quite as simple as that; Shepard was desperately pushing for another Mercury mission, while Slayton was still anxious to fly and was pulling as many strings as he could to be cleared. There was talk that the choice of pilot for MA-9 would actually depend on whether or not there would be more than one all-day mission. The *New York Times* quoted an unidentified Mercury official as saying that if only one such flight was planned, space officials were beginning to feel that it might be "prudent to entrust this flight to a pilot with Mercury experience. Commander Shepard is a known quantity and it would be a waste of acknowledged talent to finish Mercury without giving him an orbital flight since he handled the first mission flawlessly."

Deke Slayton later wrote that, by the time of this decision, Cooper had "alienated some higher-ups in NASA" to the point that they were considering not flying him at all. But to Cooper, everything seemed fine as he waited for an announcement on the upcoming flight. "None of that ever came down to me," he said of the misgivings some had whether he was the right person for the job. Wally Schirra was certainly backing his friend for the MA-9 task. Shortly after his successful flight, Schirra publicly said he was pulling for Cooper. "I hope soon to be working for Gordon," he told a *Life* reporter. "He's ready to go. I hope we can swap places on the next trip." Alan Shepard, while waiting for a second flight (and in fact covertly seeking to steal the flight from Cooper), stoically bit his lip and told reporters: "Gordo's been waiting around a long time. I must say, I admire him for being able to project as much of himself into the program as he has without being chosen for a flight up to this time." He must have privately choked on that line.

In his new position as coordinator of astronaut activities, Slayton would prove not only influential but almost autocratic in selecting crew members for upcoming flights. He looked objectively at some of the issues raised about Cooper's attitude, but in the end he decided that "Gordo was a capable pilot and could do the job, so I recommended him. There was some grumbling out of HQ, so I said: 'Either we fly him on MA-9, or we send him back to the air force now. It isn't fair to keep this guy hanging around if we're not gonna fly him.'" As usual, Slayton got his way, much to Shepard's chagrin. On 14 November 1962, Gordon Cooper was named as pilot for the mission, with Shepard as his backup. The next time the two would jockey for a flight, in the Apollo program, the result would be very different.

The MA-9 mission was planned to last longer than all of the preceding Mercury flights, stretching the spacecraft and its resources to the limit. This also meant that the spacecraft would be far heavier than those on previous missions, as it would have to carry additional amounts of consumables such as oxygen and water. Cooper said there were even discussions about removing the spacecraft's contour couch and suspending him in a net. "We could have saved a lot of weight; the net seats were really comfortable. That was a real pet project of mine, trying to get that net seat approved. We had to do all of these drop tests on everything, and it just ran behind a little bit. [In the end] we just couldn't do it." Considerable debate had centered on whether the net seats had a little too much spring, which might shake the astronaut more than was desirable. Although the lightweight seat was not used, the leg portion of the contour couch was removed, leaving only the back portion and the footrest, significantly reducing the craft's weight.

The flight of *Faith 7*, as Cooper had named his spacecraft, was originally planned for April 1963, but it fell victim to a succession of delays, and the launch date was finally announced as 14 May. As they covered the buildup to launch, the press decided that they really liked Gordon Cooper, with his boyish good looks and evident spirit of adventure. He also threw them some good lines, as when he spoke about his work schedule and what a tight squeeze it was getting into the overloaded *Faith 7*. "Climbing aboard the spacecraft is probably the hardest task of all," he chuckled in response to one question. "At first, nothing seems to fit correctly, especially the pilot."

Then, just a day before he was due to launch, Cooper buzzed the Cape in one of NASA's F-102 jet airplanes. Shocked and enraged, Walt Williams immediately called Chief Astronaut Deke Slayton. He didn't care if the launch was just two days away; he wanted Cooper taken off the flight and replaced by Alan Shepard. Shepard was told to get himself ready to fly. "Williams had long believed I was not taking things seriously enough," Cooper wrote about the incident in his memoirs. "He'd come from being chief of operations at Edwards for NASA's rocket plane program—the X-1, X-2 and X-3—and was accustomed to dealing with serious and studious research types, not fighter jocks. I guess I just rubbed him the wrong way, because much earlier he'd raised with Deke Slayton the subject of giving Shepard the nod over me." Fortunately for Cooper, Slayton talked long and hard with Williams, standing up for his fellow astronaut and pointing out why he should not and could not be replaced at this late stage. He promised to talk very bluntly with Cooper. Eventually, Williams was mollified enough to allow Slayton to deal with the matter. One can only imagine what Slayton said to Cooper after the incident, but the paint must have blistered on the walls.

He would have pointed out, quite correctly, that Cooper had come within a hair of losing his flight and astronaut career for the sake of an ill-conceived prank, while Cooper would have expressed his disgust at the last-minute tampering with his spacesuit. It was a worrying few hours for Cooper, but at ten o'clock that night he got a call informing him that Williams had finally relented. It had been a *very* close call.

On Monday, 14 May 1963, the astronaut, *Faith 7*, and Atlas 130–D were ready. As he stepped out of the gantry elevator, Cooper didn't look like someone who carried the latest hopes and fears of America on his slim shoulders. He flashed a smile at pad leader Guenter Wendt, stopped in front of the amiable German engineer, and threw him a crisp salute. "Private Fifth Class Gordon Cooper reporting for duty," he intoned. Wendt smiled and returned the salute. "Private Fifth Class Wendt standing by," he responded. This salutation harked back to a bit of mischief Cooper had conceived some two years before, on a day when NASA's Public Affairs office had allowed the press to film some prelaunch activities. It was file footage they would use when America's first astronaut—whoever that might be—was actually sent into space. Several people, including Cooper, Guenter Wendt, suit technician Joe Schmitt, and physician Bill Douglas were part of an elaborate gag, which took place as a fully suited-up Cooper left the transfer van at the foot of the gantry. While the press happily clicked away and movie cameras followed the action, the four men dutifully left the bus and strode confidently to the elevator. Cooper, carrying his portable ventilator, began looking up at the waiting Redstone rocket. Then, as he reached the gantry elevator, he suddenly grabbed the doorframe and began screaming, "No, No! I won't go!" With an incredulous press recording the hilarious scene, the protesting astronaut was unceremoniously bundled into the elevator cage and the door closed behind him. As the elevator rose, the four men broke into hysterical laughter.

Jack King from NASA Public Affairs was livid at this unexpected frivolity, but even more so when some influential media people called for Wendt and Schmitt to be sacked for their folly and for Douglas and Cooper to be immediately demoted. It all blew over, but from then on, Cooper and Wendt would always perpetuate the gag by referring to themselves as Privates, Fifth Class.

At 6:36 a.m., after a little good-natured pad banter and last handshakes, Cooper was sealed inside his spacecraft and began to check off his systems. He was ready for liftoff from Pad 14 around nine o'clock. Then there was a hold, but it had nothing to do with the spacecraft or booster. Instead it was a diesel engine on the ground, used to move the twelve-story service tower away from the Atlas rocket. The engine had never failed before, but on this occasion it was dead and would not start. As engineers struggled to fix the problem, Cooper decided to take a little nap. While he dozed, it was discovered that water in the diesel engine was the culprit. He was still blissfully asleep when word came through that a radar breakdown in the Bermuda tracking station had compounded the problems that day, finally causing a postponement of the flight until the next morning. Cooper, who had been in his spacecraft for five hours and fifty-one minutes when told of the decision, was understandably disappointed. "I was just getting to the real fun part," he said with a shrug and a palms-up gesture after his hatch was reopened. "It was a very real simulation." He was tired and drenched in perspiration, but as always he was resolute. "We'll try again tomorrow," he said.

The following day everything came together—after a while. At 3:00 a.m. on the morning of 15 May, Gordon Cooper was awakened, had a shower, ate breakfast, had biosensors stuck onto his body, and was suited up once again for his flight. He said he was still a little tired; he hadn't gone to bed until 11:00 p.m. the night before, but was hoping this would prove to be his big day. When Cooper was once again sealed inside *Faith 7*, he began to go through his checklist as part of the three-hour preflight routine. Once this had been completed and while radar and tracking technicians worked on their final calibrations, Cooper found that he was beginning to nod off. He was tired, had nothing else to do, and felt he could squeeze in a quick catnap. Meanwhile, Guenter Wendt and his team were going about things in their usual methodical way. "This was no time for talking emotion," Wendt recalled.

You have a job to do and you better not make a mistake. Reflecting is not on your checklist. They are in there, you are fighting the clock, making sure you don't make a mistake, that everything is right. There are hundreds of others, all at that moment watching things, making sure that they do everything correctly. Astronauts are in a different frame of mind. They have done so many simulations, so many practice runs, that they know they don't have anything to do until they get going. They are thinking, "For heaven's sake, let me go!" There is nothing they can do. Cooper was so relaxed, he went to sleep on me in the spacecraft!

As Cooper took his leisurely nap, the final technical, weather, and out-station reports came in, confirming the launch could proceed. Then it became time to make history.

"Gordo!" CapCom Wally Schirra's voice suddenly cut through the radio silence. The astronaut had been in a deep sleep. He was instantly awake and ready to go.

"Uh-huh," he responded.

"Hate to disturb you, old buddy," Schirra said, with a touch of laughter in his voice, "but we've got a launch to do here."

"Let's do it!" said Cooper. "Ready when you are."

Just after 8:04 a.m., the sudden roar of three Atlas engines rolled across the Cape, and *Faith 7* was hoisted into the warm Florida skies on the last and longest Mercury mission. Wally Schirra, known by the CapCom call sign "Sigma Seven" that day, could see that the launch was proceeding normally.

"Sigma Seven, *Faith 7* on the way," came the familiar Oklahoma drawl.

"Feels good, buddy," Schirra replied.

Cooper would later state in his mission debriefing that the moment of launch was smooth but very definite, and the acceleration was "very pleasant." He would also report that while he could hear the staging, he was more aware of feeling the separation. Soon after BECO (booster engine cutoff), the launch tower was jettisoned, and it arced away to the left. Next came SECO, the sustainer engine cutoff, and it too came right on time, placing the craft in orbit at 8:09 a.m. Cooper said it came with the same "glung" noise as BECO and that the spacecraft separation was not quite so noticeable in sound, "but was noticeable in the boot from the posigrades firing."

Faith 7 automatically turned around after separation, and Cooper's attention was attracted to the booster, not more than two hundred yards away. "I could read the lettering on the sides," he would later report in his mission debriefing. "I could see various details of the sustainer, the tanks, and many little details. It was silver—very bright silver in color, with a frosty white band around the center portion of it." He would keep the booster in sight for several minutes, noting in glimpses out the window that it was quite clear and distinctive and that it was slowly turning "in a counterclockwise rotation. . . . I kept it in sight the whole time and could note the Cape and Florida in the background with the booster against them. It made an ideal picture to view."

As scheduled, Cooper then turned off his autopilot and switched to fly-by-wire mode, allowing him electronic control of *Faith 7*. Next he aligned his spacecraft in orbit, facing the blunt heat shield in the direction of flight and establishing a temporary attitude at which he could fire the retrorockets should Mercury Control detect a problem that would require an immediate deorbit.

When Cooper had his first good look at the Earth from space, time seemed to stop. "It is so spectacular," he told the authors. "It absolutely is a fantastic view. Just unbelievable how pretty it is, how calm and peaceful it is." Cooper was particularly well equipped, with movie and still cameras, for the long-duration flight and later took photographs of weather phenomena, cloud formations, and stars. In fact, as soon as he reached the nightside of the Earth he began calling off the stars he could recognize in the night sky south of India. "There's Orion," he reported, "and Betelgeuse. What a beautiful night tonight." He would even transmit pictures of himself working inside his cramped spacecraft by using a special ten-pound slow-scan television camera—the first such pictures of an astronaut in space ever seen by the American viewing public. Insufficient light inside the spacecraft made the images disappointingly indistinct, but those taken of the Earth through the window were of reasonably good quality.

To the amazement of ground controllers, Cooper soon began reporting that he could see all kinds of small features on the ground, such as roads on the Arabian Peninsula, a powerful ground light set up at Bloemfontein in South Africa, railroads, and even individual houses. Cooper later recalled that many experts doubted at first that he could have seen such small features from orbit, but he would prove them wrong. "The thing that people didn't realize then is that when you are running optical tests here on the ground and you are trying to see things one hundred miles away, you are looking through a hundred miles of atmosphere. When you looking down from space, you are only looking through about five or six miles of atmosphere. So the vision from space down to Earth is certainly different over the same distance."

After his flight, he would tell astonished reporters, "I could spot houses in Asia by following the smoke down to the chimneys. I could see the dust blowing off a road, and then find the road. I found a steam locomotive puffing along in northern India by seeing the smoke and then the railroad tracks and finally the engine." Initially, Cooper's claims met with a lot of doubt and skepticism. "But when NASA went back and verified the things I said I had seen, they found that yes, there had been a train running down that particular track, and yes, there had been a truck going down that particular road. Yes, there were monasteries up in that Himalayas just as I described." At one point, Cooper even described seeing what he thought was an aircraft carrier on the St. John's River near Jacksonville, Florida. This was later checked and found to be a tugboat and barge, but the fact that an astronaut could discern large objects from orbit certainly caused a major stir of interest. The crew of *Gemini 4* would later confirm everything that Cooper had said.

Aside from Cooper's clear visual impressions, some of the clearest photographs of the Earth to that time were taken from *Faith 7*. These included breathtaking panoramas of Africa's Atlas Mountains, the Great Indian Desert, a long streamer of clouds above the brown sands of the Arabian Desert, and heavy cloud layers over China. Geologists were certainly impressed, enthralled at seeing such topographic features as the Himalayas in a manner never before available to them. As Carpenter had, Cooper also used a special camera to take photographs of the zodiacal light and of the glow of Earth's upper atmosphere.

Nearly five hours into the flight, Wally Schirra told Cooper that they were very impressed with the work he was doing. He added: "We lay a pat on the back from Walt Williams." That, Cooper thought, was especially gratifying to hear. His well-constructed orbital work schedule included nearly a dozen vital experiments. One of these involved deploying a five-inch ball fitted with two flashing 60,000–watt xenon lights during Cooper's third orbit. This test was designed to show how well an astronaut could see another craft in space and judge distances during rendezvous. Next to the biomedical experiments, designed to determine how the astronaut fared during an extended flight, the flashing beacon was regarded as the most important experiment on his flight. Cooper lost track of the small satellite for some time, but on the next orbit, as he swept over a darkened South Africa, he suddenly observed "a light coming up from below me on the ground." For a few moments he thought that something was being launched toward him, but then he noticed it was flashing and realized it was the beacon. He could see the flashing lights approximately ten miles from *Faith 7* and was able to visually track the beacon for some time.

In an experiment similar to Scott Carpenter's on his flight, Cooper also deployed a balloon that was dragged along behind the spacecraft. This was another test of the astronaut's ability to judge distances in space as well as to determine how dense the traces of atmosphere are in space and if they affected the balloon's flight. But just after deployment, as on Carpenter's flight, the balloon experiment failed.

Cooper also performed some moderate exercise regimes, in which he pulled on a rubber cord with a recoil tension of sixty-four pounds. He would later complain that in his crowded cockpit it took almost as much exercise to retrieve the cord from its stowage as it did to actually use the thing. There was also an experiment involving a new kind of dehydrated food that did not quite go according to plan. Whenever Cooper squeezed water into plastic bags containing the powdered food some droplets escaped, and he had to chase down them down before they drifted into sensitive electrical areas. Cooper also had a little PR work to conduct, sending special messages of greeting as he flew over several countries.

Flight Director Gene Kranz would later report that as the orbits began clicking by, the systems were "performing perfectly and the early Go/No Go decisions easily made." On Cooper's ninth orbit most controls were shut off to allow the astronaut to sleep, and *Faith 7* flew on in a slow, drifting roll. Renowned for his calm, confident ability to nod off anywhere, anytime, Cooper slept for a total of around seven-and-a-half hours on his flight, although he did wake once to adjust the temperature in his spacesuit. Throughout, sensors attached to Cooper's skin recorded his body temperature, and electrocardiograms monitored his respiration and blood pressure. When he finally woke fully he began his second day aloft with a breakfast of reconstituted dehydrated chicken and gravy. Later in the flight he was asked, in a message relayed from a flight surgeon, if he remembered dreaming while he was asleep. "Negative," he responded. "I slept too soundly to dream."

In his postflight debriefing, he would suggest that sleeping in space should present no problems for astronauts on prolonged missions:

One indication of my adjustment to the surroundings was that I encountered no difficulty in being able to sleep. When you are completely powered down and drifting, it is a relaxed, calm, floating feeling. In fact, you have difficulty not *sleeping. I found that I was catnapping and dozing off frequently. Sleep seems to be very sound. I woke up one time from about an hour's nap with no idea where I was and what I was doing. I noticed this again after one other fairly long period of sleep. You sleep completely relaxed and very, very soundly to the point that you have trouble regrouping yourself for a second or two when you come out of it.*

Through delicate use of controls and by drifting much of the flight, Cooper was able to conserve a surprising amount of his consumables, such as fuel and oxygen. In fact, his expenditure of these necessities was so far below the allowable maximums that Alan Shepard, performing CapCom duties, jokingly told his orbiting colleague, "You can stop holding your breath and use some oxygen if you like!" Doctors on the ground later attributed Cooper's frugal use of oxygen partly to the fact that he was the first nonsmoking astronaut to fly into space.

Cooper's flight duration exceeded that of all the other manned Mercury flights combined, but pushing the spacecraft this far had its risks. He encountered a multitude of problems, including rising temperatures and carbon dioxide levels, a slow oxygen leak, and numerous electrical difficulties. Things came to a head on the nineteenth orbit, when a plunger designed to collect his perspiration malfunctioned, pumping moisture into the confines of his spacecraft; this is probably why *Faith 7* began to lose its electrical system.

On that same orbit, the green .05G light on Cooper's control panel, which normally indicated the start of an automated reentry, suddenly lit up. This was well ahead of schedule. The light was not meant to illuminate until he'd actually begun his reentry and a force equal to five-hundredths that of gravity was being exerted on the craft. Ground technicians confirmed with Cooper that he had not somehow initiated reentry, and after looking into the situation they determined that a closing relay had rendered some automatic control equipment inoperable. During the following orbit, Cooper was instructed to turn on his automatic controls. He then reported that his attitude gyroscopes, which automatically aligned the spacecraft for reentry, were not working. This meant that all of *Faith 7*'s automated control systems had lost power. "Things are beginning to stack up a little," he calmly reported to Hawaii Control. After considering all the options, Cooper was given instructions to manually realign *Faith 7* for reentry. With a temperamental automatic control system, no gyroscopes, and other systems failing, he managed to carefully align the spacecraft using the horizon.

"It was very interesting," he would recall. "I felt trained for it, able to do it without any problem. Of course, I had to have a little bit of luck. The temperature was rising rapidly; the carbon dioxide level was coming up rapidly. I had no way of controlling or doing anything about either one of those. If I could tolerate them, then I could make my nominal landing point." With the prospect looming that he might have to make a manual reentry, Cooper remained philosophical. "We had been wanting piloted reentry anyway," he later commented. "We weren't planning on having automated reentry for Gemini. We were planning on having a computer giving us guidance, but still piloting." Once again, a test pilot's skills were required in space. If Cooper still needed to prove to anyone that he was up to the challenge, this was the moment.

John Glenn was monitoring the flight of *Faith 7* from aboard the ship *Coastal Sentry Quebec*, cruising south of Japan. When he asked Cooper if he had checked his attitude by visual means, the unflappable astronaut responded, "Roger. Right on the old gazoo." The onboard problems worsened on the final orbit when Cooper reported through Zanzibar Control that his inverters, which provided electrical current for the reentry control system, had also failed. "Have you tried the standby inverter?" asked a worried ground communicator, Jim Tomberlin.

"Roger," Cooper responded immediately. "It would not start." If the backup inverter wouldn't come on line, it meant the entire automatic reentry system was dead. Tomberlin was completely mystified by this, but a decision and a suitable response were needed in a hurry, so he made it. "Roger; then we'd better get on with the checklist now. Attitude permission to bypass."

"Roger; retrorocket arm to manual," Cooper confirmed. He would now have to initiate a manual fly-by-wire reentry and correct any oscillations without any automatic aids to help him. It was a challenging scenario with certain similarities to the life-or-death reentry of Scott Carpenter twelve months earlier. As Cooper later explained to reporter Roy Neal, he had to manually activate the jets for attitude control using push/pull rods, then visually control his pitch, roll, and yaw attitude by looking out of his window. "I had a wristwatch for timing," he said. "I had to activate each and every one of those relays, and I'd have to manually fire the retros while manually flying the spacecraft. . . . I'd have to control the spacecraft all the way through reentry. I'd have to put my drogue out manually [and] deploy my parachute manually. I'd have to deploy the landing bag manually."

At one point, while maintaining his communication with Glenn, Cooper remarked, "I'm looking for lots of experience on this flight." To which the freckle-faced astronaut replied with a lightheartedness he did not truly feel, "You're going to get it!" Based on medical advice from Dr. Charles Berry, ground controllers took the unusual step of telling Cooper to take a Dexedrine stimulant tablet to help keep him alert during the reentry phase.

With Glenn helping him to reel off the countdown, Cooper hit the retrofire button at the zero count, the three small rockets fired perfectly, and the craft began to slow for an eventual plunge back to Earth. Now the once-underestimated astronaut proved his test-piloting skills beyond any doubt, keeping a fine touch on the controls as he steered *Faith 7* through the increasingly heavy atmosphere on a gently curving path down to the Pacific. On that day, like Scott Carpenter before him, Cooper was doing nothing less than averting the loss of his spacecraft, once again demonstrating the rightful place of skilled pilots in the Mercury program.

Meanwhile, he and Glenn continued a running conversation. "Real fine flight, Gordon," Glenn assured him. "Real beautiful all the way." Cooper flew the reentry beautifully, or as he had said, "right on the old gazoo." After the radio blackout period ended, his small drogue chute was deployed, followed by the main chute. Aboard the USS *Kearsage*, anxious crew members heard the boom of his supersonic reentry and scanned the skies. They soon spotted the red-and-white candy-striped parachute and the small black craft below.

Despite the odds, Cooper's precise piloting enabled him to splash down less than four miles from the waiting carrier. His would be the most on-target splashdown of the entire Mercury program. "Mostly," he said with obvious satisfaction, "I was elated . . . to have splashed down a mile closer to the USS *Kearsage* than Wally." For his part, Schirra was elated too, not only that their concentrated manual navigation training had paid off, but that his colleague was safe—even if he had been outdone in the process. "That son-of-a-gun landed closer to the carrier than I did!" he later observed. "And he had no automatic control system at all. No horizon, reference from the horizon scanners . . . none of that worked."

Faith 7, with Cooper electing to remain inside, was plucked from the ocean by one of the carrier's recovery helicopters. Once it had been lowered onto the deck and secured, Cooper blew the hatch on command. Following some quick questions and blood-pressure readings he was assisted from his overheated spacecraft at 8:11 p.m., to the tumultuous cheers of hundreds of crewmen. He was totally soaked in sweat inside his spacesuit and had lost seven pounds during his flight, but despite feeling a little dizzy and disoriented he was in good health and spirits.

A relieved President Kennedy had watched the splashdown live on TV, and he phoned Cooper aboard the *Kearsage* by radio-telephone to convey his congratulations for what he called a "great flight." After a welcome meal, Cooper spent that night enjoying a sound sleep, and the next morning boarded a helicopter bound for a huge reception and debriefing in Honolulu. As the helicopter closed in on Pearl Harbor, Cooper was asked if he would honor the lost sailors aboard the USS *Arizona* by dropping a wreath over the site. Wanting to do a good job, but still weakened by his daylong space flight, he leaned forward at the opened door and threw out the wreath. It was fortunate that a crewman was standing behind Cooper, for he was wearing a huge flowered lei, and as he flung the wreath out the door he overbalanced and nearly followed it. A strong arm reached out and pulled him back to safety. Cooper often wondered how the navy would have explained dropping a flight-fresh astronaut onto the revered memorial.

When he finally landed and stepped off the helicopter at Hickham AFB, still wearing his giant lei, Cooper had another welcome surprise waiting for him. His old nemesis, Walt Williams, was standing expectantly on the tarmac. As the two men met, Williams cordially shoved out his hand with a warm smile and offered his sincere congratulations. Cooper later recalled that it was the only time Williams had done this, and it was particularly unexpected after Cooper's low-level swoop over the Cape had nearly prompted

Williams to yank him from the flight. He was astonished when Williams said, "Gordo—you were the right guy for this mission." A delighted Cooper felt that in Williams's own way "he was letting me know that he'd been glad, after all, to have a self-professed hotshot fighter jock at the controls."

Trudy was also waiting for Cooper, together with their daughters Camala and Janita, who, draped in huge colorful leis, had flown across from their home in Houston. For Trudy it was the first time she'd been to Hawaii since 1947. The family kissed and embraced each other warmly. That Sunday, Cooper and his family flew back to Cape Canaveral. He was greeted on arrival that afternoon by one of the largest crowds in the Cape's history, with an estimated eighty thousand jubilant people lining the parade route. He was especially thrilled to see that a 100–foot-long, 8–foot-wide red carpet, emblazoned with the initial "G" in gold and bordered by yellow tassels, had been painted onto Highway AIA, the south entrance to Cocoa Beach. Beside it was a large sign saying "Cocoa Beach's Red Carpet for Gordo."

Two days later, the Cooper family flew from Cape Canaveral aboard the presidential aircraft that normally operated as Air Force One straight into a tumultuous welcome in Washington. The presidential jet touched down at Andrews AFB, eleven miles east of the capital, where several thousand people had already assembled to greet him. In keeping with Washington's traditional greeting for Mercury astronauts (as John Glenn had found a year before), the skies were gray and there was an intermittent drizzle. But heavy skies were not enough to prevent people from assembling dozens deep along the footpaths of historic Pennsylvania Avenue, ready to show their appreciation. First, however, there was a small ceremony to perform. A beaming President Kennedy officially welcomed Cooper to the White House and presented him with NASA's Distinguished Service Medal, telling the astronaut, "You have given us a great day and a great lift." He also hailed Cooper as a worthy successor to Charles Lindbergh, who had flown solo across the Atlantic that same day thirty-six years before. In a speech later that day, Kennedy would publicly applaud the achievements of the last Mercury flight and NASA in general by stating: "I know there are lots of people who say, 'Why go any further in space?' When Columbus was halfway through his voyage, the same people said, 'Why go on any further?' And they want to stop now. I believe the United States of America is committed in this decade to be first in space, and the only way we are going to be first in space is to work as hard as we can here and all across the country, and support not only Major Cooper, but all those who come after him."

Following the White House ceremony, the Coopers took part in the much-anticipated limousine parade down Pennsylvania Avenue, now lined with cheering masses of enthusiastic well-wishers. Faces filled every window along the route, while a storm of paper, streamers, and twirling ticker tape showered down on the motorcade. Later that memorable day, Cooper addressed a joint session of Congress, the Cabinet, and the Supreme Court, receiving a two-minute standing ovation before he even had a chance to speak. Toward the end of his talk he showed the more serious side of his nature by reading out a personal prayer he had composed and taped during his flight. Sitting alongside Trudy Cooper and her two granddaughters, Hattie Cooper was so overcome during her son's recitation that she had silently wept with love and pride.

NASA's John Yardley was an engineer who led the design team for the Mercury and later spacecraft and was launch operations manager during Cooper's flight. He would subsequently discover that the length of the MA-9 mission had created unforeseen problems such as increased humidity, which caused many of the equipment and electrical failures endured by the astronaut. Before MA-9, the longest time an American manned spacecraft had been in orbit was just over nine hours, but Cooper's flight lasted nearly four times longer. Yardley said his team discovered postflight that "everything started getting wet because the humidity was soaking in, and the equipment was not humidity-proof." They concluded that moisture, possibly from Cooper's perspiration, had corroded a connection on a compact electronics box called an Amp Cal (amplifier calibration), and this had caused the problem with the automatic controls. A short circuit, possibly due to faulty insulation, had also occurred in an Amp Cal solder connection, and this had knocked out other automatic reentry controls.

According to Flight Director Gene Kranz, "Mercury worked because of the raw courage of a handful of men like Cooper, who sat in heavy metal eggcups jammed on the top of rockets, and trusted us on the ground. That trust tied the entire team into a common effort." He would later add that

"Cooper, the loner and rebel against the spaceflight bureaucracy, had pulled off a great mission and a picture-perfect entry. Gordo's test pilot mentality, coupled with the superb performance of the ground team, was a fitting finale to America's first manned venture into space."

Cooper had silenced his critics—for now. But questions about his attitude toward training and authority would soon resurface. He'd wrest one more space mission out of NASA, but in the end the instinctual trust in his own judgment that saved him in space was his undoing on the ground.

The flight of *Faith 7* effectively brought the Mercury program to an end and work began in earnest on the next program, Gemini. At least that's the way NASA saw things evolving. One man named Shepard was determined to change their minds.

Without any doubt, the Mercury astronauts were a competitive bunch. As Gordon Cooper prepared for his MA-9 flight, Alan Shepard had gone full circle in the flight rotation. He had been the prime pilot for the first, suborbital flight, and now he was the backup for what was scheduled to be the last Mercury mission.

Shepard, and Glenn for that matter, felt there was still mileage to be had from the one-man spacecraft that had served so well. The fact that he had only accomplished a fifteen-minute mission also irked Shepard, as he watched his colleagues flying increasingly longer orbital missions around the Earth. He may have been the first, but that touch of glory was beginning to wear a little thin, and Shepard started to feel that he had been left behind in the scheme of things. It was not a good situation for a restless naval aviator—he wanted to go into orbit. If getting a second Mercury flight meant going right to the top and using a little of his renowned tenacity, then Alan Shepard was prepared to give it his best shot.

He even had a spacecraft to fly—Mercury spacecraft 15B. And just like Shepard, it had been there since the beginning.

Once Project Mercury assumed official existence on 26 November 1958, NASA had begun to move quickly on the design and eventual ordering of the one-man Mercury capsule. By this time, several potential designs had already been submitted and were undergoing evaluation, including four full-scale steel prototypes (more commonly known as boilerplates) that had been constructed at NASA's Langley facility. By February 1959, the final specifications of the bell-shaped capsule had been determined and an order for twelve production models was placed with the McDonnell Aircraft Corporation, based in St. Louis, Missouri. This order would later be increased by eight. Of these Mercury spacecraft, fifteen would eventually be used on flight missions, while production models 10, 12, 15, 17, and 19 never made it off the launch pad.

The Mercury capsule assigned to Shepard's suborbital flight was Spacecraft Number 7, and quite by chance it was flown on top of the seventh Redstone rocket produced specifically for NASA. This gave Shepard a numerical starting point when he decided to attach a name to his craft, in a time-worn tradition observed by military pilots. Thus, he inadvertently began the trend of naming each Mercury spacecraft using the number seven, as fellow astronaut Scott Carpenter recalled in an interview for *Space Flight News* magazine. "Al Shepard named his *Freedom 7*," Carpenter explained. "*Freedom* for obvious reasons, *7* because it was the seventh spacecraft off the line. The press and everybody else thought that Al named it Seven, not because it was Spacecraft Number Seven, but in honor of the seven astronauts. He never corrected them, we never did, so the misconception remained, and everybody who flew in the Mercury program from then on had to follow the precedent and name his spacecraft 'Something 7'. So that's how we all got 7."

Originally, NASA planned to fly each of its seven astronauts on suborbital missions using Redstone vehicles but later trimmed this manifest to just four flights. Glenn was originally scheduled to fly the third such mission. Shepard's spacecraft, *Freedom 7*, gave him a beautiful fifteen-minute suborbital ride, and his flight was virtually duplicated two months later by Gus Grissom aboard *Liberty Bell 7*. But a jubilant NASA, eager to catch up with the Soviets, decided to abandon the remaining suborbital missions. As expected, when a new flight manifest was drawn up, John Glenn was confirmed as pilot of the first Mercury-Atlas manned orbital mission.

Earlier, on 13 August 1961, McDonnell's Mercury Spacecraft 15 had been shipped over to Cape Canaveral. The intention at that time was to configure the spacecraft for the third of the proposed Mercury suborbital flights, MR-5, then slated for launch in late August or early September with John

Glenn on board. With the abandonment of the final suborbital missions, however, Spacecraft 15 was reassigned as a temporary research vehicle for testing reaction control and environmental control systems under conditions of simulated altitude. It was eventually returned to the McDonnell plant on 17 January 1962 for reconfiguration as an orbital vehicle, having been assigned first to the potential MA-13 mission and later to MA-12. Both flights were ultimately cancelled.

The newly configured spacecraft made a return journey to the Florida spaceport on 16 November 1962, and two months later Spacecraft 15B (as it was now designated), had a fresh assignment to a possible MA-10 mission. Further modifications were made by McDonnell's technicians based at the Cape, who redesigned the spacecraft's fuel system to allow for a longer-duration flight, relocated external electrical connections, worked on the hand controller rigging procedures and the pitch and yaw control valves, and updated the interior design. It is estimated that the spacecraft, had it been sent into orbit, would have weighed around 3,284 pounds at launch.

Shepard knew that Spacecraft 15B had not only been assigned to a possible last mission but that it had also been substantially upgraded to make it capable of a prolonged orbital flight. He would find it hard, however, to convince the NASA hierarchy to let him fly it. On 11 May 1963, a few days before Cooper's mission, Deputy Assistant Administrator for Public Affairs Julian Scheer patently dismissed speculation about a further Mercury flight. "It is absolutely beyond question," he stated, "that if [Cooper's] shot is successful, there will be no MA-10." This, and increasing disinterest on the part of other administrators for an extra flight, only made Shepard all the more determined. There is no question that he felt MA-10 was "his" mission. As Cooper's backup he would have been automatically slotted in as prime pilot for an additional flight. Because the other astronauts were now busily engaged in assignments specifically related to Projects Gemini and Apollo, he and Cooper were the only astronauts still actively involved in Mercury training. Cooper, therefore, would have been named as his backup. Shepard resolutely proceeded with his plans for MA-10, evidently to the point of renaming Spacecraft 15B *Freedom 7 II* and cheekily having that logo painted on its shingled exterior, just below the trapezoidal window.

When Gordon Cooper's *Faith 7* plunged into the Atlantic after completing twenty-two orbits, Project Mercury was presumed to be at an end. Dr.

Robert Seamans, NASA's associate administrator, reiterated this at a postflight press conference. When asked if there would be another Mercury flight, as was still rumored around NASA, Seamans responded, "It is quite unlikely." Despite this, a final, official decision still had to be made. A small window of opportunity was still open for him, so Shepard began to press even harder. The first manned flights of the Gemini spacecraft were still at least a year away. Shepard argued that one more Mercury mission would not only prove beneficial, principally in spaceflight dynamics and engineering, but would also fill a considerable gap between the two manned programs at a time of increased Soviet activity. Additionally, there were still two Mercury spacecraft and boosters at the Cape available and basically ready to go.

Shepard had a good deal of support in his campaign, not only from his fellow astronauts, but also from two influential men within NASA—the director of the Office of Manned Space Flight, Brainerd Holmes, and Walt Williams. Williams had advocated flying an astronaut aboard MA-10 on a minimum-duration flight of one hundred hours and perhaps up to an "open-ended" six days, as long as the pilot was in good condition and spirits. He argued that such a flight could replace two of the early, short-duration Gemini missions, cut six months from that program, and allow the first two-man flights to orbit for a week instead of just one to three days. In an argument with a more competitive purpose, he also believed that the United States could seize the space-duration record then held by Soviet cosmonaut Andrian Nikolayev, who had completed a sixty-four-hour flight aboard *Vostok 3* the year before.

In February 1998, Shepard spoke with interviewer Roy Neal about the mission he wanted to fly:

My thought was to put me up there, and just let me stay until something ran out; until the batteries ran down, until the oxygen ran out, or until we lost a control system or something. Just an open-ended kind of mission. And so I recommended that, and they said that they didn't expect to hear anything else from me. But I remember when Cooper and his family, and the other astronauts and families were invited to the White House for cocktails with Jack Kennedy, and we stopped at Jim Webb's house first and had a little warmup there, and I was politicking with Webb, and I said, "You know, Mr. Webb, we could put this baby up there in just a matter of a few weeks—it's all ready to go. We have the rockets. Just let me sit up there and see how long it will last, get another record out of it." Well, he said, "No, I really don't think so. I think we've got to get on with Gemini." And I said, "Well, I'm going to see the President in a little while. Do you mind if I mention it to him?" He said, "No, but you tell him my side of the story too." So I said, "All right."

So we got over there and we were all sipping our booze, getting some of our taxpayers' money back drinking at the White House, and I got Kennedy aside and said, "There's a possibility we could make another long-duration Mercury flight—maybe two, maybe three days—and we'd like to do that." He said, "What does Mr. Webb think about it?" I said, "Webb doesn't want to do it." So he said, "Well, I think I'll have to go along with Mr. Webb." At least I tried.

John Kennedy's ambivalence on the matter was evident at a regular press conference on 22 May 1963. Just a week after Cooper's flight, the president was asked about the chances of the space agency conducting a further Mercury flight. Kennedy's response indicated that NASA should and would make that determination. Following the conference, Kennedy rang James Webb. After the usual greetings he came to the point; he wanted someone to make a decision about the flight. "You know who's going to make this decision, don't you?" he asked. "I think I do," was Webb's diplomatic response. It was good enough for Kennedy. "You're going to make it," he said with finality. The next day, a closed hearing of the House Subcommittee on Manned Space Flight discussed the possibility of a further one-man flight. Although the committee indicated that they had some support for one last Mercury flight, they also agreed that the decision should be left to NASA.

Things were heating up. In Washington on Thursday, 6 June 1963, Robert Gilruth, Brainerd Holmes, Kenneth Kleinknecht (manager of the Mercury Project Office), and Walt Williams gave a two-day presentation before James Webb, his deputy administrator, Hugh Dryden, and Robert Seamans, in which they expounded reasons for conducting at least one more Mercury mission. There was no real impediment, they argued—a number of spacecraft had been updated and reconfigured, and these were capable of carrying sufficient consumables such as fuel, oxygen, and water to allow for missions of several days' duration. Spacecraft 15B, which had served as the backup capsule for Cooper's flight, stood ready to make the next flight, and Atlas 144-D had been tentatively assigned to carry it into orbit should the mission be approved.

Alan Shepard and Tom Stafford, the planned original crew for the first manned Gemini flight. Courtesy NASA.

Webb weighed up all the arguments, which he admitted were quite convincing, but when he stood before the Senate Space Committee six days later, he began by stating in part: "There will be no more Mercury shots." He went on to explain that Project Mercury had satisfactorily accomplished its goals, and there should now be new priorities. All the energies of NASA and its contractors, he said, should now be fully employed in focusing on the Gemini and Apollo missions. As well as declaring that the Mercury spacecraft was in effect obsolete, Webb, it was known, had also expressed an innate fear that any accident suffered during an additional mission would seriously delay both the Gemini and Apollo programs. He was also an angry man, as Tom Stafford recalled in his autobiography, *We Have Capture*:

"Webb was furious about the way the astronauts—especially Shepard—went around him to the president. From this point on, he took every opportunity he could to let us know he was boss."

In all, Project Mercury had carried two men on suborbital flights and four on orbital missions. NASA's astronauts had spent fifty-three hours, fifty-seven minutes, and twenty-seven seconds in flight for a total of thirty-four orbits. The job had been done, and with extreme competence. Following Webb's announcement, the now-obsolete Spacecraft 15B was removed from flight status and placed into storage at Cape Canaveral. It would never fly into space.

Shepard gained some consolation when he was selected to fly the first Gemini mission, with Tom Stafford as his copilot. They began preliminary work in early 1964 and were carrying out preparatory training in the simulators when Shepard was suddenly struck down with a serious ailment that threatened to end not only his astronaut career, but also his days as a pilot. He had woken one morning feeling dizzy and unexpectedly collapsed when he stood up. At the same time he became disturbingly nauseous. Shepard was not overly concerned, thinking this was just an isolated incident. Five days later, however, he suffered another sudden bout of dizziness, vomited uncontrollably, and heard a loud, recurring ringing in his left ear. After these attacks had struck him down several times, Shepard finally realized this wasn't something he could simply tough out and reluctantly made an appointment to see the flight surgeons. Following extensive tests a panel of NASA doctors recommended that he be removed immediately from the flight rotation. Suddenly, he found himself grounded.

The ailment proved to be Ménière's Syndrome. "The problem is not considered very significant for an earthbound person," he told Commander Ted Wilbur during a 1970 interview for *Naval Aviator News,*

but it sure can finish you as a pilot. I convinced myself it would eventually work itself out. But it didn't.

Tom Stafford had told me about Dr. House, out in Los Angeles, who could perform an operation on this particular kind of inner ear trouble. At first it sounded a little risky but, in 1968, I finally decided on having it done. With NASA's permission I went out to California. In order to keep the whole business quiet, Dr. House and I agreed that I should check into the hospital under an assumed name. It was the doctor's secretary who came up with it. So, as Victor Poulis, I had the operation, and six months later my ear was fine.

Despite the successful operation, Shepard had lost his chance to fly on Gemini, and it seemed doubtful that he would ever spend more time in space than his fifteen-minute "hop" in 1961. Earlier, in order to remain part of the astronaut cadre, he had accepted an interim appointment as chief of the Astronaut Office, which made him a major force in the training and assignment of his fellow astronauts. Eventually, his never-say-die attitude would enable him to not only regain active astronaut status but land on the moon as commander of *Apollo 14*.

America's first astronaut left NASA in 1974, retired from the navy with the rank of admiral, and concentrated on a variety of business ventures. He died in his sleep on 21 July 1998 while undergoing treatment for leukemia at a hospital in Monterey, California.

Paul Haney knew Alan Shepard as well as anyone, and he told the authors he'd spoken with his old friend for the last time about a year before Shepard succumbed to his illness:

I called him at his Pebble Beach, California, home to see if he would be available to come to New Mexico to speak at a Space Hall of Fame function to honor the late Walt Williams, NASA's rough-and-tumble head of flight at Houston, and before that at Edwards Air Force Base.

The first thing that Al said was that he had a "touch" of leukemia. "Come on, Al," I said, "Nobody, not even you, has a touch *of leukemia!" Well, almost combatively, he said, yes, we'd see about that. Then he rattled off some numbers indicating progress. I wished him well. Then he asked the date of the Williams function. He flipped around in a calendar and said, "Gosh Paul, I'm sorry. I'm playing golf that day . . . with Tiger Woods!"*

Speaking about the loss of Alan Shepard on behalf of his fellow Mercury astronauts, John Glenn told a memorial gathering at the Johnson Space Center: "Alan Shepard was many things. He was a patriot, he was a leader, he was a competitor, a fierce competitor. He was a hero. More importantly to us, he was a close friend.

"America has lost one of its true adventurers," Glenn concluded. "We have lost more than a friend; we've lost a brother."

9. A Seagull in Flight

Space is for everybody.
It's not just for a few people in math and science,
or for a select group of astronauts. That's our new frontier out there.

S. Christa McAuliffe

On 13 June 1963, the world's press was buzzing with mounting speculation about an exciting landmark in space exploration. Informed reports of an imminent spaceflight by a woman cosmonaut were freely sweeping around Moscow. These reports suggested that the history-making flight might last up to eight days and that the woman would not be alone in space. This led to conjecture by Western analysts that the Russians might be ready to launch the first two-seat spacecraft or to engineer another "rendezvous" in space similar to the one flown ten months before. Cosmonauts Yuri Gagarin and Gherman Titov also fueled speculation in a *Pravda* article by declaring that "both sons and daughters of the Soviet Union" would soon be launched into space.

Relying on unofficial sources, Western newspaper articles described the woman as "good looking, short, strongly built and in her early twenties." Furthermore, it was rumored that she might be a close friend and even romantically involved with Andrian Nikolayev, the only bachelor among Russia's four known cosmonauts. Several newspapers confidently named the mystery woman as Anna Massevitch, vice president of the Aeronautic Academy of Science in Moscow. They reproduced her photo alongside their articles, and Massevitch certainly seemed to match the general description.

The following day, quelling some rumors but also increasing speculation, Lt. Col. Valery Bykovsky was launched into space aboard *Vostok 5*. A taciturn, twenty-eight-year-old air force pilot, Bykovsky successfully rock-

Cosmonaut Valery Bykovsky, pilot of *Vostok 5*. Courtesy Colin Burgess Collection.

eted into an elliptical orbit around the Earth. Eleven minutes after launch he calmly reported that he was "feeling fine."

That night, through the Eurovision system, western Europeans were treated to fuzzy images of the world's latest space explorer in his Vostok craft. Television technicians in London taped and edited the pictures before transmitting them to the orbiting communications satellite, *Telstar 2*. A receiving station in Maine gathered the signals and flashed images of Bykovsky to waiting television networks across America.

The suspense was not sustained for long. On the morning of 16 June a cosmonaut dressed in a bulky orange spacesuit walked toward the foot of a waiting silver-and-black rocket. Beneath the spacesuit, sewn onto the left breast of a blue thermal outfit, was a large dark-blue flight patch specially prepared by a couple of women in the Zvezda (Star) spacesuit research bureau. It depicted a small snow-white dove clutching an olive branch, flying against a backdrop of golden sunrays. Beneath the dove, the initials CCCP (USSR) had been strikingly embroidered in red cotton. After waving to spectators from the gantry hoist, the cosmonaut was carried up fifteen stories to where the open hatch of *Vostok 6* beckoned. With a final slow gaze out over the barren steppes and the swarm of people gathered below, the cosmonaut clambered into the confines of the capsule. Once strapped in with communications established, the cosmonaut's first duties included reporting to the chief designer, Korolev, and the launch team: "Ya Chaika" (I am Seagull), came the clear, steady voice. "I feel fine."

At 12:29 that afternoon, Moscow time, the four strap-on boosters of the SL-3 variant rocket carrying *Vostok 6* fired simultaneously. There was a shrill whistling and then a mounting roar. Observers saw a brilliant white-hot flame gush from beneath the pad and roll away in a maelstrom of red sand, kicked up by the fury of the blast. It took almost ten seconds for the rocket's boosters to reach full thrust, and then the four gantry hold-down arms pivoted backward in unison on huge counterbalances. At the same instant, the vehicle could be seen moving upward from the pad on intensely bright lances of flame.

Meanwhile, Valery Bykovsky was overhead, flying his thirty-first orbit. He cheered when he heard of the successful launch. *Vostok 6* successfully achieved orbit, and for the second time in history two cosmonauts were circling the globe in tandem. As with the dual flight of Nikolayev and Popovich, the two spacecraft would achieve their closest proximity during *Vostok 6*'s initial orbit, closing to within 3.1 miles of each other. Although the two orbits were inclined at a similar angle to the equatorial plane, the new spacecraft's orbit crossed Bykovsky's at an angle of thirty degrees. By today's standards it was not a big thing, but in 1963 it represented another huge engineering and publicity coup for the Soviet Union. "Have started carrying out joint spaceflight," the pilot of the newly launched spacecraft reported. "Dependable radio communications established between our ships. Are at close distance from each other. All systems in the ships are working excellently. Feeling well!"

Soon after, televised images of the newest space explorer swept across the world. The cosmonaut's name was Valentina Tereshkova. She liked to be called Valya.

Valentina Vladimirovna Tereshkova was born 6 March 1937 on the Priziv collective farm in the small, rural village of Maslennikovo, located in the Tutayev district of Yaroslavl region. The village was comprised of forty rough-hewn wooden huts without water or electricity, clustered around the crown of a small hill about 160 miles northeast of Moscow. Valentina's father, Vladimir Aksyonovich Tereshkov, was married to a petite lady formerly named Yelena Fedorovna. They had an elder daughter, Ludmila, affectionately known to everyone as Lyuda. Valentina might have been part of a larger family, but she was conceived after her parents had lost twins they were expecting. Her own entry into the world began at four o'clock in the morning, but there was no midwife in attendance, and her father was not up to the task. He told his anxious wife that he would go and get his mother and rushed out, making his way to the other end of the village. But he had misjudged his wife's condition, and while he was coaxing his mother back to the small wooden house Valentina was born. When the breathless grandmother finally arrived and saw that they were too late, she loudly berated her son for leaving his wife alone.

Baby Valentina was given a pet family name—Valya. Years later, Valentina's mother would reveal that she wanted to name her new daughter Rayechka. It seemed, however, that her husband had other plans. Vladimir unexpectedly went missing for several hours on the day the baby's name was to be registered. On his return he told Yelena that he had been into the district council and registered their daughter's name—as Valentina.

Sadly, as she told biographer Lady Lothian, Valya would barely get to know him. "He was a tractor operator by profession," she said of her father, "an amateur accordion player and musically gifted. He was young and handsome and was loved, not only by my mother and grandmother, but by all our neighbors in the village. Russian people call this type of young man, one who is a popular and hard-working family man, 'the first guy of the village.'" Vladimir Tereshkov, drafted into the Red Army in September 1939, was killed four months later on 25 January 1940, while fighting as a tankman in western Karelia during the Soviet-Finnish war. News of his death eventually reached the family in the midst of a fierce blizzard.

Valya was not quite three at the time, and today she retains only vague memories of her father giving her rides on his tractor. She does recall her pregnant mother sobbing for days on end when she learned of his death, but was too young to fully comprehend what had happened. "So I was only my father's darling for the first two years of my life, although I have preserved throughout my life a very special love for my father," she told Lady Lothian. "I don't recall him physically, but when I close my eyes I remember him as a long, warm shadow." Many years later, in 1988, a mass grave was discovered in Karelia. Tereshkova firmly believes her father was one of those interred there.

Vladimir Tereshkov's death made life even more difficult for the peasant family, following the birth of his son five months after he was reported missing. The boy was named after him but was better known as Volodya. Though life was hard, most of Valentina's memories are of happy times. "Like other houses in our village, ours was small but adequate," she told the authors during an interview in 2004. "But the size did not matter; I remember fondly how warm and pleasant it was for us in this loving house. Yes, we were poor, but our needs were also very simple, and we always had food, bread, and milk on the table."

In fact, Valentina loved everything about where she grew up—the Volga River, people, the animals, and the rural area. Her teachers may have given her guidance and education, but it was her mother who proved the greatest influence. Yelena protected her family from shortages during and after the war years, taught her children songs and stories from all over the world, and gave them ideals and pride in being morally strong. In 1976, Tereshkova told Soviet interviewer Alexander Romanov that her mother was "a friend, an advisor, wise through life experience, the dearest person in the world. She taught us the sense of respect for man, for labor. She taught us not to be afraid of difficulties, to find the strength to overcome them."

The local boys openly admired Valya for her pluck and a stubborn refusal to behave like other girls her age. She enjoyed riding horses bareback and loved the feeling of galloping across the fields with the wind whistling in her ears. Village life, though spartan, was family oriented and healthy.

When there was deep snow and early darkness in winter, Tereshkova said, everyone in her house and neighborhood felt safe and comfortable.

At the end of the war, without any income and needing work, twenty-seven-year-old Yelena took her family off the collective farm. They traveled to an outlying suburb of Yaroslavl and moved in with her mother. Happily, there was plenty of room in the house, and Yelena's mother was delighted to have them stay with her. Shortly after, Valentina began attending school.

Yelena was not idle for long. Desperately wanting a job to pay her family's way, she finally found low-paid work as a weaver in a textile mill. Over the next few years she struggled to pay the bills and give her children a good education. There was a regular postwar government benefit of fifty rubles for each child, which helped supplement the ration-book supplies a little, but with a loaf of bread from the market costing 180 rubles it did not last long. Nevertheless, the children grew up mindful of their mother's many sacrifices. Though they had no toys, they lived in a house filled with love, Valentina remembers, and one steeped in tradition. Difficulties eased a little when Ludmila also took on work at the mill.

By her own admission, Valentina was "an ordinary pupil" at school, but as a young girl growing up in Yaroslavl she often dreamed of becoming an engineer on steam trains. Her grandmother's house was situated by some railroad tracks, and she spent many idyllic hours sitting by them, watching the trains clattering by. She was totally enthralled by the thunderous sound and spectacle of the massive locomotives, hauling freight and passenger cars to Moscow or the Far East. She envied the engine drivers and wished she could be in their place. For their part, the engineers came to count on Valentina's presence as they roared past, and would often wave. As a young woman trapped in a repetitive, mundane life Valya yearned to travel and see a picture-book world about which she could only fantasize. In 1954, having left school at the age of sixteen, Valentina took a job as an apprentice cutter in the assembly shop at a local tire factory. It was demanding work, and the factory was always stiflingly hot, but she had obligations to fulfill. On her first payday she bought her mother a beautiful scarf patterned with flowers and some candy. When Yelena received the gifts that evening, she burst into tears of happiness.

Around this time Valentina also had a steady boyfriend, Yuli, a good-looking, well-mannered young engineer who courted her for nearly two years.

Unfortunately for the youth, he assumed too much, and once mentioned to Yelena his love for her daughter and his intention to marry her. When Valentina found out, she was outraged that he had declared this without consulting her. She told Yuli that she was too busy to marry anyone. The romance quickly soured, and ended on a chilly note.

In April 1955, Valentina resigned from the tire factory and joined her mother and older sister at the Krasny Perekop (Red Canal) textile mill. The work interested her, but she grew to hate the dust that permeated everything. "My skin suffered," she recalled, "especially my face." Valentina later blamed this dust and indifferent treatment by the plant's doctors for a series of small, paralyzing strokes suffered by her mother. Despite some misgivings, Valentina genuinely enjoyed working with machinery and learned how to operate coarse linen machines. Soon after, she joined the factory's Komsomol, the Young Communist League.

Supplemented by evening classes at trade school, a correspondence course with the Light Industry Technical School enabled Valentina to graduate as a cotton-spinning technologist. An increasingly active role in the plant's Komsomol also occupied her time at work. For recreation there was a welcome involvement with a folk music group, for which she played the *domra*, a three- or four-stringed instrument related to the balalaika.

At the age of twenty-two, in a move that would determine her future, Tereshkova became interested in joining a parachuting school at the nearby Yaroslavl Aero Club, which was recruiting new members from the mill. Her college friend Galina Shashkova, already an accomplished parachute jumper, encouraged Valentina and her college friends to join. However, in her memoirs, *Stars Are Calling*, Tereshkova recalls that, like many other girls, she was very hesitant at first. "I didn't believe in my abilities very much—it looked like this sport required courage and was created for people with great willpower and strong muscles. I liked sports; I could ski and swim. I swam across the Volga many times, but there was nothing heroic in it." In the fall of 1958, skydiving instructor Viktor Khavronin talked to the girls about the sport of parachuting and their anxieties about it. They found themselves listening with increasing interest and enthusiasm. This led Tereshkova to the door of the aero club at Svoboda Street 9. "For a long time I hesitated to go inside," she said in her memoirs, "looking at the pictures of parachutists and pilots and reading again the advertisement inviting the new members. Finally, I made up my mind and opened the door."

That day, Valentina returned home a little later than usual. Her mother quickly sensed that something was out of the ordinary and began to ask questions. "Has anything happened?" she asked. "You are so strange today, Valyusha." Valentina could not bring herself to tell her mother that she had joined a parachute club. She did not want to worry her mother, and she was still unsure about her future in this bold new adventure.

Valentina's parachute group began taking instruction almost immediately. They learned about the composition and theory of parachutes and how to pack them. They were taken to an airport to watch experienced jumpers and observed how professionals prepared for their jumps and meticulously checked their parachutes. Viktor Khavronin convinced the novices that they too could become experts, and his words gave them the confidence to continue. After several sessions of instruction and theory training Valentina's class was told they were ready for the next big step. Their first jump would be on Thursday, 21 May 1959.

That day, a propeller-driven Yak-12 aircraft carrying Valentina and other novice jumpers bumped along the grass strip and growled into gray, rain-scattered skies over Yaroslavl. She was tense but attentive. With the clamor of the engines and the air roaring past an open doorway she felt a novice's concern about missing the signal to jump. As she checked her main and reserve parachutes, Valentina chatted excitedly with another close friend, Tanya Torchilova. Meanwhile, the small green aircraft banked, climbed, and finally leveled out at fifty-three hundred feet.

To this day Tereshkova cannot explain what happened next. Perhaps it was a touch of nerves, perhaps a misunderstanding in the midst of the commotion. Whatever it was, without her instructor's command to do so she stood up and moved to the open door. She then attached the static line, as she had been taught on the ground. Moments later, to the astonishment of her instructor and the other jumpers, she shut her eyes, took a deep breath, and tumbled out into the void. For several heart-stopping moments she fell rapidly before the main chute blossomed out above her with a reassuring jerk.

Tereshkova was elated by the sensation and the beckoning sight of the ground below. "Jumping into nothingness my heart turned over and I was very happy," she later wrote. "I was reminded of how, years ago, my friends had dared me to jump from a high bridge into the Kotorosl, which flows into the Volga, and I had done so." She began tugging on the straps to control her motion and soon hit the wet grass with her knees slightly bent to prevent injury, as she had been instructed. Despite this, the landing was hard and awkward. She toppled to the ground and was dragged across the wet grass until she regained her feet and managed to collapse the parachute. "My first jump was over," she wrote. "I felt I wanted to do it every day."

The elation lasted as long as it took for a red-faced instructor to catch up with her shortly after. He was almost speechless with fury. In colorful terms he shouted that if she could not obey orders there was no place for her at the school. Contrite, but nevertheless exhilarated, Valentina promised to stick to the rules.

After the encounter with her instructor, Valentina knew that she had to reveal the truth to her mother. Yelena was fearful of and not at all pleased with this newfound interest in a dangerous sport. However, she finally accepted, grudgingly at first, that this was what her daughter wanted to do. "Several times she locked me in, [and] did not let me go to the airfield," Tereshkova reflected in a 2003 interview with Nepal's *Sunday Post*. "But I ran away from her. Even in my school days, whenever I did something my mother did not approve of, I later concluded she was right. But sometimes she was not. She feared for me when I began parachute jumping, but I do not regret having done that."

The next jump occurred a month later. For this exercise the students had to pantomime the procedures for opening their main chutes, even though they were hooked up to a static line and the chutes would automatically deploy. Then, during the third jump, they had to open their reserve parachutes so they knew what to do in the event of main chute failure. "Fear . . . did I have it when I jumped for the first time? Yes and no," she recalled in her 1964 memoirs. "I jumped myself; nobody pushed me. And at the same time, any feelings after the first jump were very peculiar. These feelings made me close my eyes. I closed my eyes during the second jump, too, when I left the plane. During the fifth and the sixth times I could keep my eyes open, overcoming the fear."

Tereshkova gained top marks in jumping and other disciplines and began to excel at the sport. Her proficiency and self-confidence grew rapidly.

Three months later she proudly received a rating of parachutist third class. Several times a week, before setting off for work, Tereshkova would make her way to the flying club for some additional practice. As keen as she was, there were still some setbacks. On one jump she narrowly avoided landing in the Volga River and could easily have drowned. Another time an instructor chided her, saying she landed "like a bear." This stung the young girl, and she practiced touchdowns with increased resolution. Within two years she had achieved a first-class rating and was instructing other novice jumpers from the mill.

Meanwhile, in 1960 Valentina was elected branch secretary of the Komsomol and became an ardent communist. On 12 April 1961, Tereshkova had just opened party proceedings in one of the plant shops when a flustered mill girl rushed in, panting with excitement. "There's a Russian man in space!" she cried. "We could hardly believe it," Tereshkova told the authors in a 2004 interview. "The news was quite unexpected and breathtaking, and the girl had to repeat the radio bulletin to us. Of course, we knew of the triumphs of other Soviet spacecraft, but to think that a man might actually be in space took our breath away, and we were excited beyond words." The meeting was quickly abandoned as everyone rushed out to listen to the factory radio. The announcer could barely contain his pride as he repeated the bulletin: Maj. Yuri Alexeyevich Gagarin had just completed a single orbit of the Earth, lasting 108 minutes. Promoted from lieutenant during his flight, he was the first human being to fly into space. Tereshkova wrote in her memoirs that she was in awe of the feat:

When we learned that Yuri Gagarin had landed safely our joy knew no bounds. All of the town's inhabitants were out on the streets. By request of the employees, that day the plant's public organizations decided to wire congratulations to Yuri Gagarin. We met to formulate the text. The chairman of the plant's trade union committee, Mikhail Kalinin, teased me. "Valya, you keep jumping and jumping, but look where Gagarin has soared!" At that time I was doing a lot of parachute jumping. "First the men will fly, and then we women!" I replied with no great confidence. I never thought then that I would be so lucky.

Later, she would read that Yuri Gagarin had also studied at a flying club, and a small seed was sown. "I was inspired by Gagarin's flight in space. I kept thinking about new developments in space research, and I became determined to try and join the space program. It would mean leaving my childhood home, and the work near my mother and family. But nothing was going to stop me in my resolve to reach the sky." Four months after Gagarin's flight, on Sunday, 6 August, there was more excitement. Tereshkova was sitting at home listening to an opera on the radio, when a broadcaster broke into the program to announce that *Vostok 2* had been launched into orbit. The pilot was Gagarin's previously unnamed backup, Gherman Titov.

According to Tereshkova, in a story later popularized in the Soviet media, she impulsively picked up her pen and wrote a letter to the Supreme Soviet in Moscow, declaring her willingness to be considered for spaceflight. The following day she showed her letter to Komsomol organizer Valentina Usova, who encouraged her to send it off. She promised to support her friend's application. Tereshkova later revealed that as time went by she wished she had never sent the letter. She was embarrassed by her enthusiasm, feeling it was foolish to imagine that a simple textile worker from a peasant background would ever be considered for a task performed only by elite male air force pilots. There is little reason to doubt that she actually sent off this letter, but it is unlikely that it had any influence on later events.

In the meantime, Valentina maintained her proficiency at parachuting. On Air Force Day, 18 August, the residents of Yaroslavl assembled on an embankment of the Volga to watch some parachutists jump. Tereshkova's mother, brother, and sister were among the crowd. It was finally announced on the public address system that it was Valentina's turn. She landed in the Volga River. Soaked, she finally approached her mother, filled with excitement. "Did you see me jump?" she asked. Yelena gave her a cold stare and replied, "If you crash, don't come home!"

Lt. Gen. Nikolai Kamanin had overall responsibility for the cosmonaut training program. In diary notes dated 24 October 1961, he mentions the flood of letters forwarded to his training center in Moscow, addressed to various agencies and the Supreme Soviet. All sought information on becoming a cosmonaut, and several were from women. Kamanin did not dismiss these entreaties as others might have done. Instead, he used them to put forward a proposal that a female cosmonaut be selected. His country now had the edge in manned space exploration, and he wanted to maintain the impetus. "We cannot allow that the first woman in space will be American," he wrote. "This would be an insult to the patriotic feelings of Soviet women. The first Soviet women cosmonauts will be as big an agitator for communism as Gagarin and Titov have turned out to be." He was adamant that suitable candidates be located and given immediate training.

The Central Committee of the Communist Party of the USSR held a meeting inside the Kremlin on 30 December. One of the main topics for discussion and resolution was Kamanin's proposal to investigate the selection of a woman to fly on an early Vostok mission. There was general agreement that there were only a few qualified women pilots, providing an extremely small candidate pool. It was considered crucial that the first woman cosmonaut be proficient at parachute jumping. With this in mind they began examining the qualifications of likely candidates.

During the first few years of the Soviet space program, propagandists put together fanciful "autobiographies" and stories attributed to cosmonauts, which resulted in the prevalence of many state-manufactured untruths. Certain inconsistencies in the biographies of prominent cosmonauts such as Tereshkova are yet to be dispelled, and the lady herself rigidly adheres to many of the choreographed fallacies. It would be nice to think, as recorded in some of these so-called biographies, that Tereshkova's impulsive letter brought the Soviet air force to the Yaroslavl Aero Club seeking her out. Unfortunately, such stories are untrue. Quite simply, there was increasing pressure for a new space spectacular, and Kamanin's recommendations helped promote the concept of a female cosmonaut. The Central Committee of the Communist Party finally approved a proposal by the Ministry of Defense to select sixty new cosmonauts in 1962. That number was to include five women. A special Mandate Commission selected the five from the group of potential candidates, all of whom were considered suitable to fly in space.

Before any commitment had been made by the Central Committee, Colonel Ovsenko of the Soviet air force had been assigned the task of finding appropriate candidates at parachute clubs. Kamanin armed him with a master list of around four hundred names, requested from aviation clubs across the Soviet Union. Of these, only fifty-eight would meet the minimal standards required for further evaluation.

Initially, Tereshkova was not even considered one of the top candidates. In fact, she was described in Ovsenko's report as "second echelon" only, but fortunately another door opened for her. Any candidates had to be politically suitable—either a party member or a member of the Komsomol—so once again Tereshkova's name bobbed to the surface. Furthermore, she came from a hard-working, proletarian family; she was young, single, self-motivated, and even quite attractive. Overall, she fit the commission's profile. In December 1961 the nervous young textile worker was called before a regional committee of the National Society for Assistance to the Army, Air Force, and Navy (DOSAAF), a civil defense organization in the former USSR . They told her that she was under consideration for a special top-secret agenda. In her memoirs, she recorded: "I understood only one thing; they invited me to start training for the flight in a spaceship. We all believed in the reality of this flight then, but no one thought that it might happen so soon."

In January 1962, Tereshkova reported to Moscow for a series of interviews and medical examinations. They were to be conducted under the strictest security; no one could be told the real purpose of her trip. Reluctantly, she told her family and friends that she was traveling to Moscow to take a qualifying medical examination for the national parachute team. Lying, especially to her mother, was distasteful, but it was something she was obliged to do. Tests and interviews were conducted on the remaining women at the Scientific Research Institute of Aviation Medicine in Petrovsky Park, on the outskirts of Moscow. "Part of the tests were the same as our [later] training program," Tereshkova recalled in a 1996 interview with Kerrie Dougherty. "Starting with the famous centrifuge. Then we had a lot of medical tests."

On 16 February 1962, Valentina Tereshkova was selected as a cosmonaut candidate. She was elated but apprehensive. Then a telegram arrived, directing her to report to Lt. Gen. Kamanin in Moscow. As before, she was to tell no one the purpose of her trip. Leaving Yaroslavl on 2 March 1962, she was greeted on arrival in Moscow by Kamanin. He told her that she was the first of the five selected women to arrive. She was then introduced to several future cosmonauts, including Andrian Nikolayev, Pavel Popovich, and Valeri Bykovsky. They were interested to hear that she was a parachutist and that she had made a total of 163 jumps before her selection.

The other women selected were Tatyana Kuznetsova, Irina Solovyova, Zhanna Yorkina, and Valentina Ponomareva. As required, Tereshkova reported to the Air Force Cosmonaut Training Center on Monday, 12 March, where she was introduced to Kuznetsova and Solovyova. They quickly became friends. Kamanin had earlier informed them that Yorkina and Ponomareva were delayed and would arrive in four weeks. Once again they met several of the cosmonaut group, but it was Gagarin who assumed the role of tour guide for the three awestruck women, escorting them through the center's facilities. As he showed them around, he patiently answered their questions. "Let me tell you, when I entered the Star City of cosmonauts, my heart was about to stop," Tereshkova later reflected in *Stars Are Calling*. "How will they meet me there? I didn't do anything great in my life; in fact, I didn't see much of a real life and these were real pilots. And two of them, Gagarin and Titov, were heroes whose names were known all over the world!" In fact, the Star City of today did not exist at the time the women reported for training; the center was simply a collection of office buildings, and the women were quartered in a rehabilitation center.

Tereshkova would soon be as famous as the first two cosmonauts she had just met. A different destiny awaited the other women about to commence their training.

At just twenty years of age, Tatyana Dmitryevna Kuznetsova remains the youngest person ever chosen for spaceflight training. Born in Gorky on 14 July 1941, she joined a local flying club and became a parachutist, later moving to Moscow to continue the sport. At the time of her selection she held several world parachuting records. She had also served as a scientist for the applied mathematics department of the Academy of Sciences.

Irina Bayanovna Solovyova was born in the town of Kireyevsk on 6 September 1937. An accomplished parachutist, she joined her local aero club in 1957. When selected, she had been a member of the Soviet national team at the 1957 world parachuting championships and held numerous skydiving records.

Born in Ryazan on 6 May 1939, Zhanna Dmitryevna Yorkina studied at her hometown's pedagogical institute, graduating as an English teacher. While at the institute she joined a nearby flying club through the DOSAAF program and became a proficient skydiver. She had applied for cosmonaut training after Gagarin's flight.

Valentina Leonidovna Ponomareva, the only married member and only pilot of the group, was born in Moscow on 18 September 1933. She learned to fly and parachute with an aero club attached to the Moscow Aviation Institute and worked at the USSR Academy of Sciences' Institute of Higher Mathematics. After Titov's flight, a ballistics expert named Vsevolod Yegorov had approached her. With professional connections to Sergei Korolev, he knew that women were about to be selected for cosmonaut training. Yegorov urged her to write a letter expressing interest to Mstislav Keldysh, then vice president of the Soviet Academy of Sciences and the chief theoretician of Soviet cosmonautics. "I didn't take him seriously and told him this was science fiction," Ponomareva reflected in 2000 during an interview conducted in Moscow by Bert Vis and Rex Hall. "Nevertheless . . . I wrote the application. Keldysh signed this, and I went to the medical commission. We underwent the tests at TsNIAG. When someone failed a test, she was simply sent home, after crying a little. As soon as a candidate failed, she was told. But I successfully went through all the tests."

Her husband, Yuri Ponomarev, was a fellow student from the Moscow Aviation Institute (MAI). Although later also selected as a cosmonaut, he would never fly into space. They had a son, Aleksandr, born in March 1958. A year to the day after the first manned spaceflight, on 12 April 1962, Ponomareva and Yorkina arrived at the center to belatedly commence their cosmonaut training. As Ponomareva told interviewer Slava Gerovitch, there were some early challenges to be faced and readjustments to be made when they arrived at the center and were enrolled as privates in the Soviet air force: "We found ourselves in a military unit, in which we became an alien part, with our different characters and different concepts. Our commanders had great difficulty dealing with us, since we did not understand the requirements of the service regulations, and we did not understand that orders had to be carried out. Military discipline in general was for us an alien and difficult concept."

At first, the women formed something of a mutual support group. This solidarity helped them through the initial stages of their training, particularly when Ponomareva and Yorkina joined them. Unlike the male cosmonauts, however, they knew that they were all competing for what would probably be a single flight opportunity, and this made them fiercely competitive. They also realized that they were all under great scrutiny simply because they were women. Their male counterparts, openly derisive when discussing the women, were keeping watch on them. The men knew that none of the women had military flight backgrounds, although Ponomareva had learned to fly through a civil defense air club. Gherman Titov was one such outspoken skeptic; he did not believe at first that women belonged anywhere near an airplane and insisted that only men could carry out all the tasks involved in spaceflight.

Not long after, their training began in earnest. As Tereshkova recalls, they were introduced to many of the torturous devices that would all too soon become familiar to them. They were shown the heat chamber, where a trainee had to carry out a set number of complicated tasks, and then the centrifuge. Finally, they saw the isolation chamber—a training facility that basically shut a candidate away from the sights and sounds of the world. Future cosmonauts would have to spend several days here in complete isolation. In this "cabin of silence," Tereshkova says that Gherman Titov passed the time by reading out loud the poems of Pushkin, Pavel Popovich sang Ukrainian songs, and Valery Bykovsky drew cartoons.

The women were soon spending time in the centrifuge, the thermochamber, and the decompression chamber. It was daunting stuff, and Tereshkova would later write that when she was being whirled around in the centrifuge it felt as if her blood had turned to mercury. With her usual determination she stood up well to the unsettling loneliness of the isolation chamber, designed to study the psychological effects of prolonged isolation on cosmonaut trainees. She would later describe the training as "very complicated."

During this period the women were escorted through the Vostok plant, accompanied by some male cosmonaut trainees. Here, Tereshkova was introduced to the mysterious chief designer, Sergei Korolev. She was not sure how Korolev would receive the women, but he soon put them at ease. Tereshkova later wrote that for such an important person he turned out to be a fairly unassuming man and possessed a keen sense of humor as well as being very sociable. He knew all about the women, their names and their backgrounds, and discussed with them the progress they were making in training. As the conversations continued, enthusiastic young workers constantly approached Korolev with blueprints and questions. All too soon he stole a glance at his watch, and the women sensed it was time for them to thank the legendary designer for his time and depart.

All of the male cosmonauts had served with the Soviet air force. In order to maintain the status quo all five women were enlisted as honorary privates in that branch of the forces. Each received basic cockpit training under Col. Vladimir Seryogin (later killed in a tragic aircraft accident along with Yuri Gagarin), which would qualify them as flight-trained passengers in Ilyushin 14 turboprop transports and two-seater MiG-15 trainers. Though they took the controls to gain a feel for flying, none would be permitted to fly solo. In this way, Tereshkova completed twenty-one flights, for a total of sixteen hours and forty-two minutes, and described Seryogin as "a kind and patient teacher, a great master of his profession." At the end of 1962 the women were each given a commissioned rank as junior lieutenant.

The training schedule remained arduous, but Yuri Gagarin would later admit that he was full of admiration for Tereshkova after watching her at practice and in classroom exercises. He knew it was tough to master rocket techniques and study spacecraft designs and equipment, but he felt she tackled the job stubbornly and devoted much of her own time to study, poring over books and notes every evening. Any concerns he harbored were melting away, and he once wrote, "Valentina has amazing abilities; she is doing very well in flight training, celestial mechanics, and many other things. And she's modest. She's not afraid to say, 'This is something I don't know about; help me.' And we help her. She may be thin and weak looking at first glance, but she has great strength, energy, and willpower. She's a real Russian woman!"

In August 1962, an excited but apprehensive Tereshkova was transported to the Tyuratam Cosmodrome, where she witnessed a rocket launch for the first time, in the company of the other women cosmonaut finalists. She returned to Moscow thrilled by the experience.

At one stage of the training it seems that Zhanna Yorkina performed poorly during a three-day flight simulation in a Vostok mockup, and was even said to have fainted. She was asked to stand down from training for a while. In later years, Tatyana Kuznetsova would reveal that she also left the program temporarily in its latter stages "because of illness." Other sources say her withdrawal was due not only to poor health but also a less than satisfactory performance in the centrifuge and pressure chamber tests. Despite this, both women would remain part of the female cosmonaut detachment. These problems meant that the list of flight-eligible female candidates for *Vostok 6* was reduced to just three—Tereshkova, Solovyova, and Ponomareva.

One thing Tereshkova took to eagerly was her advanced parachute training with Gagarin's trainer, Nikolai Konstantinovich Nikitin. Considerably more difficulty was associated with this program, as it not only included parachute descents into the Black Sea, but the land and water jumps had to be performed wearing a cumbersome, heavy spacesuit laden with life-support systems to emulate cosmonaut ejection procedures then in place.

While the women trained for their half of the dual flight, the male cosmonauts trained for theirs. Valery Bykovsky and Boris Volynov were in prime training for the *Vostok* 5 mission, although it had not been decided which of them would fly. Ordinarily, the State Commission would sit three days before any launch to make a final determination. As the remaining three women were undertaking simultaneous training with Bykovsky and Volynov, the five often went on exercises together. "Practically, we were in one group," Volynov recalls. He was witness to one hazardous training exercise in which the trainees, wearing their full spacesuits and burdened down by parachutes and survival kits—nearly three hundred pounds in all—had to leap from an aircraft flying at three miles' altitude and splash down near a waiting recovery ship. As he told interviewer Bert Vis, he was aboard one of these vessels when Tereshkova made her jump:

When she was on board the ship after finishing the training she opened her spacesuit and inside there was a heat-resistant suit with a pocket in it, in which we kept our identity cards. She took out a lilac branch and gave it as a present to the captain of the ship, who had just saved her from the sea. Well, the man was a great, experienced sailor . . . tanned, looking severe. But when he received that flower from the woman who was taking off her spacesuit . . . who had just jumped from an aircraft and had splashed down, fulfilling a great task, and then gave a flower . . . it was so touching. There were tears in his eyes . . . it was so emotional.

On an earlier jump Tereshkova had landed in what was described as an "abyss," so Nikitin made her perform a series of twenty-six more jumps, each progressively harder and more demanding. Altogether she made a total of forty-four jumps during her training, with delays in opening her parachute ranging from five to sixty seconds. Four of these were made wearing the special training spacesuits. When interviewed by researcher Rex Hall in Coventry in September 2004, both Tereshkova and Bykovsky revealed that part of their parachute training also involved simulated drops strapped into Vostok ejection seats wearing full spacesuits. They said it was a harrowing exercise, as the heavy seats had to be manually shoved from the doorway of a slow-flying aircraft.

As Tereshkova's training progressed, other cosmonauts began to notice that Nikolayev, fresh from his triumphant dual spaceflight, was a constant visitor at her sessions. A bachelor, he always seemed to be there to encourage the trainee, to talk her through difficult studies, and to attend her simulation exercises. They were often seen deep in conversation in the library. Many of the center's residents began hinting at a blossoming romance between the two cosmonauts.

Valery Yazdovsky, later selected as a cosmonaut, worked at the Korolev Design Bureau. As he told interviewer Bert Vis in a 2000 interview at the training center, Korolev asked for his help in training the women in the Vostok's systems: "I conducted lessons with them as an OKB-1 specialist. In order not to frighten this group after the completion of the lessons, I proposed to organize the examinations as if they were a job interview. There was no examination board as such. My assessment was that the best-prepared person was Valentina Ponomareva. She had graduated from MAI and had flown aircraft. I gave her the highest marks."

After Yorkina's resumption of training, and with Kuznetsova temporarily eliminated from consideration, only four of the women took their final exams late in 1962. In his diary notes for 29 November, Kamanin reflects that the best performer in these tests was Ponomareva, who came out on top in both practical and theory assignments. However, the women were being assessed just as much, if not more, for their propaganda value, and the decisions were strongly affected by the prejudices of the era. In his assessments of the four candidates he expressed concern over Ponomareva's "unsteady morals"—independence, outspokenness, and a perceived overconfidence in her own abilities.

As Kamanin recorded in his diary, Irina Solovyova was a little too self-assertive and private for his liking. Though she was morally and physically tough, she had picked up a reputation as a loner and lacked visible "activity in social tasks." Zhanna Yorkina had found much of the training hard going and was rated as the "weakest" of the four candidates, although she had gained a lot of ground on the other women. Kamanin was prepared to be a little benevolent with Yorkina, saying that she would make a good cosmonaut one day. Tereshkova, on the other hand, had rated quite highly in her tests and was deemed to be "a model of breeding." Furthermore, Kamanin added, "We must send Tereshkova into space first, and her double will be Solovyova. Tereshkova, she is a Gagarin in a skirt."

Following the tests, the candidates were sent on a two-month break to a resort in the Urals and did not report back until 10 January. They were told that a final decision on which of them would fly would not be made until three or four days before the actual mission. Final details of the flight were still under discussion two months later. Mission planners were undecided whether they should simply send a woman into space on a single mission or two women at the same time on separate Vostok flights, emulating the feat of Nikolayev and Popovich. Another option was to send up a male cosmonaut for five to seven days and then a woman for a flight of around two days. It was known that Kamanin favored the second option and was ready to send two of the women into space on another joint "rendezvous" mission.

The issue was finally resolved at a meeting of the Central Committee on 21 March, when it decided on the mixed flight—a male and female cosmonaut flying simultaneous missions. The committee recommended that the flight take place no earlier than August. Given this, Kamanin consulted with Korolev, and they mapped out the finer details. A male cosmonaut would be launched on a mission lasting eight days; a woman would join him in orbit toward the end of his flight. Both would land two days later. There was only one major problem; Korolev's engineers pointed out that the two Vostok spacecraft designated to the mission, both of which had already been assembled, had parts that would soon expire. They were designed to be flown no later than May or June. On 29 April the Central Committee concurred with their recommendation to bring the launch dates forward.

Subsequently, things were set in motion. The State Commission responsible for naming crew members to *Vostok 5* and *6* made its preliminary selections on 11 May. Valery Bykovsky was appointed prime pilot for the first mission, with Volynov as his backup and cosmonaut Alexei Leonov in a support role. For *Vostok 6* they selected Valentina Tereshkova as prime pilot, with Irina Solovyova and Valentina Ponomareva as her backups. These selections became official on 4 June. It was further announced that the ra-

Valentina Tereshkova and Valery Bykovsky, who never shared a spacecraft cabin but flew in space at the same time. Courtesy Colin Burgess Collection.

dio code name for Bykovsky would be Yastreb (Hawk) and for Tereshkova, Chaika (Seagull).

Though he knew that fame would attach itself to him because of his forthcoming spaceflight, Valery Bykovsky also realized that Tereshkova would capture most of the media attention focused on the joint mission. This sat well with him, as he was essentially a quiet, laconic man who preferred a low profile. Bykovsky was born on 2 August 1934, in Pavlovsky-Posad, an ancient textile-manufacturing town located thirty-five miles east of Moscow. Many of the town's craftsmen practice the intricate art of fabric printing, and the town is widely known as the origin of large, vibrantly colored woolen shawls and kerchiefs worn by Russian women. The future cosmonaut was the second of two children born to Klavdia Ivanova and Fyodor Fedorovich Bykovsky. At the time, his older sister, Margarita, was aged three. Their father was seeking to better himself after a hard life as a mine and railway worker. Although it is not clear when he was recruited, Fyodor began working for the powerful KGB and graduated to a good position with another, unnamed government agency. Details are scarce, but it seems he was a good provider for his family.

When Valery was four, World War II erupted, and the family traveled to the port town of Kuibyshev on the Volga River. Barely a month later they made a second move, to the town of Syzran, another major port and rail center at the confluence of the Volga and Syzran rivers. Here Valery developed a deep fascination for ships and the sea, and dreamed of one day becoming a sailor. He would often draw battleships on pieces of paper and make small models of sailing vessels. In 1941 the family was on the move yet again, when Fyodor was posted to the Iranian capital of Tehran. They would not return to Russia until the summer of 1948, at which time they settled in Moscow.

While completing his seventh grade at school, Valery boldly informed his parents that, like his father, he wanted to attend naval school. Fyodor told his son that he could not contemplate such things until he had finished tenth grade and received his school-leaving certificate. Valery was disappointed but knew his father was right in saying that a full education was vital, even for those who wanted to become sailors.

Valery grew up playing volleyball, basketball, soccer, hockey, and *gorodki*, a Russian version of skittles. In his early teens he also took up cycling, fencing, and swimming. While in tenth grade, still determined to join the navy, a chance meeting changed his life forever. One afternoon, as he prepared to go home, Valery was told that a spokesman from the Komsomol, a flight instructor, was giving talks on aviation careers in the school's committee room. He made his way there and listened intently as the man addressed a small but eager gathering of students. At the end of the talk Valery boldly stuck his hand in the air and asked if he could enroll at the local flying club, which brought forth similar cries from other students. The delighted speaker took their names, but told them they would have to pass strict admission and medical tests to gain entry.

His earlier dreams of becoming a sailor now behind him, sixteen-year-old Bykovsky could think of little else but enrolling at the flying club. He and his friends often discussed this, but he soon noticed that their early enthusiasm was beginning to wane. On the day he attended the medical and admission tests only one other boy came with him. To Bykovsky's joy, he passed, and flying lessons soon began at the Moscow City Aviation Club. On graduating from secondary school in 1952 he also gained his sports pilot's license. Then, two weeks after his graduation, Bykovsky enlisted for primary flying studies in the air force's school for fighter pilots.

Small of stature but lean and athletic, Bykovsky was a keen and determined student, thoughtful and industrious. Eventually, he flew several different aircraft types from a number of airfields, and certification records show that he took to these changes with great competence. He made his first solo flight on 27 September 1952, flying a Yak-18. In making his comments and recommendations, Bykovsky's instructor wrote that he should be permitted to continue his jet fighter training. Graduation came from the Kachinskoye Air Force School in 1955, with Bykovsky earning excellent grades in flying and combat training.

The next step was an assignment to a regiment at which many of the pilots were veterans of the war, eager to share their experiences with young recruits. Bykovsky listened attentively to the stories and advice of these battle-hardened men. When he flew with them he learned a lot about air battles, interceptions, and shooting. He took to the skies almost every day and studied aeronautics in his spare time. Four years later he became a senior fighter pilot and a certified parachutist with seventy-two jumps to his credit.

In 1960 Bykovsky was selected in the first group of cosmonauts, and the following year he witnessed the first launch of a person bound for space. The cosmonauts had witnessed dogs and human dummies being sent up, but this time—for the first time in history—it was a man, and a friend. None of them, Bykovsky would later state, had any idea back then what enormous consequences Gagarin's flight would have.

Despite his aviation career and arduous studies, Bykovsky found time to meet and wed Valentina Mikhailovna Sukhova. Their first child, also named Valery, was born in 1963. Like his father he would later become a pilot, but in 1986, at the age of twenty-three, Valery was killed in a tragic flying accident. Shortly after takeoff from an airfield in Afghanistan, the Tupolev Tu-134 in which he was copilot collided in mid-air with an Antonov transport plane, killing everyone aboard both aircraft. A second child, Sergei, would be born on 12 April 1965—four years to the day after Gagarin's spaceflight.

The fifteen-month training period was almost at an end, and soon it was time for Tereshkova, Bykovsky, and their backups to prepare themselves for the long journey to the launch site at Tyuratam. There, on 6 June 1963, the launch of *Vostok 5* suffered the first of several delays caused by technical problems and a failure in the command radio line. It was rescheduled for 11 June, but concerns about expected high solar flare activity caused yet another postponement. That day, the cosmonauts relaxed quietly at a riverside beach house. Korolev and Tereshkova spent several hours sitting on a second-floor balcony discussing her readiness for her own flight.

On the morning of 14 June, Titov assisted Bykovsky into his bright orange spacesuit. Then, accompanied by his backup, Volynov, Bykovsky boarded a bus for the drive out to the launch pad. After a short speech thanking those who had brought him and his spacecraft to readiness he climbed into the gantry elevator, waved, and was carried up to a platform where the *Vostok 5* spacecraft was waiting. Watching from a nearby observation area, Valentina knew the next rocket launched from this place would carry her into space and the history books. She later admitted to being a little nervous at the thought. "Yuri Gagarin noticed it," she recalled, "but instead of saying something that would make me calm, he sympathized with me: 'I understand you. It's hard to be the first.'"

The countdown proceeded, but a number of problems caused rolling delays. The first occurred when a string lanyard, used to arm the ejection seat prior to hatch closure, snapped when it was pulled. This information was not immediately relayed to Korolev, but as the countdown progressed technicians decided the problem should be investigated. The chief designer was informed, and he ordered the hatch reopened, instructing Bykovsky to exit until the problem was rectified. The countdown then proceeded until thirty minutes before the revised liftoff time, when a gyroscope in the rocket's upper stage failed. There was talk of postponing the flight until the next day. Eventually, the faulty unit was replaced, and the countdown resumed. As the day wore on and time ticked by, the well-being of the cosmonaut began to be a concern. Apart from a brief interval, he had been strapped into his spacecraft for several hours, and his bladder was sending him an urgent message. "On the launch pad for five hours, awaiting liftoff, besides all the usual and solemn impressions, I was dominated by a normal human sensation," he later reported. "I had drunk too much tea and was waiting eagerly for the moment when I could open the spacesuit. This sensation impaired all other emotions!"

Volynov, meanwhile, was sitting patiently in the transfer bus. Dressed in a similar spacesuit, his job was to be at hand if Bykovsky had to be replaced. It was a tense time for the backup pilot, as he told Bert Vis:

I drank water just like him, I ate the same, just like him. [Bykovsky] took his seat two hours before the [scheduled] launch. I was sitting in the bus . . . it was extremely cold. The standard ventilator that was installed in the spacesuit went out of order. It was supposed to be used only for a short time. They brought some . . . compressed air . . . and they used [it] to ventilate the spacesuit. But when two hours had passed there was still no launch and I was sitting in the bus. Then, the doctors and specialists came and said, "Don't relax, soon we'll change. Bykovsky will not fly—you will fly!" I sat on the bus for another thirty minutes, and then the doctors said, "Okay, now we start the replacing. Be ready to fly!" I was going through the flight plan.

It was a complicated situation. I was not allowed to go out. [After] thirty minutes more . . . there were slight malfunctions in the radio connections. That's why we came to the conclusion to change everything. All the time they kept telling me "Thirty minutes!" but all in all it was five hours. And finally, [still] with small disorders in the radio connections, they decided to send him into space.

Bykovsky's rocket finally thundered off the launch pad at 2:59 p.m., Moscow time. As he ascended, the cosmonaut radioed that everything was fine. Soon *Vostok 5* had settled into an initial elliptical orbit of 112 by 146 miles, and Bykovsky tried to orient the spacecraft so he could see the third stage. As he revealed in his postflight debriefing, Bykovsky now realized his work had begun:

"After ten minutes of wasteful attempts to orient the capsule to see the stage the air pressure in my orientation tanks was down to ten atmospheres. The spacecraft moved very slowly under manual orientation. I needed to conserve five atmospheres of orientation tank pressure for retrofire. To turn the spacecraft perpendicular to the direction of motion, or towards a star or moon to check the orientation took eight minutes."

As soon as a transmission slot became available, Tereshkova was able to speak with her colleague from the mission control center: "Yastreb, Yastreb [Hawk, Hawk]," she called. "Do you recognize my voice? A warm greeting to you. . . . I congratulate you on a good beginning." She could see Bykovsky on a television monitor, and he smiled as he replied: "I am waiting."

Later, Tereshkova met some Russian press representatives at the launch pad. According to the Soviet news agency, Tass, she was "wearing an elegant blue linen dress and stiletto heels." The press release said her shoes "looked quite out of place on the coarse concrete of the pad." With Bykovsky safely in orbit a launch date of 16 June was confirmed for *Vostok 6*.

Two days after Bykovsky's launch, on 16 June, Tereshkova was awakened at 7:00 a.m. She and Irina Solovyova had spent the previous evening drinking tea with Sergei Korolev and discussing the flight and the work to be done in space. Tereshkova had slept well, in the same bunk that Gagarin occupied the night before his own flight. For the next half hour the two women exercised and then enjoyed a hot breakfast. Both underwent a final medical examination before donning their spacesuits, after which they were transported to the pad. In his role as launch CapCom, Yuri Gagarin accompanied them on the short bus ride.

As she prepared to make her way up to the spacecraft level, Tereshkova embraced and kissed Gagarin. She repeated the farewell with Titov and then with the man who would later become her husband, Nikolayev. Korolev offered some encouraging words before she entered the gantry elevator. Doctors monitoring Tereshkova's pulse rate noted that it soared to 140 as the elevator climbed. Soon after she was settled into her ejection seat inside the spacecraft, and the hatch above her head was closed. "I was not nervous any longer," she remembered later of that moment. "There was just the expectation of something unknown and wonderful ahead."

Vostok 6 lifted off at 12:29 p.m. For the first time a woman was on her way into space. Almost to herself, Valya cried out, "I'm off!" She later reflected in her memoirs that "The music of launch begins with the low sounds. I hear the roar that reminds me of the sound of thunder. The rocket is shaking like a thin tree under the wind. The roar grows, becomes wider, more upper notes are distinguished in it. The spaceship is shivering." As *Vostok 6* climbed ever higher into the sky, dwindling into an intense, starlike dot for spectators below, Tereshkova felt the trembling vibration associated with launch and a steadily increasing heaviness in her arms and legs. She also experienced an unseen weight pressing down on her chest. "It becomes hard to breathe," she reported. "I can't move a single finger." Every time she thought the pressure had reached its limit it continued to mount, but she remained focused on the task ahead. "Somewhere in the star-filled height flies the lonely spaceship controlled by Valery Bykovsky," she later reflected in her memoirs. Then, she heard some reassuring words: "'Chaika, Chaika, everything is excellent, the machine is working well.' I shake with surprise. The voice of Yuri Gagarin sounds right near, as if he's sitting next to me, as an instructor in the right seat of the plane. I answer not at once; maybe because of the cheering words of my friend; maybe the flight to orbit is over, and the pressure disappeared, as it melted under the warm wave spreading in my body. Breathing became easy."

Soon an excited voice was broadcast from the orbiting spacecraft. "Ya Chaika, Ya Chaika [I am Seagull]! I see the horizon—a light blue strip, a blue band. This is the Earth; how beautiful it is. Everything goes well." She later told a group of reporters that her first sight of Earth from space was "overwhelming": "It was breathtakingly beautiful, like something out of a fairy tale. There is no way to describe the joy of seeing the Earth. It is blue, and more beautiful than any other planet. Every continent, every ocean, had its own distinct beauty. The Earth was visible very clearly, even though the craft was traveling at [five miles] a second. Africa shone out in yellow and green, Australia was fringed in an opal color. Unfortunately, every time I went over Europe it was sheathed in cloud."

By three o'clock that afternoon the two cosmonauts had established radio contact. Dubbed the "cosmic couple" by the press, they were soon exchanging comments and information about conditions aboard their craft, as well as communicating with ground controllers. The same equipment that had allowed Soviet scientists to study Bykovsky in orbit was now brought into play for Tereshkova. Television images of her were beamed to the world, smiling happily as a pencil and small logbook drifted in front of her helmet. Telemetry data was good, and Tereshkova seemed to be adapting easily to weightlessness.

However, all was not going as well as it seemed. During her first orbits of the Earth, in a worrying parallel with the illness-plagued flight of Gherman Titov, Tereshkova experienced feelings of nausea. Similar sensations still affect a good number of space travelers today. "When trying to get used to the absence of gravity," she later related, "I couldn't do it at once. It was too unusual for me." Fortunately, the feelings soon passed. After checking her medical telemetry and talking with her, flight controllers reassured themselves that all was well.

According to news reports of the time, during her first afternoon in space Tereshkova also provoked the controllers' concern when she failed

Valentina Tereshkova, the first woman in space, during her *Vostok 6* flight in 1963. Courtesy Colin Burgess Collection.

to respond to their repeated transmissions as well as to urgent calls from Bykovsky. Whether because of illness or fatigue, she had apparently nodded off and slept so soundly that ground controllers were unable to wake her. Eventually, Bykovsky's repeated calls woke her, and communications were restored. She was contrite and apologetic. "There was very little time to sleep," she explained decades later. "Once I did sleep a little longer and missed a call. I admitted to having been asleep."

On her fourth orbit Tereshkova enjoyed a light-hearted chat with Premier Khrushchev. In concluding he told her to be of good cheer, that the Soviet people were proud of her feat, and expressed wishes for a successful completion of her flight. In response, she said she was deeply moved by his attention and fatherly concern.

Despite the sounds of communications and *Vostok 6*'s operating noises, Tereshkova knew her spacecraft was surrounded by a profound silence. It created in her a distinct longing for the familiar things of life back on Earth. "The further away a spacecraft drifts," she once reflected, "the more you start to miss the sounds of nature—rainfall, for instance." Following her conversation with Khrushchev, Tereshkova made preparations to sleep—this time according to the flight schedule. By ten o'clock that night Bykovsky had completed fifty-four orbits and Tereshkova twenty-three. She had flown one more orbit than America's most-traveled astronaut at that time, Gordon Cooper. Tereshkova woke early the next morning, Moscow time, and spent fifteen minutes performing some light physical exercises. She then washed her face and hands using a dampened cloth and ate breakfast.

In a later interview with Kerrie Dougherty, Tereshkova disclosed that her flight unavoidably allowed her mother to finally discover the truth about her activities. "It was top secret. My mother, [like] the mother of Yuri Gagarin, first knew about it with the rest of the country. It was a very big surprise. One can understand what a mother feels like in this kind of situation." She had explained her lengthy absence from Yaroslavl by telling her mother that she had to complete advanced parachute training in Moscow. Yelena had accepted this. Though uncomfortable with the danger, she was nevertheless proud and wanted her to do well. Her daughter wrote home regularly, and on two occasions made brief visits to Yaroslavl to see everyone. Tereshkova hated being untruthful about what she was really doing in Moscow. Before the flight she had written ten letters saying she was busy but well and happy. A friend had promised to post one each day to her mother; however, one of the later letters was delayed, only arriving in Yaroslavl on the day of the flight.

As Tereshkova revealed to Dougherty, friends told her mother that on television they had seen a woman in space who greatly resembled her daughter. Yelena dismissed this as idle gossip. "She said, 'No, my daughter is just a parachute jumper. She could not be aboard this spaceship!'" Tereshkova related. "My mother was absolutely certain that I could not hide anything from her, but then she saw my photo, and when she heard my message to her—because I had a special message I transmitted to her—then she recognized her daughter on the television screen!" Though proud of this remarkable achievement, Yelena is said to have been deeply hurt by the deception. It would be some time before she finally forgave her Valya for what she considered to be her lack of faith in her own mother.

As with the dual flight of Nikolayev and Popovich, Western journalists trying to gain facts about the mission were frustrated by reams of propaganda emanating from government-run news agencies. With no evidence to the contrary, they were forced to make conclusions based on unreliable, unsourced information. This arrangement suited the Soviets; they could revel in the published rumors of a space rendezvous spectacular and later deny what had been reported.

Meanwhile, the flight continued. In mid-afternoon Moscow time of her second day in space, Tereshkova was soaring above her native Yaroslavl and over the Volga. She later recalled looking so hard for her birthplace that her eyes ached, but the area was shrouded in white clouds. She recorded gaining a momentary glimpse of a thin silver stripe, a familiar contour of the ancient city, but also said that it may have just been wishful thinking on her part. She did take a few moments to send greetings to all the mothers in the world and told her own mother not to worry, that everything was going well.

To this day, Tereshkova insists that after her brief bout with nausea on her initial orbits, a single instance later in the mission when the food made her vomit, and headaches brought on by the sensors attached to her head, she otherwise felt fine for the rest of her flight. On the second day of her mission, however, the flight controllers were not so sure. In the television images they were receiving Tereshkova looked tired and weakened and, more importantly, she was not keeping up with her flight program. That day, she failed to carry out one of the flight's most important objectives—manual orientation of the spacecraft using the attitude control system. The controllers were aghast and deeply concerned that she had not attempted this procedure; if the automatic systems failed in flight, manual orientation would be the only way Tereshkova could safely return.

Despite these worrying developments, *Vostok 5* and *6* remained in orbit until 19 June, by which time Bykovsky had exceeded Nikolayev's ninety-four-hour spaceflight record. Bykovsky had successfully completed his program of experiments, which included using a special camera to film the Earth's horizon, observing the sun's corona, and even studying the growth of peas in weightlessness. On that day, following further instruction from the ground, Tereshkova did finally perform the manual orientation test of her spacecraft, keeping it in the correct position for a quarter of an hour. "I still remember how the powerful spaceship was docile in my hands," she later described. "When the spaceship works well, there is a feeling of cooperation." She did not, however, manage to carry out any of the biological experiments that she had been assigned. She later stated that she had been unable to reach them in the spacecraft.

As *Vostok 6* began its forty-eighth orbit, the spacecraft was oriented for reentry by means of a solar sensor located in the service module. Tereshkova was supposed to describe the operation of this sensor to the ground as well as the sensations of reentry. She did neither, to the frustration of the ground controllers. The braking rocket fired as scheduled, slowing the spacecraft, which was then separated explosively from the service module. The spherical *Vostok 6* craft now began a fiery ballistic return to Earth, shielded from the immense heat of reentry by a protective ablative coating. As Tereshkova recalled in *Stars Are Calling*, the pressure pushed her back in her couch, and though it was difficult to keep her eyes open she took note of what was happening to her space capsule: "I notice the dark red tongues of the flame outside the windows. I'm trying to memorize, fix all the feelings, all peculiarities of this descending, to tell those who will be conquering space after me. My mind is working calmly and logically. With a loud roar, the spaceship bumps into the dense atmosphere. The noise grows with every second; it already reminds me of the thunder of hundreds of drums, the final part of some outrageous heroic symphony."

Four miles above the ground, bolts securing the pilot's hatch were severed explosively and the hatch, situated above her head, was jettisoned. Two seconds later ejection rockets fired, catapulting Tereshkova and her contoured seat out of the craft. After a parachute descent to thirteen thousand feet the seat was also discarded, and Tereshkova continued her descent under a separate parachute. The abandoned spacecraft's parachutes also deployed at that altitude, bringing it to the expected heavy landing—one deemed too heavy for the Vostok pilots to safely sustain. As she descended, Tereshkova raised her faceplate for a better view of the landing area. Below her was a large field, with a lake nearby; she became concerned that she might actually land in the water. However, there was a strong wind blowing, and it carried her away from the lake. Then, despite orders not to do so, she looked up at her red-and-white parachute and was struck in the face by a small piece of falling metal, leaving a small cut and bruise on her nose.

Soon after, at 11:20 a.m., Moscow time, Tereshkova touched down in a wheat field in the southern Urals, 390 miles northeast of Karaganda in the Altai region. The first spacewoman had clocked up seventy hours and forty-one minutes in orbit and had flown one-and-a-quarter million miles. On the ground, workers on a collective farm had watched in fascination at the sight of the spacecraft, the ejection seat, and an orange-clad figure descending beneath large parachutes. They cautiously made their way to the site, joined soon after by other workers who had been building a bridge over a nearby river. Meanwhile, Tereshkova was removing her spacesuit and changing into a more comfortable track suit. She then began gathering everything together—her spacesuit, parachute, and ejection seat—and tried to carry the accumulated weight to the capsule over a thousand feet away. Soon the people who had been working in the field arrived and helped her carry the seat, which she could not manage single-handed.

Tereshkova later reported that her next action was communicating her location to the rescue team. Following this, she maintains that she asked one of the farmers to drive her to a nearby village so she could make a telephone call to the Kremlin, during which she gave a short report to Khrushchev. Then, according to her interview with biographer Lady Lothian, Tereshkova returned to the field to wait for the rescue airplane, enjoying the warmth of the summer day despite some soreness from the injury to her face. An hour after the landing a medical team parachuted down from a small aircraft. Among them was renowned parachutist Lyubov Maznichenko, who became one of the first to greet Tereshkova. (Maznichenko is the blond woman featured in a famous photograph of Tereshkova whom the Western press misidentified for many years as Tereshkova's backup, Irina Solovyova. In the photograph the two women are shown seated between the spacecraft and Tereshkova's ejection seat, with a small crowd of curious onlookers in the background.)

Tereshkova was in fine spirits. She had even eaten a traditional gift of black bread and salt, as well as some fermented cheese, cakes, and milk given to her by the workers. By way of thanks she gave all her remaining space food to these people. Both acts would later anger doctors who had planned to check her food consumption against her physical condition.

One orbit later, *Vostok 5* also returned to Earth. Bykovsky had set a new space endurance record of 118 hours and 57 minutes. Plagued by technical problems and a lower-than-expected orbit, he had been ordered to deorbit and land. Temperatures in his service module had reached uncomfortably high levels, and after the flight he said he was having difficulties with the waste management system: "When I was told to return to Earth on the eighty-second orbit, I received conflicting messages via the telegraph. Finally, Gagarin confirmed that I should return that day. The solar orientation for retrofire worked correctly, and the TDU braking engine fired for thirty-nine seconds."

In an unfortunate repetition of what had happened to Gagarin and Titov, the restraining strap that kept the two spacecraft sections together prior to reentry failed to disconnect. The combined service module and capsule gyrated wildly until the heat of reentry burned through the faulty strap. Bykovsky coolly described this violent event as "disorderly":

There was a powerful explosion when the cabin hatch blew off, and I was ejected from the capsule in my seat two seconds later. I landed between two trees in a steppelike region. I was first approached by a man on a horse, then an auto drove up. The local people helped me out of my suit. Soon over a hundred spectators had gathered. Then An-2 and Il-14 search planes flew overhead. I couldn't reach them on the radio. I was driven in a Volga automobile to the spacecraft, which had landed [a mile to a mile and a half] away from me.

Though further Vostok flights were still planned at that time, Bykovsky's landing would in fact mark the end of the successful program, during which five men and a woman had logged over 381 hours in space.

Tereshkova returned to a rapturous welcome in Moscow, and together with Bykovsky was later awarded the titles of Pilot Cosmonaut, Hero of the Soviet Union, and the Order of Lenin. At the official welcome ceremony in Red Square, the crowd roared with laughter and then applauded when Russian premier Nikita Khrushchev playfully pushed Andrian Nikolayev close to Tereshkova, and the two cosmonauts embraced briefly. She then addressed an international women's peace conference before setting out on an exhausting goodwill tour. In October 1963 she accompanied Yuri Gagarin on a trip to Mexico and Cuba, and later that month received a standing ovation at the General Assembly of the United Nations in New York.

The chairman of the Senate Space Committee, Clinton Anderson, was among those who believed the Russian propaganda that trumpeted Tereshkova's flight as a faultless success. He expressed keen disappointment that NASA had no women in training. "The Russians have proved to us," he told one reporter, "that you don't have to have twenty years of test pilot experience before you can handle one of these capsules." He did not know that Tereshkova, and her spaceflight predecessors, were little more than passengers buckled into a seat aboard an automated spacecraft that needed far less piloting than a Mercury spacecraft. Nor did he know that Tereshkova, the first nonpilot to fly Vostok, had performed the flight test objectives of her mission so poorly.

Nikolai Kamanin, after reading through the flight reports, noted of Tereshkova in his diary that "she became noticeably fatigued, ate little, and slept a lot. Her work capacity was less than satisfactory." A postflight medical team under Professor Vladimir Yazdovsky also concluded that Tereshkova had not performed well during her flight. She angrily denied this, stating that her only problems had been fatigue and a lack of sleep and that overall she had felt fine. However, Yazdovsky's report was quite damning, mentioning illness that Tereshkova suffered on the thirty-second and forty-second orbits, her greatly diminished appetite, vomiting, and weak cardiac activity. Concerned by the report, Korolev insisted on an audience with Tereshkova and Bykovsky on 13 July to discuss these issues. He was far from pleased with her performance and wanted a full explanation. He would also later write: "Solovyova and Ponomareva were better prepared for the first flight than Tereshkova. I am deeply convinced that on future flights they will prove to be better than Tereshkova, but neither of them can compete with Tereshkova in the ability to influence crowds, arouse sympathy among people and to appear before an audience. It is these three qualities that made Tereshkova the first woman in space."

Tereshkova still tersely maintains that she "cannot understand the reasons for spreading fabrications. There was the lie that I nearly died during the flight and came back very ill. . . . I cannot understand why people denigrate and distort achievements in this way. I could not have completed my program if I had been ill."

"The couple wishes to make as little show of the event as possible, and have asked only one or two reporters and a photographer to be present," declared the Hungarian news agency, MIT. The agency was reporting on the upcoming wedding of Andrian Nikolayev and Valentina Tereshkova on 3 November 1963. The wedding was meant to be a secret, with the ceremony held in one of two "Palaces of Marriage" in Moscow. Simple registry offices, they had been created two years earlier in an attempt to win couples away from the more traditional style of church weddings, now frowned upon by the state. Although the time and place were not announced, the so-called secrecy did not prevent several hundred Muscovites from crowding into midcity Griboyedov Street. They packed nearby balconies and windows, hoping for a glimpse of the happy couple. It was freezing cold, but nothing could contain their excitement.

Tereshkova wore a simple white, knee-length dress and flower-dotted veil, while the groom was decked out in a black suit and silver tie. A shower of flowers greeted them as they pulled up outside the registry. Cameras then whirred and loudspeakers played Strauss waltzes as the couple signed a red-leather register and exchanged rings. A beaming Soviet premier had the honor of giving the bride away. Even Korolev was able to attend the public ceremony, although for reasons of national security he had to remain an anonymous guest.

When he first learned that the two cosmonauts wanted to wed, Khrushchev is said to have asked them to wait, fearing that it might appear they had been ordered to marry and have children just to test the effects of spaceflight radiation on reproduction. Others have suggested the opposite—that the wedding was ordained and organized by the state for the sake of publicity. These sources imply that though the cosmonauts were friendly there was no real romance but since Khrushchev found the idea appealing he applied pressure through Nikolai Kamanin for the couple to wed.

Hints at the truth may lie in Kamanin's diaries. In them, he explains that he was meeting with Nikolayev as late as 7 October to try and persuade the cosmonaut to commit to a date, as the wedding would be the subject of a government decree. Nikolayev was evasive, avoiding direct answers, and refusing to commit to a date. By 29 October, Kamanin was getting ten phone calls a day about the matter from important government officials and in frustration called Tereshkova and Nikolayev. He gave them a direct order—set a wedding date. Finally, at 2:30 p.m. the next day, the couple gave in and agreed to a wedding date only three days later.

Following the marriage service, Khrushchev and his wife Nina were among the guests who greeted the newlyweds at a lavish champagne-and-caviar reception in a former czarist palace in the Lenin Hills. He proposed the first toast to the couple and led cries of "Gorko, gorko" (bitter, bitter), a traditional wedding chant offered to encourage the couple to overcome the bitter taste of wine with the sweetness of a kiss. The newlyweds happily obliged. Yuri Gagarin was quite lyrical when he proposed his own toast, suggesting that the marriage would be "as lovely as the crisp autumn twilight filtering through the great glass windows." Altogether, twenty-one toasts were proposed and drunk to the bride and groom. The sun was setting over the reception house when the premier reluctantly pleaded business obligations and left. The newlyweds stole off soon after. As a wedding gift, Khrushchev presented the delighted couple with a new seven-room apartment at Kutzovsky Prospect 30132. Such luxuries were normally reserved for the most senior members of the Communist Party. Interestingly, it is said that the apartment could be divided into two separate sections should the couple ever wish to live apart.

At 2:00 a.m. on Monday, 8 June 1964, Yelena Andrianovna was born by Caesarean section. "There were medical reasons for having a Caesarian," Valentina explained, "but not to avoid pain." The first child ever born to parents who had both been in space was perfectly healthy. Discussing her daughter's birth, occurring only twenty-eight weeks after the well-publicized marriage, Valentina provided a clever disclaimer: "She was born at seven months, and I believe it was because I had been traveling and working exceptionally hard."

Despite motherhood and her demanding role as a traveling diplomat, Valentina wanted to fly in space again. To this end, she began attending the Zhukovsky Air Force Engineering Academy in order to advance her education and gain a college degree.

In 1967 it was widely (and correctly) rumored that the Soviet Union was preparing to fly to the moon and that cosmonauts were in training for the flights. Somewhat prematurely, Tereshkova boldly declared during a visit to Cuba, "The moon team has already been picked. Major Gagarin is head of it, and I am on it!" Years later, the names of those in training for the Soviet lunar flights were revealed. Ironically, if a cosmonaut had been the first human to circle the moon, that person would quite possibly have been Valery Bykovsky. He and Nikolai Rukavishnikov were deep in training for such an undertaking, rehearsing their flight in a Zond simulator at Star City and also in the capsule that eventually flew the unmanned *Zond 4* deep space mission, launched 2 March 1968. Despite her confident prediction, Tereshkova's name did not appear on this list or on any future crew manifest. She was now considered a valuable Soviet propaganda tool, and the flight aboard *Vostok 6* was always going to be her first and last.

On 27 March 1968 Yuri Gagarin was killed during what should have been a routine test flight with Vladimir Seryogin. Tereshkova was distraught. She and Gagarin had become close friends who helped and stood by each other, and now her "beloved brother," as she described him, was gone. Alexei Leonov identified Gagarin's remains at the crash site, and the nation went into deep mourning. Following a cremation service, his ashes were placed into a niche in the Kremlin Wall. Tereshkova, still deeply upset, was one of the four cosmonauts who ceremonially carried Gagarin's urn to its final resting place. She is adamant that her colleague's death was a major factor behind her forced retirement from active status with the cosmonaut team. Pressure was applied on her to give up too-risky parachute jumping; in fact, she was not permitted to either fly an airplane or jump. Tereshkova requested that she be put back on active duty and tried to persuade people of influence that she was a living person and not just a museum piece. "It was quite natural that we would lose some friends, that there would be some sacrifices," she reflected, "but they would not allow me to go back into training."

Her well-publicized marriage, once supposedly a match made in the cosmos, would also fail. There is no way to know if Khrushchev, taking advantage of his cosmonauts' immense global popularity, had prearranged it. But when he fell out of political favor, the marriage also began to crumble. The space heroes drifted apart and would not even stand beside each other in cosmonaut group photographs. "I do not find it easy to speak about that period of my life," Tereshkova admitted to biographer Lady Lothian. "It is too sensitive. Our marriage came to an end in 1977. Real life breaks many illusions about love. Perhaps women are too romantic, too emotional? But if we were not, life would be pale and dull."

In October 1969, Tereshkova and her one-time backup Irina Solovyova graduated with honors from the Zhukovsky Air Force Engineering Academy, but Tereshkova's ambition to fly in space again was never realized. Shortly after her graduation the female cosmonaut detachment was officially disbanded. In 1976, now bearing the rank of colonel, she also earned a candidate of technical services degree. From 1968 to 1987, Tereshkova represented the Soviet government at numerous women's organizations and events, and held the position of chair of the Soviet Women's Committee. She was a member of the Central Committee of the Communist Party and also deputy of the Supreme Soviet of the USSR. In 1979, she married for a second time, this time to eminent surgeon Yuri Shaposhnikov. He later became director of the Central Institute of Orthopedics and Traumatology. Those who know her say he was the one true love of her life.

In 1989, the former cosmonaut was appointed chair of the Union of Soviet Friendship Societies and elected to the new Congress of People's Deputies. She became the vice chair of the Russian Agency for International Cooperation and Development and vice president of the International Federation of Democratic Women.

The fall of Communism brought on a number of changes for Tereshkova, a committed member of the Communist Party. Life was not easy for her during the transition period, and even now she pines for many of the old familiar ways. "Yes," she admitted in a 2002 interview with the Finnish newspaper *Helsingin Sanomat*. "I am truly sad that the Soviet Union has collapsed."

On 5 May 1995, Tereshkova was promoted to the honorary rank of major general in the air force reserve. Two years later she turned sixty, and her compulsory retirement from the air force took effect on 28 April 1997, under Presidential Decree No. 429; the official Ministry of Defense order for her retirement was dated two days later. When asked how she had occupied herself since leaving active flight status, she told Dougherty: "I am still working as an instructor at the Cosmonaut Training Center, training others for spaceflight. I have always been with the center, and never stopped these activities. I have some extra jobs—what we call voluntary jobs, as I am not paid for these. But I get my salary from the Cosmonaut Training Center." She was also asked if she still saw the four women with whom she had trained: "Yes, we are still living in the same houses," she replied, "we are still friends. We have led our different lives with our families, our children, our grandchildren. But our—let's say our space 'sisterhood'—is still alive. We all still live in the Cosmonaut Training Center and are part of it. Two of us are now retired, and Irina and myself are working there."

In 1994, the Russian federal government appointed Tereshkova to head its Center for International Scientific and Cultural Cooperation. She was also elected to chair the Presidium of the Russian Association for International Cooperation—an international organization that has forged cultural and friendship associations with fifty-six foreign countries.

Tereshkova's daughter Yelena went into medicine, and today is a qualified orthopedic surgeon. In 1995 Yelena and her husband, who is employed by Aeroflot, made Valentina Tereshkova a grandmother, with the birth of their son Alyosha. Sadly, Valentina's second husband died of cancer in June 1999. As she told *Moscow Times* reporter Marina Uvarova for the 6 April 2001 issue, it was a devastating time for her: "He was an unbelievably gifted, strong, and decent person. With him I did the impossible—I achieved happiness. I lived with my husband, Yuri Georgievich Shaposhnikov, for nearly twenty years. He was a talented surgeon, an amazing man, honest and kind. On many occasions he would go to an operation at night, rush to the hospital and call. Yuri devoted his entire life to prolonging the lives of other people. I had the words, 'You burned up by illuminating others' carved on his gravestone."

Among the vast list of honors bestowed upon Tereshkova is a crater on the far side of the moon named for her and Asteroid 1671, which was given the name of her call sign, "Seagull." She is also the subject of a bronze bust unveiled in Moscow's Space Heroes Alley to commemorate her sole flight into the cosmos. In October 2000, the London-based International Women of the Year Association named Tereshkova its "Greatest Woman Achiever of the Century." They had previously honored the former cosmonaut by naming her "Woman of the Year" in 1984.

Tereshkova still looks fit enough to fly into space again. Before John Glenn achieved his second space flight in 1998 at the age of seventy-seven, she was asked if she would be ready to make a similar return journey. "Always!" was the immediate response. Elegantly dressed at all times and wearing her hair brushed back much as she did at the time of her space flight, Valentina Tereshkova remains a dignified and popular figure wherever she travels. The years may have softened her handsome oval face's features, but her gray eyes still maintain that mischievous and inquisitive sparkle the world grew to love back in 1963.

Valery Bykovsky also remained in the space program, but unlike Tereshkova was allowed to prepare for future flights. Together with crewmate Nikolai Rukavishnikov he trained as the commander of a proposed manned circumlunar mission planned for launch around March 1969. The program was later abandoned after immense technical difficulties and the successful flight of *Apollo 8* around the moon. In 1976 Bykovsky made a second spaceflight as commander of *Soyuz 22*, with flight engineer Vladimir Aksenov. Two years later, together with East German researcher Sigmund Jähn, he completed a successful Interkosmos mission aboard *Soyuz 31*. This third flight gave him a total of twenty days in space. Remarkably, Bykovsky still holds a unique record. His 1963 flight aboard *Vostok 5* remains to this day the longest solo spaceflight by any astronaut or cosmonaut.

Following Tereshkova's historic flight in 1963, the four women selected with her had continued general training. According to Solovyova, both Korolev and Kamanin were keen to fly two women aboard *Voskhod 5*. They were even divided into crews early in 1965, with Ponomareva listed as commander of a prime crew, teamed with Solovyova. Yorkina and Kuznetsova were the backup crew. In that period, according to Ponomareva, there was talk that the mission would last fifteen days and include an EVA performed by Solovyova. However, these plans came to an abrupt end after Korolev's unexpected death in hospital in January 1966.

Ponomareva and the other backup women kept up their training for three more years, but there were mounting problems from within the training center, as she recalled in a 2002 interview with Slava Gerovitch. "After Tereshkova's flight the commanders of the center wanted very much to get rid of us," she said. "But the fact that we were regular officers presented an obstacle to such efforts. It was not so easy to get rid of us. Later, however, they found a way, but this first time they failed." Ponomareva recalls that it was a time of mixed feelings for the women. On the one hand, there was hope, she said, on the other, skepticism. It was clear that there was very little support for women in cosmonautics. "The main task—priority—was fulfilled," she told Gerovitch, "and men would handle the rest": "Kamanin came to us and advised us to write a letter to the Central Committee that we wanted to serve our country by making a space flight. We wrote that we had been training for a long time. I'm not sure if this letter was the reason, but we were summoned to the Central Committee at the Old Square and were thanked for our services. Shortly afterwards (in October 1969) the group was disbanded. Maybe it was for the better that the group was disbanded, that we stopped training, but there was of course disappointment."

It was not until nineteen years later that Svetlana Savitskaya became the Soviet Union's second woman in space. In 1982, in what seems to have been a deliberately orchestrated attempt to upstage NASA, she flew into orbit just ahead of America's first woman in space, Sally Ride. Two years later she also became the first woman to make an EVA, during her second space flight. This took place just eleven weeks before Kathy Sullivan was scheduled to achieve that honor on shuttle flight STS-41G. Despite their denials, it seemed Russia's space planners were still playing the do-it-first game. Sullivan's spacewalk was scheduled for three hours and thirty minutes. Savitskaya's spacewalk lasted for three hours and thirty-five minutes—one might say a remarkable coincidence.

Another ten years would pass before Yelena Kondakova became the third (and to date last) Russian woman to fly into space. Three female cosmonauts in thirty-one years; and it is ironic that the other three women who have lived aboard a wholly Russian spacecraft came from Great Britain, France, and the United States—Helen Sharman, Claudie André-Deshays, and Shannon Lucid, who worked extended missions aboard the *Mir* space station. Sharman and André-Deshays were launched to *Mir* on top of Russian rockets, while Lucid arrived as part of a NASA space shuttle crew. Three other women, Japan's Ryoko Kikuchi and NASA astronauts Wendy Lawrence and Bonnie Dunbar, also trained for possible flights to the space station aboard a Russian Soyuz spacecraft, but ultimately did not fly those mis-

Vostok cosmonauts reunited: Valentina Tereshkova and Valery Bykovsky at an autograph convention in Coventry, England, in 2004. Courtesy Colin Burgess Collection.

sions. Seven Western women (including Lawrence and Dunbar) also entered *Mir* for short visits as part of space shuttle ferrying crews—some of them visiting more than once. Yelena Kondakova was eventually given a second flight opportunity, flying in May 1996 aboard space shuttle *Atlantis* on mission STS-84, which docked with the *Mir* space station.

By the end of the millennium, no further spaceflights had been made by any Russian women, and only one female candidate was in training—Nadezhda Kuzhelnaya, selected to join the cosmonaut unit in 1994. Sadly, a decade later, with little prospect of occupying one of the very few seats available on a Soyuz mission, she resigned on 27 May 2004 to become a Tu-134 pilot with the airline Aeroflot. For more than two years, until candidate Elena Serova was selected for training in October 2006, there were no female cosmonauts other than an American space tourist.

It seems clear then that Tereshkova's flight was simply a one-off for propaganda purposes and that the Russians never had any real intent to inject equal opportunity into their space program. In terms of inspiration, however, the flight of Valentina Tereshkova opened many doors, allowing other women to follow in her wake. In realizing the legacy of her flight, we are also drawn to tragedies along the way. Four women perished on space shuttle missions while realizing their own dreams of spaceflight: Christa McAuliffe, Judy Resnik, Kalpana Chawla, and Laurel Clark. "I am aware that it is very difficult for space pioneers to go along a path that hasn't been explored yet," Tereshkova said in a 2003 interview on this subject. "Practically, we had to sacrifice our daily lives for the sake of it. It took our lives sometimes, too."

Today, Valentina is obviously enjoying a less frenetic and more relaxed ambassadorial lifestyle. In October 2004, at a convention held in Coventry, England, she and Valery Bykovsky enjoyed an often playful rapport, laughing and poking fun at each other, happily jousting like many old friends do. "Ah, my little space Chaika, she is the love of my life," a smiling Bykovsky told the authors through an interpreter, a mischievous twinkle in his eye. "We were united by spaceflight, and we are still close to each other." Tereshkova, feigning surprise at his words, quickly admonished her colleague with a gentle shove and a retort in Russian that caused both of them to chuckle. She subsequently pointed out that when they first met he was "just another fighter pilot," while she was an accomplished parachutist. "Maybe so," he observed, still smiling, "but a good fighter pilot should never have to learn how to use a parachute!"

Watching this easy amiability and the obviously well-oiled routine, one is left with the impression that despite any achievements they may have realized after their dual flight, they both know that their names and lives are forever linked because of the events of 1963. And both are more than comfortable in that knowledge.

Valentina still dreams of flying into space again, but as the years slip by that precious dream is rapidly dimming. "Those who have already been in space yearn with all their heart and soul to hasten there again and again," she once wrote in her memoirs. "With every single day passing, time leaves my flight in the past. Occasionally the wind will whisper something from the tops of the tall pine trees, and then everything becomes silent. In such minutes I remember the most bright and wonderful experience in my life: the flight into space."

10. Stepping into the Void

Often the test of courage is not to die, but to live.

Vittorio Alfieri

It was an astonishingly simple plan and one quite breathtaking in its audacity. A gamble for glory that would earn acclaim and tremendous banner propaganda for the Soviet Union but that would subsequently attract justifiable criticism when the full facts were revealed. Words such as *reckless*, *hazardous*, and *ill-conceived* would later be attached to the first manned flight of the Voskhod spacecraft, yet that flight created history at a time when the world was breathlessly anticipating each new space spectacular.

In December 1957 a group of talented young Soviet designers, most of whom had recently graduated from technological institutes in Moscow and Leningrad, was assembled in the planning section of design bureau OKB-1. Their mission was to begin detailed studies of manned orbital flight and to develop spacecraft capable of carrying and sustaining future space travelers. Chief Designer Korolev jokingly called these designers his "kindergarten." Many of them, including a serious young scientist named Konstantin Petrovich Feoktistov, would later become pioneering cosmonauts.

Feoktistov, a name of Greek origin that translates as "loved by God," was then in his late thirties, held a degree in rocket design, and also maintained considerable authority in this group. Konstantin, or Kostya as he was affectionately known, was born on 7 February 1926, the son of bookkeeper Petr Feoktistov and his wife, Mariya. The family lived in the industrial city of Voronezh, in central Russia. It was his older brother, Boris, who first introduced ten-year-old Kostya to the wonders of space travel by bringing home a book called *Interplanetary Travel* by Yakov Perelman, which they often read together. Kostya made a vow that he would one day fly to the moon, meticulously designing his rocket and planning the voyage in a tattered copybook. According to his boyhood calculations, he estimated that he would be making his first space journey in 1964. He was not far wrong, but he would never get to share his experience with Boris. His beloved older brother was lost battling the Nazis in 1941.

His father was also fighting on the front line, so when Konstantin turned sixteen he also volunteered for duty as a combat scout in a neighborhood reconnaissance platoon during the fiercely fought defense of Voronezh. He managed to complete several dangerous missions before he was captured by a German patrol. They dragged the sixteen-year-old boy to a brick courtyard filled with other captured Russian soldiers, where two bodies were already sprawled on the ground beside a freshly dug trench. He realized with horror that he was about to be executed.

"The German toyed with his Parabellum [Luger] gun while asking me questions," Feoktistov would later recall. "I was playing it dumb but his playing with the gun was getting on my nerves, and I thought it would be nice to kick the gun out of his hands. My face must have betrayed my thoughts because the German pulled the trigger." Left for dead in the body-filled trench with a bullet wound to his throat, Feoktistov waited until darkness fell several hours later. He then managed to crawl out of the pit, reach the Voronezh River, and make his way across, rejoining his unit. Soon after he was taken to an army hospital where his mother eventually located him. When he was fit enough they traveled away from the front to the Uzbek city of Kokand, where he was rehabilitated. Mariya Feoktistova took a job as a railroad traffic regulator, while Konstantin and other school friends from tenth grade helped bring in the harvests of corn and red beets between studies.

Konstantin ended his secondary education in 1943 at the top of his class and applied for training at the Moscow Aviation Institute. His enrollment papers took a long time to arrive, and much to his disappointment, when he finally arrived in Moscow he was told that his application was a month too late. Feoktistov then applied to the E. N. Bauman Moscow Higher Technical School, from which he graduated in June 1949. Over the next two years he worked as a factory engineer in Zlatoust and at various research institutes. In 1951 he worked for the first time with Sergei Korolev, but Feoktistov said that he "didn't like him in the beginning. He gave one of those inspirational speeches which seemed to me to be full of clichés."

Konstantin then undertook postgraduate engineering studies in Moscow. Four years later, in 1955, he won a magisterial degree in the technical sciences and at the age of twenty-five was subsequently reunited with Korolev, when he became a member of the chief designer's OKB-1 design bureau. "Eventually, he won my respect and he was the right man for the enterprise he ran," Feoktistov told Korolev biographer James Harford. "We had a very good relationship."

An ardent worker at OKB-1, Feoktistov would, like many others, occasionally incur the wrath of the perfectionist designer when things weren't going right. It seems, however, that he could be just as hard-nosed. *Novosti* press agency correspondent Alexei Gorokhov characterized Feoktistov as "an amazingly obstinate man . . . who seemed to be unable ever to raise his voice. This obstinacy sometimes almost drove Korolev mad. It was nevertheless Feoktistov's team which was given the task of drawing up tentative plans for a manned spacecraft." Being much older than his colleagues and possessing a degree in rocket designing, Gorokhov says that Feoktistov "enjoyed incontestable authority in the group." It is telling that, despite political pressure, the future cosmonaut refused to ever join the Communist Party.

OKB-1 engineer and later cosmonaut Oleg Makarov once said of Feoktistov that he "set the pace. He has a bright head. I think he saw the spaceship before it was built. We younger engineers learned from him." By May 1958, Feoktistov and his team were working steadily on some early designs of what was known as Object OD-2: a craft intended for spaceflight and manned occupancy. That month he presented his team's concepts to Korolev, and on that day the world's first manned spacecraft was born. "He was highly inspired by them," Feoktistov recalled of Korolev's reaction.

Korolev studied the many proposals and settled on what he was felt was the best option. He subsequently ordered Feoktistov's team to deliver full design reports. All other ideas and prototype designs were abandoned, and he issued instructions that everyone was to concentrate their efforts solely on the design of the manned spacecraft, which became known as Vostok (East) after a suggestion by members of the design team. Two months later, the proposals were back with Korolev, and he gave the go-ahead to build the Vostok spacecraft. Less than three years later, Yuri Gagarin flew a Vostok into orbit, becoming the world's first space traveler. On the day before the flight, Feoktistov extensively briefed Gagarin on the operation of the spacecraft, and the cosmonaut flew carrying instructions from Feoktistov on how to manually override the Vostok's automatic systems if necessary. "He was a real, meticulous scientist," Gagarin would later write.

Other flights took place, and triumph followed triumph. While the American manned space program toiled along at a steady pace, the Soviets seemed invincibly superior with their Vostok spacecraft and rocket technology. But the achievements were mostly politically driven. Ostensibly creating space spectaculars with ease, they were actually stretching their technological capabilities to the limit. This seeming success, and the propaganda associated with these flights, would actually cause overwhelming difficulties, as the incumbent Soviet premier, Nikita Khrushchev, continually exerted pressure on Korolev and his design bureau to pull off ever bigger space triumphs. Korolev was of the opinion that he and his engineers could turn the Vostok capsule into a two-pilot spacecraft, but this seemingly did not satisfy Khrushchev or his urgent need for international recognition and prestige. The Americans were relentlessly pressing ahead with their two-man Gemini program, and Khrushchev wanted to maintain the lead. It is therefore likely that he insisted on a flight carrying three men in a single spacecraft, and his formal demand was relayed to Korolev.

The six Vostok missions had been a great achievement for the Soviet space program. They had established that it was possible to fly into space, perform basic functions while in orbit, and return the cosmonauts to Earth safely. These were test flights and apart from the purely propaganda-based selection of Tereshkova, were flown by cosmonauts recruited from the Soviet air force. Air force chiefs had made it very clear that they were totally against any plans to include scientists and doctors in the early cosmonaut selections, which would take away precious seats from the pilot cosmonauts. However, with plans for a highly automated, two-person Voskhod (Sunrise) spacecraft on Korolev's drawing boards as an intermediary precursor to the Soyuz craft, crew requirements changed. Korolev had determined that there was no real need for a two-pilot operation, and it was time to think of selecting and flying space researchers, scientists, physicians, engineers, and even journalists into space. His visionary concepts did not sit well with the military's hierarchy.

Nevertheless, Korolev's views did not go unheeded. The medical services staff of the Soviet Ministry of Defense, initially opposed to the idea, was forced to screen up to thirty possible "observer" participants and the same number of physicians in April 1964. From these, six received approval to begin cosmonaut training. At the same time, thirty-six candidates were also being screened after their selection by the Academy of Sciences and the Ministry of Health. Fourteen passed the medical board examinations, and this number was eventually reduced to ten by the evaluation committee. They finally announced that five were suitable for cosmonaut training. This number would include Institute of Medical-Biological Problems physician Boris Yegorov, then aged twenty-seven.

Boris Borisovich Yegorov was born in Moscow on 26 November 1927, the son of Boris Grigoryevich Yegorov, a leading brain surgeon and politically a very influential man. His father's power probably had quite a bearing on his selection for the Voskhod program. His mother Anna Vasilyevna was a prominent eye doctor, who died when Boris was just fourteen. During his eighth grade at school Boris developed an interest in technology, especially radio engineering, and began rebuilding broken radios. He also built his own crude eight-valve television set, and fashioned a telescope from some old field glasses, which he used to study the stars and planets. Upon leaving school, and despite his interest in instrument technology, he decided to enter a medical institute.

After graduating from medical school Yegorov eventually became interested in research applicable to space medicine, and particularly enjoyed conducting research on the vestibular apparatus in the inner ear. Joining a Moscow-based research institute, he was part of the postflight medical support team for the first two cosmonauts in 1961 as well as one of the first investigators of Gherman Titov's illness in space—later attributed to Space Adaptation Syndrome.

In August 1961, shortly after Titov's flight, the department head of Yegorov's research institute approached the young physician. Academician Oleg Gazenko, the chief of space medicine in the Soviet Union, was a man of considerable influence. He asked Yegorov if he would consider taking part in a planned space mission in order to conduct medical tests on its crew. The prospect delighted Yegorov, and he agreed to be nominated. After his selection as a Voskhod flight candidate, Yegorov was one of the seven trainees who arrived at the Cosmonaut Training Center in June 1964. This number was later trimmed to five when two candidates dropped out for medical reasons. As a full member of the Academy of Medical Sciences, Yegorov's father had become well acquainted with Sergei Korolev and so championed his son as the best candidate for the position of spaceflight physician.

Korolev looked over the young man's résumé and discovered that he actually had quite strong qualifications. He read that the younger Boris Yegorov, a civilian, had worked for some time at the air force's Institute of Aviation and Space, he had been associated with medical tasks involving the first cosmonauts, and his work in this area had been glowingly acknowledged by Gazenko. The only problem for Yegorov, who was otherwise in excellent health, was his poor eyesight. However, it was reasoned that since he would have nothing to do with piloting the craft, he could potentially fly on the upcoming Voskhod mission.

Fourteen OKB-1 bureau candidates had also been nominated on 17 May, of whom eight were deemed medically qualified. Ultimately, only one candidate from this group was nominated for spaceflight training. Then it was discovered that the sole candidate had suffered a spinal injury that would not only preclude him from parachute training but would also restrict him to spaceflights of less than a day's duration. Korolev personally intervened, and in this way Konstantin Feoktistov was cleared to become a cosmonaut candidate.

When it was decided that Vostok would be reconfigured into a three-person craft, Feoktistov is said to have declared that it was an unsafe concept and that he wanted no part in such a hazardous scheme. He is also said to have confronted Korolev with his views, but the chief designer had a small inducement up his sleeve that he had worked out with his political superiors. He told Feoktistov that if the bureau could find a way to cram three people into the spacecraft, an engineer from OKB-1 would be chosen to occupy one of its seats. There was no reason, he said, why Feoktistov should not be the one to occupy that third seat.

Korolev already knew that the proposed mission would require very little of the crew. Apart from the pilot commander, they would be virtual passengers during the flight. In fact, there would be no space in the cabin to conduct anything but nominal experiments. The three cosmonauts would be jammed together like stellar sardines. To save space and weight, it had already been decided to eliminate ejection seats from the design, which caused the engineers much consternation. This meant no escape would be possible for the crew if the dangerous launch phase went awry, and it also meant the cosmonauts were in for a particularly heavy landing at the end of the mission: instead of parachuting from the spacecraft, the crew would have to land inside. For this reason, a second spacecraft parachute was added, and the crewmembers were seated in shock-absorbing couches. To further minimize the impact and risk of injury on landing, a small but powerful rocket, designed to fire momentarily just before touchdown, was attached to the base of the spacecraft.

An uneasy decision had also been made to dispense with spacesuits. The Vostok spacecraft was essentially a good design, but not so good that fundamental safeguards such as spacesuits and ejection seats could be discarded. Feoktistov had to weigh the excitement of an opportunity to fly in space against the potential danger of flying in a spacecraft whose safety had been fundamentally diminished. But as he told researcher Bert Vis in 1992, it was a long-held dream of his to journey into the cosmos:

I was the leader of the team dealing with design and construction of the spaceships, including the Vostok. After the first unmanned tests of the Vostok, I proposed to transform it to a piloted craft as quickly as possible. I proposed this to Korolev. Since these were rather scary decisions that had to be made, I thought it would be necessary that I would be the first to fly on this new craft. He was absolutely opposed to this . . . but he still felt that he owed me something. And I have always thought that my flight on the Voskhod was a reward for my work on Vostok. We started work on Vostok in 1958, and in June 1964, he let me get started training for flying with the Voskhod.

Documentation and recorded oral histories seem to indicate that Feoktistov had no real reluctance about redesigning the Vostok spacecraft once he had the incentive of being one of the crew. In fact, Feoktistov appears to be the one who first proposed the risky idea of eliminating the spacesuits and ejection systems. Despite Korolev's promise to get him one of the three seats on the spacecraft, Feoktistov's selection was far from certain. When Nikolai Kamanin heard that the design engineer was a candidate, he is said to have flown into a rage. He wanted to know how anyone could put a man into a space ship "if he is suffering from ulcers, nearsightedness, deformation of the spine [and] gastritis?" In fact, doctors had supplied Kamanin with documents stating that Feoktistov was unfit for space flight. Then Korolev began to exert some pressure of his own and had the deputy minister for health sign a favorable medical certificate. Air force detractors including Kamanin then had to back down, but with great reluctance. Feoktistov was now guaranteed a seat. He told Bert Vis that his cosmonaut training began on the tenth of June and was completed early in October: "They didn't have to teach me all the technical sides, of course—I invented it all myself. I'm a technical intellectual who never does any exercise, so I had to spend the whole four months in physical training!"

On 9 October 1964 the Soviet State Commission formally but covertly announced the assignment of three cosmonauts, including Boris Yegorov and Konstantin Feoktistov, to the first flight of a Voskhod spacecraft. All aspects of the flight and crew would remain top secret until the launch took place a few days later. Commanding Feoktistov and Yegorov on the mission, then planned for a single day but with provisions for two, was a pilot from the first cosmonaut group selected in March 1960: Vladimir Komarov. A native of Moscow, Komarov was a thirty-seven-year-old air force fighter pilot who had once been assigned to the Soviet Air Force Research Institute. He was born in Moscow on 16 March 1927 to Mikhail Yakovlevich Komarov and Kseniya Ignatevna (Sigalaeva) Komarova. His father was not a highly educated man and supported the family by working as a janitor, porter, storekeeper, and laborer. Komarov grew up in a large rambling apartment complex on Third Meschchan-Skaya Street, which was home to twelve families. Although it was crowded and the children had only a small grassy area in which to play, everyone got along well.

Vladimir Komarov was nearly seven when he first became interested in flying, a result of listening with his family to the real-life drama of the stricken ice-breaker *Chelyuskin* in February 1934. The vessel had become stuck in pack ice during a scientific expedition in the Chukotsk Sea at the far eastern tip of Siberia. The 104 men and women on board took refuge on the ice field, where they were stranded as their ship was slowly crushed. While anxious Russians followed the drama on their radios, six Soviet pilots repeatedly flew dangerous retrieval missions to the area and eventually rescued all of the trapped explorers. For their bravery, the pilots were the first to receive the supreme award of their country, the title of Hero of the Soviet Union. It was their story that apparently first spurred the young Muscovite to dream of one day becoming an aviation hero himself.

Every summer, Vladimir would travel to his grandmother's house in the small suburban village of Filino, north of Moscow. "I liked that picturesque spot with its pine tree forest, the little lakes and the small Klyazma River," he would later recall. "It was a nice place to have fun." He was at his grandmother's house on Sunday, 21 June 1941, a date he said he would always remember. His father Mikhail, then employed as a fitter and turner, was supposed to join them that day but did not arrive. His family sensed something serious had happened. It turned out they were right. In the wake of the surprise Nazi invasion, he had been called up and immediately mobilized. The mass mobilization of men of all ages quickly swept through the entire area, creating a sudden and dramatic shortage of labor. Vladimir and other children from the village began doing all they could for the collective farm, carrying vegetables to the storage area, harvesting, and threshing grains. They were hard times, but Vladimir grimly realized that he was part of a massive coordinated effort, and he quickly came to understand the principles of responsibility and teamwork.

When the family finally returned to Moscow two months later, life was at a low ebb in the city. Not only were Nazi troops approaching the city, but there was also a serious shortage of fuel and food supplies. Despite these hardships, Vladimir still entertained his youthful dreams of graduating from an aviation academy, acquiring engineering skills, and then becoming a test pilot. His schoolwork later resumed with ninth-grade special classes, and he became more determined than ever to become an aviator. After graduating in 1942, Komarov sought admission to the ten-year school at the First Moscow Special Air Force School and was given a conditional enrollment at the age of fifteen. "But," he later said with pride, "when I passed all my exams and went through a medical checkup, they admitted me as a full member of the school."

Then Mikhail Komarov was killed. His family was apparently never given any details about when or where he lost his life fighting the Nazi invaders. Immediately after his enrollment at elementary flight school, Komarov's group of student pilots was told they were going to be evacuated to Siberia. It was another hardship for Vladimir, particularly as it meant leaving his widowed mother, but he was steadfast in his desire. "We were growing up at a time when, I believe, practically every boy was dreaming of becoming a flier," he would later declare, "and it looked like my dream was coming true."

However, things would change. At war's end the air force school was shut down, so Komarov took his flying ambitions to the Soviet army, where he continued his preliminary flight training. "All of us . . . went to the third Sasovskaya Military Aviation School for preliminary instruction of pilots." He later attended the Chkalov Higher Air Force School in Borisoglebsk and the Serov School in Bataisk. At the Chkalov school he displayed outstanding ability. One day his commander presented him with a certificate, on which was written: "He loves aeronautics, flies boldly, confidently, not stopping at what was achieved. In flight he is hardy, quick witted, and has initiative. He endures the altitude well, and does not have flight accidents."

Graduating in July 1949, Komarov then spent five years as a fighter-pilot lieutenant in the Caucasus military district. "Life is many-sided," Komarov reflected of this period. "We flew, we lived with studies, with work, but we also had our own, personal life. I was acquainted with a young girl, Valya. We decided to get married." His "Valya" was Valentina Yakovlevna Kiselyova, and the young couple became husband and wife on 21 October 1950. They would have two children—a son they named Yevgeni and a daughter, Irina.

In 1954 Komarov attended the Zhukovsky Air Force Engineering Academy, where three years later he received an engineer's diploma. Late in 1959 he became a test engineer at the Ministry of Defense's Central Scientific Research Institute and was working there when he was invited to apply for the first group of cosmonauts. He was the only member of the group with any test-piloting experience.

When Yuri Gagarin first met Komarov, his initial impression was that "he was older than the rest of us." Komarov was indeed at a disadvantage during the regime of physical training, but he excelled in the academic classes, despite missing six months' training in mid-1960 because of a hernia operation, which ruled him out of an early Vostok mission. Then, two years later, in a development reminiscent of Deke Slayton's situation, doctors discovered a problem with Komarov's heart. After a long session in the centrifuge, doctors had carried out some routine checks on Komarov and detected a minor abnormality, called an extrasystole—a small independent contraction of the heart. He was immediately removed from training, and for a time his cosmonaut career seemed to be at an end. In contrast to Slayton, however, Komarov's doctors eventually decided that a slightly irregular heartbeat would not affect Komarov's performance in space, although they were reluctant to pronounce him fully fit for flight. Yuri Gagarin would later write that there were "grave doubts about his future participation in space flights. It required great persistence on his part to convince the doctors." Komarov was eventually cleared to return to flight training, but he had come very close to washing out of the program.

After all his problems had been resolved the highly regarded Komarov was soon under consideration for an important mission. His tenacity in the face of a career-ending problem had paid off handsomely for him. "I will never forget the day when I was appointed commander of the Voskhod crew," he would later recall. "A dream had come true."

In the days leading up to the mission, clues emerged that a new spectacular was being planned. The government newspaper *Izvestia* began to hint that Marina Popovich, wife of the *Vostok 4* pilot, was a possible leading contender. The newspaper said that she had "finished the higher aviation school and has now left for a distant business trip." When Pavel Popovich was asked by an *Izvestia* reporter if his wife was preparing for new flights, he cryptically responded: "The machine with which she is now familiarizing herself is considerably more powerful than the one in which she set a new speed record two months ago."

Leading Soviet space expert Norair M. Sissakian, number two in the Soviet bioastronautics program, was also conspicuously absent from the opening of the International Astronautical Congress in Warsaw, "for very important reasons," *Izvestia* reported. Soviet press reports began teasing readers with reports that a new space shot was imminent and that cosmonaut number seven had the initial "K."

The anonymity of the cosmonauts participating in the flight had unintentionally humorous consequences. According to Yegorov, one night they had decided to take a walk before sleeping. On their return to the security gate the lone guard challenged the three men, saying they could not enter because "the cosmonauts are sleeping. They must not be disturbed." Komarov began to explain, but the guard abruptly cut him short, saying, "My

The three-cosmonaut Voskhod crew: Konstantin Feoktistov, Vladmir Komarov, and Boris Yegorov. Courtesy Colin Burgess Collection.

orders are to let no one in!" Finally, Feoktistov stepped forward and fortunately the guard recognized him. "Can't you see we are the men who live here?" Feoktistov said, and the apologetic guard quickly let them in.

At 10:30:01, Moscow time, on the morning of 12 October 1964, Voskhod took to the skies. Incredibly, only seven months had passed since the three-person flight had been approved. Controllers watched anxiously as an improved version of the R-7 launch vehicle thrust higher into the sky, not only carrying the first three-man crew but also the first scientist and the first physician to travel into space. "We were expecting some extraordinary sensation," Komarov later wrote of the launch. "But there was nothing special. The rocket shivered a bit, there was a slight noise."

Feoktistov was understandably more amazed at the raw power of the lift-off and ascent into orbit. "I guess we all have a bit of our ancestors' nomad instincts in us," he commented. "For me, it was a thrill to ride that beast."

Thankfully, the 11,700-pound Voskhod reached its correct orbital velocity without incident, entering an elliptical orbit of around 110 by 253 miles above the Earth. It was time for the propaganda machine to burst into life. "Three men launched into space!" came the *Tass* announcement, boasting that the craft was sent aloft "by means of a powerful, new launch vehicle." *Pravda* went one better with a large mocking headline, "Sorry, Apollo!" *Tass* reported that the stated aims of the flight were "to test the new multi-seat piloted spaceship; to check the capacity for work and interaction during spaceflight of a group of cosmonauts consisting of specialists in different fields of science and technology; to carry out scientific, physical and technical investigations on conditions in spaceflight; to continue the study of the effects of different factors of spaceflight on the human organism; and to carry out extended medical and biological research in conditions of a long flight; with the help of instruments on board the spacecraft and with the direct participation of a scientific worker, a cosmonaut and space doctor."

It was all a powerfully managed propaganda coup. For one, Voskhod was certainly not "a long flight" as indicated in the *Tass* release. Moreover, very little "extended research" would actually be carried out, and only minor studies were made of the effects of spaceflight on human organisms. Indeed, only a handful of scientific investigations could be conducted by three people crammed into a capsule intended for one, and the cosmonaut in the center was almost on top of the other two men. Instead of wearing bulky spacesuits, each of the crew was dressed in light steel-colored woolen suits and blue jackets, with white headphone caps. They were also carrying knives in the unlikely event they had to fend for themselves in the forests of Siberia after landing.

Feoktistov was pleased to finally be in space. "I had many enemies who did not want me to make that flight," he told a *Russian Weekly* reporter in 1998. "Once we took off, I remember thinking, 'That's it! No one can get me off this spaceship now.'" Despite the cramped conditions, all seemed well on board at first, although there were problems with the temperature and humidity within the cabin. Both were much higher than planned, which added to the discomfort in the confined area. Nevertheless, as they completed their first orbit the crew reported that they were "feeling fine." Thereafter, every ninety minutes, as Voskhod crossed the Soviet Union, the cosmonauts appeared on television. The pictures were grainy and filled with static, but the viewers were too excited to care. The cosmonauts could be easily identified from the televised images, and when one announcer asked a stern-looking Feoktistov to smile, he looked at the camera and gave a wide grin.

Each of the crew had specific tasks to perform. As the pilot-commander, Komarov was in charge of the control and operation of the spacecraft and would handle communications with the ground. Feoktistov was responsible for all visual and photographic tasks, and Yegorov was to conduct an array of medical experiments. One of his first tasks involved taking blood pressure and respiration readings from his colleagues.

The crew may have said they were feeling "fine," but this was far from true. Feoktistov and Yegorov were both suffering from space sickness, although Yegorov's illness was more severe. He reported feeling giddy and ill on the second orbit and had a poor appetite. The illness was apparently at its worst on the fifth orbit, but Yegorov reported feeling better after a refreshing sleep. Each of the crew talked with a delighted Khrushchev during their third orbit, then settled down to a small meal, although Yegorov mostly abstained from eating. During the fifth and sixth orbits, Feoktistov recorded observations on cloud cover and carried out photometric studies. Meanwhile, Yegorov was actually the busiest of the three men, conducting physiological studies on the state of the cardiovascular system and the functioning of the vestibular system. This included taking blood samples from himself and Feoktistov. Though his medical experiments were minor by comparison with those on later, more spacious flights, the simple fact that a doctor was conducting them in space added greatly to the publicity the flight was generating. Arrays of biological samples had also been carried on board, including cells, bacteria, algae, plants, and seeds.

Vladimir Komarov would later discuss some of his impressions of the flight in an article called "For You My Motherland" in the Russian-language newspaper *Trud*:

The polychromatic rainbow of the aureole was replaced by impenetrable darkness [as] our ship entered the dark side of the Earth. This is not the first time, and we became accustomed to such transitions, the more so because we heard stories about the wonderful play of colors, about the change of day and night in space from those whom we escorted in flight before.

A field of lights of large cities floats below. The ship flies over the European part of the Soviet Union. Konstantin Petrovich puts his hand on my shoulder. I take myself away from Vzora [the porthole] and, comprehending his inquisitive glance, I silently nod my head: "Yes, it is our Moscow."

Boris Yegorov was also moved to record his impressions when looking down at the Earth: "We had left behind the Soviet Union. The ship was now in the Earth's shadow. Before us spread out a magnificent view of the horizon and a second brilliant layer of the atmosphere a hundred kilometers above it. There stretched a luminous layer in yellow-white colors with stars shining through it. The layer of brightness was clearly visible in the moonlight. The stars in the jet-black sky sparkled like diamonds." As the flight progressed, Feoktistov performed a photographic observation of the Earth and conducted experiments on the behavior of fluids in weightless conditions. Meanwhile, Yegorov continued his observation of the condition and behavior of the crew as well as some minor research into pain and tendon reflexes.

As Yegorov later explained, the crew also enjoyed some lighter moments: "During our meal we took our weightless food not just with our hands but tried catching it with our mouths. It was a sort of a hunt we made not just for fun, but also to analyze weightlessness. It was very funny though, and we were laughing throughout the whole meal. When the medical equipment came down and floated in front of us, we called it our 'Sputnik.'" Toward the end of their third orbit, while Voskhod was flying over the Atlantic, Yegorov witnessed another unique phenomenon: "Columns of yellow light hundreds of kilometers tall, rising at right angles to the black horizon, towered above a layer of brightness that shrouded the Earth. They fringed the entire visible horizon for about two thousand kilometers. We were so entranced, we did not at once realize the nature of the phenomenon. We were observing the Aurora Australis. In the cosmic stillness, the columns of yellow light stood motionless, like an immobilized fire, and through them the stars glittered coldly."

Feoktistov told the authors that there had been many other memorable moments and sights during the flight. The most exciting experience was obviously "the feeling at the stage of the journey where the ship was taken into orbit." Commenting on the unforgettable things he had seen, he listed "the rising and setting of the sun; the observation of layers of brightness above the horizon before the ship would leave the shadow of the Earth, and the fast moving, recognizable but very unusual colorful map of the Earth's surface." He also agreed with his companions that the auroras were easily the most spectacular sights they witnessed during their flight.

The crew of Voskhod began preparing for reentry less than twenty-four hours into the mission. Without much hope of success, Komarov requested a further twenty-four hours in orbit, declaring that the crew was involved in many interesting experiments and studies. "Feel fine," he reported. "Wish to extend observations; seeing much of interest and beauty." His request did not get past Korolev, who lightheartedly quoted *Hamlet* in denying the crew further time in orbit. "There are more things in heaven and Earth, Horatio . . ." he told Voskhod's commander. The crew was disappointed but confirmed they would stick to the schedule and return that day.

Reentry went without a hitch, and as they approached the ground under parachutes the new braking rockets functioned perfectly, easing Voskhod to the ground. Feoktistov would later describe it as a "feather-bed landing," while Komarov went even further in his praise: "There was a rustle underneath us and a scratching against sand." Korolev was well aware of the dangers to the crew of landing inside their capsule and nervously paced the control room floor until news reached him of the crew's safe return, 193 miles northeast of the town of Kustany. Voskhod's mission had carried the crew on a sixteen-orbit flight lasting twenty-four hours and seventeen minutes. The chief designer knew it had been an incredibly perilous undertaking. He was both relieved and perhaps a little surprised that it had all gone so well. "Is it really true that the crew has returned from space without a single scratch?" he repeatedly asked controllers.

A bigger surprise was in store for the three crew members. Still celebrating their safe return, they were expecting to receive the telephoned congratulations of Soviet premier Nikita Khrushchev. No such call came, and the crew was surprised to hear that they were to return immediately to the launch area at Tyuratam before flying back to Moscow. They would soon learn that certain events had taken place during the twenty-four hours they were in space. Khrushchev was no longer in a position of power. He had been deposed following a special meeting of the Central Committee of the Communist Council of Ministers, and a trio of leaders had now assumed leadership. The three men now in power were President Nikolai Podgorny, First Secretary of the Central Committee Leonid Brezhnev, and Chairman of the Council of Ministers Alexei Kosygin. Khrushchev's happy telephone call to the crew in orbit had actually been his last public act. When the three cosmonauts returned to a triumphant welcome in Moscow they were congratulated by two of the three new leaders, Brezhnev and Kosygin, who met them at Vnukovo airport. Korolev's Shakespearean quote had been lent an added poignancy by these events.

In spite of the many hazards associated with their flight, the three men had performed remarkably well. Vladimir Komarov had confirmed with this flight that he was a highly qualified and consummate cosmonaut pilot. Eventually selected to pilot the first Soyuz mission, he would become the first cosmonaut to fly in space twice. Though he would make two successful launches, however, he would only return alive from his first mission. He had survived the risky Voskhod flight, but his luck with untried spacecraft would not last. Thirty months after his Voskhod mission, he would perish while test-flying *Soyuz 1*.

After his spaceflight, Boris Yegorov remained active in spaceflight medicine, but he would never fly in space again. His sudden fame seems to have changed his life completely. As Russian space writer Alexander Sabelnikov explained, "after a successful flight he left his family, a wife and a small son, and hurled himself into a sensational marriage to the famous Russian movie star Natalya Fateeva." The second marriage did not last, and Yegorov went on to marry yet another movie star. This union also failed. In all, he was married four times, a record that would not have impressed mission planners. His medical career, however, did not suffer. He became the director of a biomedical research institute but died of natural causes at the comparatively young age of fifty-seven, on 12 September 1994.

Konstantin Feoktistov fully intended to fly in space again, but it was not to be. He fought bitterly with Kamanin to be allowed to fly the second Soyuz flight, but was rejected. In the meantime, he worked on many future manned space project designs, including developing the technology needed for rendezvous and docking and setting the dimensions of the Soyuz spacecraft. He also worked on the designs of the Salyut space stations, which he hoped to visit in person, but once again was thwarted. The fact that he had just divorced his second wife apparently also placed his future as a cosmonaut in doubt. He worked instead on the missions as a flight controller.

At last, in 1980, Feoktistov managed to get himself assigned to a crew, along with his old colleague Oleg Makarov, for a mission to the *Salyut 6* space station. "I think it's high time I went up again to take a look for myself," he said at the time. But just days before launch, he was pulled from the crew for medical reasons. In the meantime, Feoktistov's characteristic outspokenness was not making him many new friends. When he made one public criticism too many about the direction of the space program, he was forcibly retired. Feoktistov formally left the cosmonaut team in 1987 and the Russian space program in 1990; he then began teaching at a technical institute. He is no longer recognized on the streets of Moscow but says he has very few regrets in life, apart from always having longed for a second flight.

"The Voskhod venture opened a door to outer space," the then seventy-five-year-old told a Russian TV journalist in 2001, "and I hoped to walk through that door once again on a serious and longer space mission. It would be very interesting to do this. Besides, I would do some research. It would be useful from the point of view of what is the upper age limit for cosmonauts and anyone, in general, who would want to fly. I would go right away, without thinking twice about it!"

"Life has, however, rewritten my plans."

In 1990, Korolev's deputy, Vasily Mishin, spoke frankly about the many hazards associated with the flight of the first Voskhod in an interview by G. Salakhutdinov for *Ogonyok* magazine. "Was it risky?" he responded, when asked about the dangers.

Of course, *it was. It was as if there was, sort of, a three-seater craft and, at the same time, there wasn't. In fact it was a circus act, for three people couldn't do any useful work in space. They were cramped just sitting—not to mention that it was dangerous to fly.*

But in the West, they drew the conclusion that the Soviet Union possessed a multiseat craft. It would never have entered anyone's mind there that we would send a crew into orbit without the appropriate means of rescue.

It was good that everything turned out all right. But what if it hadn't?

The fortunate success of the risky gamble might have made the Soviet spaceflight planners pause and take the time to build on their experiences. Instead, they pushed on with an even more daring Voskhod mission.

Originally, both the Americans and the Soviets planned to conduct the first-ever extravehicular activity (EVA)—more commonly known as a spacewalk—in a careful, progressive process as part of their gradual space testing. It was known that such an operation outside of the relative protection of a spacecraft would be an extremely hazardous undertaking, and the Soviets even planned to send animals on spacewalks before risking humans. However, when NASA openly announced its plans for a stand-up astronaut EVA from its new Gemini spacecraft sometime in 1965, the Soviets decided to trump their rivals yet again with another spectacular space first. In order to achieve this, the designers would need to adapt what was already a technological dead end, the Voskhod spacecraft, and further delay development of the new-generation Soyuz spacecraft. The timing and method chosen were not ideal, and once again a logical, progressive program was being sacrificed for the sake of propaganda. Yet the achievement would ultimately be no less impressive.

The Voskhod spacecraft had not been designed with EVA in mind, so the designers faced many severe limitations in making such activity possible. For instance, the main hatch could not be opened and closed in space, and unlike the Gemini spacecraft the main cabin could not be depressurized and repressurized. At first glance, it seemed an unachievable task, and the designers and mission planners were understandably concerned at the haste required. It would certainly have made more sense to wait for the next-generation Soyuz spacecraft, which would be even more practical for EVA purposes than its Gemini counterpart. But political pressure did not allow this luxury; somehow, Voskhod would have to be adapted to the task.

If it was not possible to depressurize the main cabin or open the main hatch, the designers concluded that a second chamber would be needed for a spacefarer to enter, seal the chamber from the crew compartment, and then open to the vacuum of space. Though it was possible to create an extra side egress hatch, adding a permanent second chamber to the side of the ball-shaped Voskhod was never an option. The designers knew it would either tear off under aerodynamic stress during launch or cause enormous instability during reentry. The air lock would therefore need to be temporary, an extra "room" created only when needed and discarded soon afterward. The only practical way to do this quickly was to create a thin-walled, inflatable air lock that would unfold after launch like an accordion, then jettison after use. At over eight feet long, their final design was longer than the diameter of the spacecraft's crew cabin. When the side hatch was opened, the only materials standing in the way of a fatally explosive decompression in the spacecraft would be two layers of thin rubber and a fabric covering.

There was no room inside the spacecraft structure to accommodate the pressurized air needed to inflate the air lock. Instead, it was stored in canisters attached to the outside of the spacecraft. There was also no easy way to connect the spacewalker to the spacecraft's life support systems; the cosmonaut would have to rely on a limited oxygen supply contained in a backpack. It was all horrendously risky, and Kamanin was deeply concerned that the flight was being prepared with far too much haste and too many engineering shortcuts. He was right. The flight would probably have been launched even sooner if Khrushchev had not been deposed in the middle of the mission planning.

The EVA spacesuit, at least, was a modification of previous spaceflight equipment and built on experience. Extra protective layers were added to regulate heat, and a liquid coolant layer was also incorporated to prevent overheating. To allow mission controllers to monitor the spacewalk, and for obvious propaganda reasons, a television camera would also be carried on the flight to record and relay the historic events.

It was quickly evident that there would not be room for three cosmonauts on a mission with so much extra equipment. This restriction would actually allow planners to claim yet another space "first," as the Soviet Union could fly the first-ever two-person mission. When the flight concluded, they could then boast of flying two- and three-person crews while the Americans still only had solo flight experience. It would be a close-run thing, however, as America's first two-person mission would launch only five days later.

Unlike the first Voskhod mission, where it had been possible to include nonpilot cosmonauts in the crew, this demanding flight required two top pilots. A group of six cosmonauts was selected to train for the mission in mid-1964, including five from the original 1960 intake. None of the six had flown in space before. Soon, two candidates rose to the top of the list: Pavel Belyayev was chosen to command the mission, and Alexei Leonov was selected to conduct history's first EVA. They knew the risks, but as Belyayev would say, "A program is a program, and I was responsible for its fulfillment."

At thirty-nine years of age Belyayev was the oldest of the first group of cosmonauts, and one of only four already in their thirties. Leonov was a full nine years younger than his mission commander. As well as being the oldest members of the first cosmonaut group, Pavel Belyayev and Vladimir Komarov were the only Air Force Academy graduates. As such, unlike the other cosmonauts, neither was required to attend the Zhukovsky Air Force Engineering Academy. Belyayev had the most flying hours of them all and was also a veteran of World War II. As the highest-ranked military officer in the first selection group, he was made their commander and retained the position until Gagarin was entrusted with it. Leonov, in contrast, was a lieutenant—the lowest-ranking member of the team.

Pavel Ivanovich Belyayev was born on 26 June 1925 in the village of Chelishchevo in the Vologda region of the USSR. Situated between Moscow and St. Petersburg, Vologda is a region of lakes and forests that is very popular with tourists. Ivan Parmenovich Belyayev, Pavel's father, was a physician's assistant who had been disabled in World War I. Pavel was one of six children, and since his mother Agrafina worked on the collective farm he soon learned to look after himself. "After school," he later told journalist Mikhail Vasilyev, "I usually took my shotgun and hunting gear and made for the woods. When I was in fifth grade I was already regarded as a professional hunter." His self-reliance was a useful personal trait: neither of his parents would live long enough to see their son fly in space.

While in sixth grade at school, Pavel's father presented his son with a single-barrel shotgun, and he was soon hunting by himself in the nearby forests. An excellent rifle marksman, Pavel dreamed of earning a living as a hunter when he was older, but at the age of twelve he and his family moved to the Kamensk-Uralsky region in southwestern Siberian Russia. The area was well known for mining and the manufacture of steel and aluminum, so not surprisingly Pavel's first job at the age of sixteen was in a pipe factory as a lathe operator, helping the war effort. He then read in the local newspaper that applications were being accepted to a special air force school in Sverdlovsk. He decided to apply but was told there was no room in the hostel for students from out of town. He later tried to volunteer for service in a fighting ski unit, but was rejected because he was too young. In May 1943, at age eighteen, the impatient youth finally received his call-up papers and was assigned to the Stalin Naval Air School at Yeisk on the Azov Sea in 1944. He graduated from the school in 1945, in time to take part in aerial combat against the Japanese in the Far East. He remained stationed in the Pacific until 1956, when he attended the Air Force Academy. While in the Far East, he met and married Tatyana Prikazchikova, and they soon had two daughters, Irina and Ludmila. On graduation in 1959, he became a squadron leader with the Black Sea fleet air regiment. One year later, he was a cosmonaut.

Both Komarov and Belyayev had medical issues that kept them out of the running for the earliest missions. In Belyayev's case, he double-fractured his left leg just above the ankle during parachute training in the first year. By the time he recovered he was over a year behind the others, and some wondered if he would be allowed to resume cosmonaut training. "There was no guarantee that I would remain in the detachment," Belyayev would later reflect. "Everything depended on the skill of the doctors . . . I am very grateful to them." Luckily, he had impressed the trainers with his ability to withstand very high g-forces, seemingly with no effect. They had also noted his calm demeanor when he quickly extinguished a potentially hazardous electrical fire in the isolation chamber. Belyayev pushed himself hard with a demanding exercise program and therapy regime, proved that his leg had fully healed by resuming parachute training, and was allowed to stay in the cosmonaut team.

In 1963, Belyayev helped to choose and train the next group of cosmonauts. By this time he had become very well respected by his colleagues, although they noted that he was generally very reserved and quiet—unless the subject was aviation. "I find speaking rather difficult," he once said, "and in addition I don't like talking. Perhaps this is because I am not a good talker. . . . Above all, I am a pilot." Gagarin described his colleague as "a man of great willpower and self-control even in the most dangerous situations, with a logical mind capable of profound self-analysis." With his age increasingly against him, Belyayev hoped that his first flight assignment would come soon. In the meantime, he continued to hunt and fish.

The eighth of nine surviving children, Alexei Arkhipovich Leonov was born in the Siberian village of Listvyanka on 30 May 1934 to a miner and electrician named Arkhip and his wife, Yevdokia. As a boy he loved to draw. "I started drawing before I could write," Leonov would later recall, "before I even knew the alphabet . . . long before I began going to school." Unfortunately, as he grew older, his artistic subject matter became increasingly violent: Leonov, like the other boys of his generation, was profoundly affected by the war his country was fighting. "We experienced the hardship, grief and suffering that war leaves in its wake," Leonov wrote. "There were all kinds of shortages. . . . We can never forget this." Like so many youths in the village, his school shoes had holes in their soles, and his schoolbag was an old gasmask carrier. When wounded Soviet soldiers returned home to Siberia, often missing limbs, Leonov would draw them. His work was considered good enough to be put on display at the local hospital where many of the injured were being treated.

Unfortunately, the family not only suffered from external dangers but also from hostile fellow villagers. When Leonov was only three years old, his father was imprisoned for alleged anticommunist activities. The charges were untrue, Leonov maintains, but nevertheless his family was forced to move out of Listvyanka, while their neighbors stole everything they owned, including the clothes they were wearing. It was a tough lesson for such a young child, and the experience left Alexei with a somewhat ambivalent attitude toward authority.

He was a mischievous child—one who "liked to pull at girls' pigtails," he admits. However, his pranks never rose to the level of bullying. In fact, he was more often the target of bullies than the instigator. On one such occasion, he was happily floating in the village pond when some older boys decided to target him. Leonov had never learned to swim and so had tied knots in his wet trouser legs, filled them with air, and was using them as a flotation device. When the boys yanked the trousers away and swam off, Leonov immediately sank. In those frantic few moments, he learned how to swim. "I swung my arms," he later recounted, "and swam ashore." If he hadn't worked it out in a hurry, he might have drowned.

In 1948, as a result of a Soviet government push to entice families to move to the area, Arkhip Leonov's job was relocated to Kaliningrad, on the Baltic Sea coast between Poland and Lithuania. Young Alexei quickly fell in love with the area, and seascapes became a subject he would paint often in later life. "I love the sea," he would later state. "It holds a strong attraction for me." Yet it was a tense few years. The area, with its strong Prussian culture, was being filled with Soviet citizens who had recent war memories and a hatred of anything related to the despised Germans. Despite this, the mixing of very different European heritages may explain how Leonov grew up to be one of the most diplomatic and well-rounded cosmonauts of his era, able to charm communists and capitalists alike, and indeed eventually transform himself from one to the other.

Leonov seriously considered becoming a professional artist for a while, but when one of his older brothers decided to become an air force mechanic, Leonov was intrigued. He later wrote: "I probably would have become an artist, if I had not been captivated by . . . the colors of the eternally magnificent sky. I became a flier to be closer to it." The decision to enter aviation was also in part financial. Leonov had taken the entrance examinations for an art institute, but they did not provide housing for their students; the air force did. In July 1953, he began attending the first of two air force schools, and in January 1955 flew for the first time with an instructor. Just four months later he made his first solo flight. In 1957 he graduated from the Chuguyev Air Force Pilots School in the Ukraine after a demanding course of both academic work and practical flying skills. He had not given up on his artistic ambitions and continued to attend evening classes in the subject while in the air force. He also met a student teacher, Svetlana Pavlovna, at the base where he was stationed, and they married in 1959. That same year, Leonov was given a very prestigious assignment: to serve as a fighter pilot in East Germany, patrolling the borders of the Soviet sphere of influence. Just as he was relocating, however, he also began another selection process—for consideration to be among the first group of cosmonauts. While serving in East Germany, he was informed that he had been selected. One of his few regrets was that he would no longer have the opportunity to visit all of the German art museums he had planned to see.

Leonov became an assistant CapCom during Gagarin's mission and second backup for Bykovsky's *Vostok 5* flight, but his first few months as a cosmonaut had not gone so smoothly. He had performed poorly in his initial centrifuge tests and was almost dismissed from the program. He managed to convince the trainers that he had been suffering from a bad cold at the time of the tests and to his relief was allowed to retake them. But it was never an easy process for him. "The medical barriers were becoming increasingly difficult," Leonov would later recount, "and the doctors increasingly exacting. This meant that our pride was at stake: a true pilot cannot bear the thought of knuckling down to doctors who might ground him at all costs. . . . The doctors are unbendable. They won't accept excuses. . . . To keep one's altitude meant going through fire and water—which we did."

During his first year of training Leonov was involved in a near-fatal accident, which proved to be just the first in a string of car-related incidents. He and his wife were being driven by a third person on the main route from central Moscow to Star City, in icy conditions, when the car skidded off the road and onto the surface of a large frozen pond. The vehicle was too heavy for the ice sheet, and it began to sink. After climbing onto the roof, Leonov was able to rescue his wife and the driver by pulling them out of the freezing water. The pond was subsequently nicknamed "Lake Leonov."

As Gagarin got to know his fellow cosmonauts better, he decided that Leonov was similar in character to Titov in many ways: both men worked fast, using their initiative and subjective thoughts, in contrast to some of their slower, more studious colleagues. And like Gagarin, Leonov was a favorite of Korolev. He could easily have been one of the first cosmonauts to fly, but he had been held back from consideration for early Vostok missions because he was simply too tall for the ejection seat. With his outgoing nature, Leonov was certainly a lively addition to the cosmonaut team, although he was prone to make mistakes out of haste. Gagarin summarized his colleague's character as "strong and impetuous . . . capable of furious activity showing decisiveness and daring." Leonov certainly impressed his parachute instructors, one of whom said that "He stood out from the group of cosmonaut trainees for his strong will, rapid mastery of new skills, and ability to adapt to extraordinary circumstances." With a winning combination of physical fitness and parachute skills, Leonov was a natural to join Belyayev on *Voskhod 2*.

No sooner was Belyayev chosen as commander than his selection came under serious scrutiny. There were concerns about his health, and it was reported that he had performed poorly in altitude chamber tests, even though faulty test equipment may have been to blame. It was also discovered during training that Belyayev had an occasionally irregular heartbeat, a condition very similar to the one that caused NASA to ground Deke Slayton. However, Vladimir Komarov had also been diagnosed with this condition and had flown a problem-free Voskhod mission, so it was decided that this concern should not affect Belyayev's command of *Voskhod 2*. Leonov, in fact, had joined with Komarov to personally appeal Belyayev's case and

The *Voskhod 2* crew discuss their upcoming mission. Mission commander Pavel Belyayev is on the left, while Alexei Leonov, who would soon become the first person to walk in space, is seated on the right. Courtesy Colin Burgess Collection.

keep him on the flight. For his own part, Leonov was given no such leeway—his health needed to be excellent in order to handle what could be a strenuous physical ordeal. To prepare, he spent many hours in the gym every day, as well as running, cycling, and skiing, to keep up his fitness level and agility. It was the most demanding exercise regime any cosmonaut had ever been through.

Leonov began training with the new spacesuit and the other EVA hardware in the summer of 1964. Pavel Popovich's space-tested Vostok spacecraft, specially adapted with the new air lock, was used for the tests. When Leonov first saw it, the air lock looked to him like a "strange pipe." When he was eventually able to train in the new cabin design, he continually bumped his head on a panel in the cramped layout. He could only trust that its designers knew what they were doing. Early in the mission training, Nikolai Kamanin had noted with interest that Yevgeny Khrunov and Leonov actually made a better team than Belyayev and Leonov. With lingering doubts about Belyayev's health there were even thoughts of making a change in command for the mission. However, Khrunov was also training impressively in EVA techniques, and it was deemed essential that he remain as Leonov's backup for the task. By this time, the Belyayev-Leonov team had bonded well.

"We understand each other remarkably well," Belyayev would later comment on their partnership. "At times it seems as if we read each other's thoughts." Leonov was in complete agreement. "Pavel was a man I sought to emulate. I took an immediate liking to him, and I learned a great deal from him."

Leonov soon progressed to training using a full-sized wooden air lock mockup in a Tupolev airplane flying zero-G parabolas. Over and over again he practiced awkward maneuvers such as putting on his spacesuit and exiting the spacecraft. "This was our flying laboratory for a while," Leonov explained. "We had to learn how to work in weightlessness. At first, everything seemed difficult." The most efficient way to enter the tiny air lock was found, as was the optimum amount of push needed to leave it altogether. Leonov tried his best to learn to move smoothly, without unwanted spinning, but there was only so much that could be practiced during the short parabolas. As Leonov later told the authors, there were aspects of the spacesuit and spacecraft design that could never be satisfactorily tested before the flight.

As part of his training, Leonov was also kept in an isolation chamber for an entire month, then immediately upon his removal made to fly in a jet and perform a parachute jump. This unusual regime was devised to help him adapt to floating into the emptiness of space after being strapped in the confines of a spacecraft. Such conditioning seems overdramatic now, but at the time no one knew how a human would handle floating freely above the Earth. Leonov didn't mind—he'd been allowed to sketch while in isolation.

Like the crew, the spacecraft also needed to be readied, and it was decided that an unmanned test of the air lock in orbit would be prudent for such a risky mission. It would also provide an opportunity to test the new spacesuit. On 22 February 1965, *Kosmos 57* was launched. The bland mission name conveniently hid its true purpose from the outside world. As mission controllers watched on remote television images, both the air lock and the empty spacesuit pressurized as planned, and the outer hatch opened and closed. The bizarre spacecraft worked—at first. As the designers celebrated the success of the tests, however, all telemetry from the spacecraft suddenly vanished. Tracking radar systems picked up dozens of signals: somehow, the spacecraft had activated its self-destruct system and exploded. The cause was eventually traced to faulty signals sent from the ground stations.

It was an embarrassment to the mission planners, and it also posed a problem. Though the air lock had worked in space, the final mission objective—determining how the spacecraft would reenter with an air-lock hatch on its side—had still not been tested. Despite this, Brezhnev pressured Korolev to go ahead with the Voskhod flight without further unmanned tests. But Korolev stood his ground: another test was indispensable. A Zenit reconnaissance satellite that was due to launch was hastily fitted with the equipment, and its successful reentry cleared the way for a manned mission.

Korolev, still unhappy with the pressure being exerted to hasten the program, did not want the crew to go into space with any misconceptions. A few hours before launch he warned the two cosmonauts that it was never possible to test everything on the ground and that they should make their own decisions based on what was safe to perform in space. Leonov spent his last free moments "in an anxious state," repeatedly going over the plans for the flight in his head.

Launch day—18 March 1965—was a cold one, and the two cosmonauts made the journey to their waiting rocket in wind and snow. "The steppe was completely white as far as the eye could see," Leonov would later recount. It was not Leonov's first trip to the pad on a launch day, but normally he was on the bus tape-recording the conversations and songs. This time, although he was singing as heartily as ever, he was the one in a spacesuit and the center of attention. Belyayev may have been the mission commander, but all eyes were on Leonov. He would be fulfilling the purpose of the mission, and thus making history. It was not easy for the crew to get into their spacecraft: they had to carefully climb through the compressed, folded air lock to reach their couches. But at last, all was ready; the mission could begin.

The launch was one of the most beautiful anyone could recall. As the rockets ignited, flames vaporized the surrounding snow, creating a glittering glow of colored gas. Inside the spacecraft, things were not quite as pretty. During ascent, a number of alarms rang out in the crew cabin. Fortunately, the spacecraft made it into orbit in good condition, and the flight was allowed to continue. The EVA was planned to occur as soon as possible after launch. That way, it would take place over the best ground stations, and the untried technology would have less time to malfunction. As soon as the spacecraft was in orbit the air lock was inflated, and Leonov began putting on his backpack. Belyayev was also dressed in a full pressure suit: they weren't taking any extra chances. Initially, Leonov did not adapt very well to weightlessness, something that could have delayed the EVA attempt. "Leonov seems not to have been able to adjust to weightlessness immediately, and was under some strain," Gagarin wrote when discussing Leonov's heart rate. Apparently, he soon recovered, and the EVA preparations continued.

"Many people have asked me what thoughts I had during those minutes when I was about to step outside the craft," Leonov recounted in 1971. "Perhaps, I had none apart from the most needed ones . . . not to allow even one hasty step." Once Leonov had floated headfirst into the air lock and attached his seventeen-and-a-half-foot safety tether, Belyayev closed the inner hatch behind his companion. As Leonov later recalled, he was now alone, with only the thin tunnel walls to protect him. "At once I adjusted the necessary pressure in the spacesuit. Once again I tested the tightness of the pressure helmet and the light filter, then checked the supply of oxygen." Having carefully examined his suit for any leaks, all seemed in good order. He was ready. The pressure in the tunnel was reduced to zero, and Leonov prepared himself for the giant leap.

As *Voskhod 2* swept around the Earth on the second orbit of the mission and Belyayev maintained contact with his comrade by radio, Leonov opened the outer hatch. First, he poked his head out of the opening, and then his shoulders. Looking down at the Earth, Leonov was initially surprised at first by how flat it looked. Only when he looked to the far horizon could he see the curvature of the planet, something he attributes to the spacecraft's low orbit.

As precious seconds passed, Leonov knew he had now reached the point where he had to totally trust the hardware and let go of the spacecraft. Exiting the air lock was surprisingly easy—in fact, Leonov says that giving himself only the tiniest of pushes he floated out of the hatch "like a cork from a bottle." This maneuver left him with the uncanny feeling that he had actually kicked the spaceship away from under him. Instinctively, he flapped his arms as he floated away. "I threw open my arms like wings," he would later say of the moment. He felt as if he was on a children's swing, arcing up into the air. As he later told Yuri Gagarin, "The line which was my connection to the spacecraft . . . slowly played itself out to its full length. The slight jolt had caused an insignificant change in the direction of the craft."

At first, the sunlight blinded Leonov, and he squinted as he adjusted to the bright conditions. He would later compare the brightness to a welder's arc. He felt the sun "bright and very hot" on his lips where the helmet filter did not reach and soon felt himself begin to sweat. He quickly recognized several land features on the Earth below, although most of his EVA would take place over the Pacific Ocean. He was rotating, a movement that he could only partially arrest by twisting on his safety tether. "I pulled it quite energetically," Leonov recalls. To cinematographer Vladimir Suvorov, watching television images live in the control center, Leonov's movements looked "clumsy, angular, and even somewhat convulsive. It seems that the spacesuit impedes his actions."

Leonov realized with pleasure that the fears of the psychologists—that humans would recoil in terror at the abyss of space—were groundless. "I not only failed to sense any such barrier," he would recall, "but even forgot there would be one. There was no time to think about it. I simply forgot myself, so enamored was I of the unusual view around me. . . . In all dimensions, the universe is boundless." Instead of terror, Leonov felt an "indescribable freedom."

Floating free in space was never a concern for Leonov, neither psychologically nor in terms of space sickness. But he quickly learned that it was not easy to control his movements in weightlessness. "Some people want to explain 'swimming' in space with swimming in water. In water you feel something supporting you; in space there is no such feeling. It seems as if you are simply flying near the spacecraft." In fact, he sometimes lost track of his orientation, meaning "I had to get my bearings from the moving ship or the sun . . . when it was fully extended, the tether was another good guide."

Meanwhile, far beneath him, the Earth was slowly rotating—a beautiful and transfixing sight. "I looked at it and was completely charmed by it. The Earth startled me with the richness of its colors . . . it was like flying over an enormous colored map." The effect was enhanced by the deep silence Leonov was experiencing. No sounds other than his breathing and heartbeat intruded on his euphoria, as the magnificent views passed below. As he slowly twisted and turned, Leonov was also impressed by the view of the exterior of the Voskhod spacecraft. "If you could only see how fantastic it looks in space! The portholes look like extremely big eyes, and the antennae like gigantic tentacles . . . like some fantastic creature out of science fiction films . . . bathed in the stream of sunlight." The spacecraft shone with a golden glow that Leonov has tried for years to capture in oils but has never been able to reproduce to his satisfaction.

As part of his work program, Leonov had attached a movie camera to the end of the air lock in order to record his spacewalk. It was scheduled for retrieval at the end of the EVA. He also planned to take still photos of Voskhod using a camera mounted on the chest of his spacesuit. But when he reached for the shutter switch, attached to his leg, he found his suit was so inflated he could not reach it. Leonov tested the suit a little more. He found he could bend his arms and legs a little, but it was hard work, as the suit would fight him to return to its ballooned shape. His hands and feet would also not extend all the way to the tips of the spacesuit when it was in this condition: he was floating within the suit, rather than wearing it, and he did not like the lack of control. "I got pretty tired," he would recall. "It was uncomfortable to work in; my movements were restricted. The gloves were not as elegant and comfortable as those we normally wear . . . the pressure suit resisted changes in the form of my body, flexing of arms and legs, so it required an effort to work."

As he floated without any practical means of controlling his direction, Leonov bounced off the Voskhod spacecraft five times as they orbited the Earth together. "I had to fend off the spacecraft with my hands," he remembered. "I returned headfirst with my hands outstretched to protect my head from hitting. . . . A thought flitted across my mind, that I might strike the craft with my pressure helmet's visor." Inside the spacecraft, Belyayev felt small jolts as Leonov came into contact with the exterior, and he used his attitude thrusters to hold Voskhod steady. He also kept up a running commentary with the ground, describing the effect of Leonov's activities. "I clearly felt everything that Alexei was doing to the ship," Belyayev would recount after the flight. "I felt it when he pushed, and heard his boot ring as it scraped against the side of the spaceship. And even when he ran his hand over the side of the craft, I heard a kind of rustle."

At one point, Leonov recalled, "my body went into a complicated twist . . . there was no way to stop my rotation. The line became twisted and my velocity dropped." He was learning what others after him would come to learn: that it was hard to maintain a practical orientation in space or control one's movements. "A flexible tether," he would add, "is to some extent an element of support—but a minimal one." After ten minutes of floating, the time Leonov could spend enjoying the view had to come to an end. "To my great regret, the time allotted passed very quickly," Leonov would state in his postflight report. At the conclusion of his spacewalk he carefully removed the movie camera and prepared to enter the air lock feet first. As he began to squeeze back in, he came to a sudden stop. He could no longer fit inside; his spacesuit had inflated to the point where he was trapped outside the spacecraft. "The way we had done it in our training sessions . . . didn't work at all," he would later say of the life-threatening incident. As he continued to struggle and squirm into the airlock, his spacesuit began to overheat. "It was getting hot," he would recall. "I could feel a stream of sweat running between my shoulder blades and over my eyes. My hands became wet, my pulse quickened. . . . It was quite a physical effort." If he could not find a way to get back in, history's first spacewalk would end in his death.

He decided to try going in headfirst, thinking that it might be easier that way. As he jammed himself into the small air lock and tried to squirm in further, his body began to heat up to the point where he was in danger of passing out. His eyes were stinging with sweat, his helmet visor was fogged with perspiration, and he could no longer see clearly. It was becoming increasingly obvious to him that he was not going to fit into the narrow tube, but he subdued the panic that others might have felt and allowed his knowledge of the suit and spacecraft to help him at this critical time. He had to make a risky decision, one with no guarantee of success. In order to reduce the size of the swollen spacesuit, he would begin to bleed the pressure from it, using a regulator valve. There was no other solution, and he reached for the valve.

The suit pressure dropped, and soon Leonov reached the limit of what he knew was considered safe. Once again he tried to squeeze his body into the tight space, and once again he failed. He had little choice—he lowered the pressure still further, down into the danger zone. He knew that in doing so he could have been affected by "the bends"—a dangerous condition that sometimes kills even the most experienced deep-sea divers. He could also have starved himself of oxygen, especially as his heart was already beating at a shockingly fast rate. Though Leonov's actions could have incapacitated or killed him, the alternative was to stay outside the spacecraft forever. Luckily for Leonov, it worked. With the last of his strength, he was finally able to squeeze into the air lock and close the hatch behind him. "I really had to strain and put all my efforts in order to complete this return to the spacecraft," he would later relate. "I floated in the lock exhausted, drenched with perspiration." In fact, he had sweated so heavily that he could feel the fluid sloshing about inside his suit.

His trials were still not over. Leonov was supposed to be turned the other way in the air lock, and so he squeezed himself around to a feetfirst position with great effort. Safety regulations stressed that he was not supposed to open his helmet visor until he was in the spacecraft cabin, but he had little choice if he was to remain conscious. Now close to total exhaustion he repressurized the air lock, opened his helmet, and floated passively for a moment, catching his breath and recovering from the ordeal. "My farewell with space was a bit drawn-out," Leonov would later joke. Only once he had recovered a little did he open the inner hatch and enter the relative safety of the crew cabin. He had been outside the spacecraft cabin less than twenty-four minutes.

Belyayev would later remark that his exhausted crewmate looked like a man who had just been reborn, as if he had come back from a new world. He clapped him on the shoulder and shouted, "Great show! Good for you!" The round-faced, perspiring Leonov, bright red from overexertion, probably looked more like a newborn baby than he cared to admit. "Sitting in my couch I felt streams of perspiration on my forehead and cheeks, which trickled into my eyes. The only thing I wanted to do was wipe the sweat from my face as quickly as possible."

At first, Leonov found it hard to clear his stinging eyes and had to raid the first-aid kit for tissues in order to wipe them. Soon after, as required, he had the task of writing down his immediate impressions of his EVA. "For writing down all that I saw and what work I had carried out in that twenty-four minutes, I needed about an hour and a half," Leonov would state. "It was better to be back inside! The craft is like your own house, and in the cabin was my friend Pavel. I think that it is a bit early to compare outer space with a place for an entertaining stroll . . . without the many months of all-around training, I would never have managed to cope."

After the hazards of the spacewalk, Leonov and Belyayev hoped that nothing else could go awry. But they were wrong. After the air lock had been discarded the spacecraft hatch did not completely seal, and the spacecraft pumped oxygen into the cabin to compensate for the leak. Maintaining the pressure was eating through their oxygen reserves at an alarming rate, and the hatch ejection had also made the spacecraft roll uncomfortably. In the purer oxygen atmosphere, the cosmonauts worried that any spark could start a catastrophic fire. "If I were to be asked whether all this was easy or difficult," Belyayev would say later of the flight, "my reply would be that this was no easy job. I think that Alexei would confirm that."

Luckily, the flight was only planned to last just over one day, and the cosmonauts could alter the cabin oxygen levels somewhat during this time. In spite of these tense conditions, Belyayev and Leonov were able to relax enough to eat a meal. They were even able to sleep a little, although with the dangerous cabin pressure it was hard to completely let their guard down. "We violated the program as far as sleeping was conducted," Belyayev would later admit. With the spacewalk carried out early in the flight, there were few other assigned tasks to perform until reentry, planned for the seventeenth orbit. One of the experiments they did perform, to check their color perception in weightlessness, was of particular interest to the artistic Leonov. Using a special chart, they both found that they had lost about a quarter of their ability to determine color differences. In the meantime, Leonov was also becoming irritated by the attitude of the mission controllers. "They were dismayed that I had made the decision on my own to release pressure from the suit," he would explain. "We were not supposed to make any decisions without consulting ground control." Quite rightfully, he had been far more interested in saving his own life than following the rules.

The dangers of the mission were not yet over. The automatic guidance system was supposed to align the spacecraft for reentry using solar orientation, but it failed, and the planned automatic reentry burn did not take place. "Voskhod 2's movements did not tally with those scheduled," Leonov would write. "We became worried." Once the information was radioed to him Belyayev began preparing the spacecraft for a manual reentry on the following orbit. He later recalled that it was a complicated task. "As a fighter pilot, I had made no small number of landings in modern high-speed planes. However, the speed of our spaceship could not be compared . . . piloting a plane and controlling a spacecraft are not the same. I was instructed to carry out a manual orientation of the craft and ignite the retrorocket engine at the calculated time." The two cosmonauts soon found that it was almost impossible to reach the necessary controls while strapped into their couches wearing their spacesuits. The only way they could accomplish this in time was to leave their seats and use the controls to carry out the manual orientation, then strap themselves back in before the firing to ensure that the spacecraft's center of gravity was correct. This delayed the moment at which they were supposed to manually fire the reentry engine by forty-six seconds. They were going to come down way off course, in an inhospitable part of the Soviet Union. "We somewhat overflew our mark," Belyayev would later admit. In addition, Leonov failed to make it back completely to his seat in time: the spacecraft was not centered when the braking engines fired.

Voskhod 2's reentry was anything but smooth. The instrument module failed to separate from the cosmonauts' cabin, eventually breaking free only when the fires of reentry burned the connections away, and the spacecraft had swung around alarmingly. Voskhod was therefore unable to follow a reentry path in which the spacecraft body could provide lift. Instead, it plunged deeper and deeper into the atmosphere on a steep descent path. "We were like a giant meteorite with a huge tail," Leonov would later say of their return through the atmosphere. "A fiery storm was raging outside the portholes. Drops of metal were falling on the glass panes, like drops of rain on a car windscreen." It was a dramatic scene that Leonov would later reproduce to great effect in a vivid oil painting. However, there was no time for casual observation since they were being subjected to a crushing 10-g force, and blood vessels burst in their eyes as they struggled to remain conscious.

The spacecraft eventually came down in a snow-covered Siberian forest, over a hundred miles north of the city of Perm. It had been a truly hazardous reentry, but survivable—and only because the pilots had taken over. Leonov later wrote that this mission had "shown convincingly that whatever the degree of automation in a spacecraft may be, humans retain the directing and organizing role in controlling the craft; . . . no cybernetic device can replace human intellect and intuition."

In the final moments of descent their parachutes caught on some treetops. "The spaceship came down between two fir trees," Belyayev would explain.

"It landed in deep snow and began to settle under the burden of its own weight." Despite being well off course and miles from any assistance, some good fortune had found them. If the sensor on the bottom of the spacecraft had hit a tree, it might have been fooled into thinking the spacecraft was touching the ground. The retrorockets would have fired prematurely and without effect, and the cosmonauts would most likely have been severely injured in a crushingly hard landing. As it was, they were wedged against a tree, and the spacecraft hatch was blocked. It would loosen, but not slide away. Using their combined force, the two cosmonauts rocked the hatch back and forth until finally it dropped off into the deep snow. They could now get out of the spacecraft, but without the hatch the interior would soon be as cold as the surrounding forest.

"The snow was thick," Belyayev later recalled of his first look around. "Everywhere there were huge trees." The two cosmonauts had survived one of the riskiest missions in the short history of manned space flight. Now they had to survive something that had confronted humans for thousands of years: the challenge of staying alive outdoors in treacherous conditions, dozens of miles from safety. They had no idea whether mission control knew where they were, even approximately. Fortunately, both men were hunters and skiers, and Leonov's Siberian childhood had also taught him a lot about how to survive in such places.

Their charred spacecraft, still hot from reentry, gradually melted the snow beneath it and finally settled onto the forest floor. The snow was literally up to the cosmonauts' chins when they emerged, which meant that any prospect of lighting a fire was remote. A noticeable amount of perspiration had built up in Belyayev's suit, but for Leonov the situation was even worse. The buildup of sweat from his EVA had pooled uncomfortably in the lower legs of the spacesuit, reaching up to his knees, and was growing rapidly colder. The cosmonauts knew they would freeze to death if they did not get rid of this moisture. In order to survive, they removed all of their clothing in the freezing snow, poured out all of the moisture they could from the legs of their suits, then wrung out their clothes. They also attempted to pull the parachute down from the trees for the small amount of warmth it might offer.

They were lucky. A few hours after their landing, a civilian aircraft spotted the spacecraft and passed details of their location to the authorities. The landing area was rugged and covered with trees, so there was no safe place for a rescue team to put down nearby. The nearest place they could land was five and a half miles away, but the rescuers would still have to make their way into the forest and cut down some trees to allow the helicopter to land. It wasn't going to happen that day. Belyayev and Leonov had to spend a miserable night in the cold forest in subzero temperatures, taking turns to stay awake. "The cold was intense," Belyayev would later recall; they almost froze to death. To stay warm, they drank some vodka they had smuggled on board the spacecraft.

The next day the specialist rescue team of parachutists and doctors finally reached them. They told the cosmonauts that a clearing was still being made to land a helicopter closer by. They were to stay put for the time being. The cosmonauts spent a second night in the forest with the rescue crew, this time in the relative comfort of tents, before working their way to the rescue site on skis. It was many more days before the spacecraft itself was recovered. Belyayev later joked that it was the first spacecraft ever recovered with bear claw marks on it.

As usual, details of the mission's more dangerous moments would be kept quiet. The flight had delivered the required propaganda victory, and Tom Stafford recalls the effect it had on the astronauts, just days before the first Gemini flight:

I remember Gus, John, Wally, and I watched in the crew quarters just a few days before launch, hearing that the Russians had upstaged us again—they had just a little clip of him floating around and said everything was fine. It was only later on, as I got to know Alexei, that we found out he nearly got killed on that mission. They had a lot of malfunctions, landed way out in the boondocks in the Ural Mountains. It was a day before they could even get a helicopter in to drop skis to them to get out of there. But, they said at the time, everything was fine!

Naturally, the Soviets made the most of the propaganda victory the flight gave them, and the two cosmonauts and their wives were quickly sent on a ten-day tour of East Germany. Yet all did not go smoothly. Symbols of Soviet pride could also be targets of anti-Soviet feeling. During a detour into the French-controlled sector of West Berlin, the cosmonauts gave a talk about their spaceflight to a theater audience. Hundreds of demonstrators gathered outside and booed their arrival, holding up placards listing the number of people who had died trying to escape over the Berlin Wall. When the cosmonauts later tried to leave in a limousine, demonstrators splattered red paint on all four of the side windows and the windscreen. The paint could not be cleared off sufficiently for the car to be driven, so the shaken VIP guests were bundled into a modest gray sedan, uncomfortably squeezed between security officers, and raced at high speed with a police escort back to the Eastern Bloc. On this occasion, the propaganda attempt had backfired.

Things were kept under tighter control within the Soviet Union. During the joyous celebrations in Red Square that followed the flight Leonov made a speech from atop Lenin's mausoleum. In it, he said, "The time is drawing close when people will move on from orbital flight . . . and will go to the moon." It would soon become a personal goal of his and one that would consume his career for years to come. That year, he would tell journalist V. Mikhailov: "I think I know the moon. I have painted several canvases of its surface as I see it. I am attracted by the moon, for this is the closest neighbor of our Earth, which humans will visit in the nearest future. I dream of it being accomplished by the men of our detachment. If I am very lucky, I will get this assignment." Belyayev was similarly passionate about the future. "You know, at the end of my flight, I felt that this was not the end. I knew that I would fly again." Both Belyayev and Leonov would be thwarted in their ambitions.

Other Voskhod missions were planned after *Voskhod 2*, including some intended to last up to five days and perform three EVAs. But arguments between Korolev and the air force over these missions' details forced the plans to be postponed in early 1966. When Korolev died that year, the plans became even less focused. Eventually, it was decided that efforts should finally be redirected to the new Soyuz generation of spacecraft. The Vostok design had been stretched as far as it safely could, and beyond. It was time to assign it only for unmanned missions and complete the designs for the next generation of spacecraft. This would take time, however. More than two years would pass before the next Soviet spacefarer was launched. Tragically, it would be three and a half years before a cosmonaut would return alive from a space mission. *Voskhod 2* was the end of a long run of risky but incredible successes for the Soviet manned space program. The hiatus proved to be a frustrating time for the cosmonaut team, and one in which some of the team's senior members took a politically risky step.

In October 1965, Belyayev and Leonov joined Gagarin, Titov, Nikolayev, and Bykovsky in composing a letter to Soviet premier Leonid Brezhnev. In it, they wrote with surprising honesty about the future of the space program. They began by listing the many firsts of the Soviet space program, then added, "In the past year, however, the situation has changed, The U.S.A. have not only caught up with us, but even surpassed us in certain areas . . . this lagging behind of our homeland in space exploration is especially objectionable to us, the cosmonauts." The letter went on to list lack of funding as one of the reasons, but then stated what the cosmonauts saw as the main reason. "Unfortunately, in our country there are many defects in planning, organization, and management of this work. How can one speak about serious planning of space research if we do not have any plans for cosmonauts' flights? The month of October is coming to an end, there is a little time left before the end of 1965, but no one in the Soviet Union knows whether there will be a manned space flight this year, the task for the flight, or the duration." The cosmonauts called for a clear, focused national plan for space flights. Whether they received a reply from Brezhnev is unknown. What is clear is that the problems they listed were not fixed.

Pavel Belyayev briefly trained for the first manned Soyuz flights and in 1966 was named as one of seven cosmonauts in line to fly to the proposed Almaz military space station. Though he continued to train as a possible Almaz commander as late as 1969, continuing health concerns forced Belyayev into reassignment in an administrative position, effectively grounding him. He became chief of staff at the Cosmonaut Training Center in 1967 and as part of his duties helped select future cosmonaut groups. When he met astronauts Dave Scott and Mike Collins that year and told them he hoped to make a flight around the moon in the near future, he must have known it was wishful thinking at best.

It seems that Belyayev did not overly impress his military superiors in the performance of his administrative duties. Kamanin expected more military discipline from him and did not think he was ready to assume a position of greater responsibility. Kamanin thought that in the meantime Belyayev would be better suited to representing the cosmonaut corps on one of the many overseas visits they were being asked to undertake. He did not know it at the time, but Belyayev was not well. He was hiding a painful ailment that would officially ground him when it was finally disclosed in the spring of 1968. Understandably, Belyayev was distrustful of the cosmonaut center doctors. After all, they had nearly cost him his place on the *Voskhod 2* mission. So when his stomach began troubling him, he did not tell them at first. But finally it was too much to ignore, and he was hospitalized in December 1969. It turned out to be stomach ulcers, which required an immediate and extensive operation. According to Kamanin, two-thirds of Belyayev's stomach and part of his intestine needed to be removed.

Belyayev underwent surgery, but would never fully recover from the operation. His condition deteriorated alarmingly, and tests revealed he had contracted peritonitis, an inflammation of the abdominal wall. There was little the doctors could do, and he died in hospital on 10 January 1970, the first spacefarer to die of natural causes. With the deaths of Gagarin and Belyayev, Leonov had lost his two closest friends.

The organizers of the state funeral faced a dilemma—should Belyayev be given the same state funeral with a burial in the Kremlin wall that Gagarin had received? The Kremlin hierarchy eventually decided that his funeral should be more muted, and Belyayev's body was interred at the Novodevichye Cemetery in Moscow, a place reserved mostly for Russia's elite. Not being buried in the Kremlin wall allowed him to have a far more elaborate grave marker: it includes a life-size statue of the cosmonaut dressed in his spacesuit, holding his spacesuit helmet. Kamanin arranged for Belyayev's wife and children to receive a generous pension, an apartment, and permanent access to Star City's healthcare facilities. Tatyana Belyayeva eventually began working at Star City herself, as hostess of the center's Cosmonaut Museum. She still lives there today, and their children live close by.

Astronaut Michael Collins later revisited his memory of the man in his book *Carrying the Fire*. "In Belyayev we found a kindred spirit," he would write. "He exuded not only good humor but also an air of quiet competence. . . . I liked him, and I would have flown with him."

After his flight, Leonov was placed in charge of training the early Soyuz crews in EVA techniques. But bigger plans were being made for him. After the *Voskhod 2* mission, Gagarin had said that "One can now entrust Leonov with a far more complex task, and I am sure he will be equal to it." Leonov was indeed soon given a new task, and it would prove to be the most difficult and complex the Soviets' manned spaceflight program had ever attempted. In 1966, he was chosen as one of the future crew commanders in the group training for missions to the moon. Korolev had once said of Leonov, "I believe this man deserves the greatest trust." Now he was being assigned a role that could have made him the most famous and accomplished spacefarer of all time.

Both Leonov and Popovich were serious contenders for the first Soviet moon flight. In 1967 Leonov undertook splashdown training, as there was the possibility of a water landing at the end of a lunar mission. By then, the focus was less on making the first lunar landing than on beating the Americans with the first circumlunar flight. At this point, Leonov believed that he was the favorite to make the first flight. In early 1968 he trained intensively on a simulator designed to test the spacecraft reentry profiles when returning from the moon. That year he also graduated from the demanding courses of the Zhukovksy Air Force Engineering Academy, which meant he was far more technically proficient than in his early days as a cosmonaut. But this proficiency would have some unintended side effects.

As Kamanin noted, Leonov and the eleven other cosmonauts then training for lunar missions were perhaps getting too familiar with the L1 lunar spacecraft and the Proton rocket that would launch it. That is, they had learned that both had many weaknesses, particularly the L1 spacecraft, which was not a very appealing flying machine. Extremely cramped, it was about half the size of the Soyuz spacecraft. The more the cosmonauts learned about the hardware, the less they liked it. After one unmanned launch attempt, during which the rocket shut down prematurely and the abort system fired the spacecraft cabin to safety, Leonov insisted on journeying immediately to the touchdown site. He wanted to see with his own eyes whether such a launch abort would be survivable.

As further unmanned tests of the L1-Proton combination were conducted and numerous problems remained unremedied, the schedule fell more and more behind. The Russians' first successful unmanned test flight around the moon did not come until mid-September 1968, meaning there would barely be time to slip in a manned flight before an American lunar mission. Leonov and cosmonaut Oleg Makarov were probably the prime crew for an attempt, although Mishin and Kamanin preferred Bykovsky for the role of commander.

By this time, Leonov's attitude was causing a great deal of concern. Kamanin had noted that the cosmonaut sometimes seemed more preoccupied with a forthcoming exhibition of his art than with his cosmonaut duties. Kamanin concluded that Leonov was suitable for a position as a cosmonaut detachment deputy commander but still lacked the discipline to be a command officer. In June 1968 his doubts were confirmed when Leonov crashed his Volga car in a late-night accident in Moscow, with a group of Italian visitors riding in the back. Leonov was supposed to be off the streets and in bed by 11:00 p.m. He was also supposed to be using a chauffeur and only meeting with foreigners when authorized. To have crashed his car with an unauthorized group of passengers was seen as inexcusable, and Leonov was severely reprimanded. It was also noted that this was his second car accident in just a few months. The following month, Leonov found himself passed over in the selection process for commander of the cosmonaut detachment commander, a position he had coveted. Unhappy with the decision, he openly disputed it and continued to lobby for the job, even threatening to leave the cosmonaut corps if he had no chance of promotion. He was eventually informed that he might receive a promotion in the future—but only if he made no more mistakes like the car accident.

In the meantime, Leonov also actively campaigned to keep himself in the forefront for any circumlunar attempt and was given the task of commanding the lunar cosmonaut group. Two further unmanned test missions were planned and then Leonov hoped to be next, on a flight planned for early 1969. He'd even come up with an evocative name for his spacecraft—Rodina, loosely meaning "Motherland," a word with stirring overtones in Russian culture.

As the lunar training progressed at a frantic pace, Mishin and Kamanin finally relented. On 28 September they decided that Leonov was indeed their prime choice for a lunar mission. Yet, in a supreme irony, caused by Leonov's own carelessness, the cosmonaut had another car crash that very evening. While driving near the Moscow suburb of Shchelkovsky, he hit a bus with his Volga. Kamanin was furious at his sloppiness, and told Leonov to take three days off work and seriously consider his future as a cosmonaut. Three car accidents in four months was unacceptable, and Kamanin wondered if he would be as sloppy when piloting a spacecraft. Leonov was allowed to keep his position as the prime pilot for a lunar mission, but was not trusted enough to drive a car—he was banned from driving for six months.

The first unmanned test flight of the circumlunar spacecraft was launched successfully. It flew around the moon and made it back to Earth, but the spacecraft cabin had depressurized in flight. This caused the parachute system to deploy early, resulting in the spacecraft smashing into the ground at great speed. The failure was traced to a minor engineering matter—a small rubber gasket had failed. Unfortunately, it was enough to put the manned LI program on hold. Leonov had just lost his chance to be the first person to fly to the moon. He was understandably furious. He felt the problem was easily fixable and was quite prepared to fly on the next flight, using the same hardware. But it was not to be. When *Apollo 8* made a successful flight to the moon on behalf of America, Kamanin wrote in his diary that "the holiday is darkened with the realization of lost opportunities and with sadness that the men flying to the moon today are not named Valery Bykovsky, Pavel Popovich, or Alexei Leonov." The team was so demoralized that they were given a three-day break from training. In his memoirs, Leonov wrote that "everything was slipping through my fingers. I could see my dreams going up in smoke."

But Leonov's chances of flying to the moon were not over. Despite the setbacks, the Soviets had another possibility: the enormous NI rocket. Originally designed to take a manned spacecraft to Mars, it was almost ready for its first test flight. It had the biggest thrust on liftoff of any booster ever designed, before or since, including America's Saturn V rocket, which could muster less than three-quarters of the NI's overall thrust. It was an ambitious monster rocket that incorporated some ingenious engineering designs, which in some cases have still to be bettered. It had been in development for a decade, and on 21 February 1969 it would get its first chance to fly. If it worked, the Soviets planned to use it for a manned lunar landing mission. The rocket, however, had never been test-fired on the ground. In a risky decision driven mostly by economics, designers were hoping to learn how well the rocket worked from test flights alone.

As Leonov eagerly watched from a safe viewing area, the first flight of the N1 rocket began well enough. With a thunderous roar it eased off the launch pad and headed skyward. It may have looked impressive, but the flight was doomed; within seconds of launch the control system mistakenly began shutting down the first-stage engines. The enormous rocket reached a height of sixteen miles before arcing back toward the ground. The engineers had hoped for a better first flight. However, problems were not unexpected, and so they did their best to learn from them and prepare for a second test.

Training for the possible lunar mission continued for Alexei Leonov with a small team of cosmonauts now under the command of Valery Bykovksy. Leonov, Bykovksy, and cosmonaut Yevgeny Khrunov were the only three training specifically for a landing, and they would sometimes stand at the very top of the N1 tower in order to gain an appreciation of how it felt when the massive rocket swayed on the launch pad. The spacecraft intended to top this monster booster had been designed in an incredibly costly parallel program to the circumlunar efforts, and the hardware was only now beginning to come together. While the Americans had been racing ahead in their unmanned and manned testing of their lunar landing hardware, the Soviets had been trying to develop five different types of manned spacecraft at once. They only had the slimmest of chances of catching up in a lunar race, and Leonov had already decided the chances of beating the Americans were over. But still he trained. It was just possible, if all the testing had gone exactly according to plan from that moment on, that he would have made such a landing at the very end of 1970.

Part of Leonov's lunar mission training included using a specially modified Mi-9 helicopter, which he repeatedly landed with the engines cut. It was an extremely difficult maneuver, but one that he mastered. He also used a special simulator that practiced touchdowns on an artificial lunar surface, giving him experience in quickly selecting a safe landing site. During parabolic flights he even practiced moving in a spacesuit in the one-sixth gravity field of the moon and supplemented this experience by walking in a spacesuit on a simulated lunar terrain.

If the program had been successful, Leonov would most probably have been the first cosmonaut on the moon. During a trip to Hungary in 1966, he made no secret of his country's lunar ambitions. "Cosmonauts are preparing for such a journey," he told the press. "I would very much like it if a Soviet man went to the moon first. . . . I believe we shall soon witness a person landing on the moon." When Leonov made similar comments to Japanese reporters in June 1969, he got into trouble once again with his superiors. On this occasion, he made the mistake of saying that a manned lunar spacecraft was in preparation and moon rocks would be returned to Earth by early 1970. Publicly giving a deadline created a huge stir in the foreign press, and Kremlin officials angrily called Kamanin and told him to keep his cosmonauts in line. Yet somehow Leonov still kept his position and continued to train for a lunar landing mission.

The second launch of the N1 was set for 3 July 1969, just weeks before the Apollo 11 mission. It was a night launch, and Leonov and Bykovksy were again at the viewing stand a few miles away to witness the event. Once again the gigantic rocket began to lift off—then violently exploded. The ground shook and onlookers were showered with hot metal as a ferocious blast wave swept past them. Bykovsky would later say that the explosion shattered every window for fourteen miles around. The blast was later estimated as being a sixtieth of the strength of the atom bomb that flattened Hiroshima, and the devastation to the launch pad was total. No people were injured, but the area around the pad was littered with birds killed in the shockwave. The rocket's flight had lasted a mere twenty-three seconds.

His dream of a lunar mission was fading with every passing day, but Leonov continued to train, now once again for a possible L1 circumlunar flight. In particular, he practiced hard in simulating a manual descent through Earth's atmosphere when returning from the moon. He witnessed another successful launch of an unmanned L1 test flight in August 1969. This time it made a near-perfect flight around the moon and a safe return to Earth. Plans were formulated for a manned flight the following April, and Leonov actively lobbied Communist Party chiefs to keep this program going. But it was too late: not wishing to appear a runner-up in the moon race, Brezhnev vetoed the idea. Leonov would never leave Earth orbit, and to date no other cosmonaut has done so either. Having finally proved the technology needed for humans to fly around the moon, the Soviets would abandon it as an expensive but lost cause. "It was a devastating personal blow," Leonov wrote in his memoirs. "I felt I had wasted the best years of my life."

In late 1969, acknowledging that he would never leave Earth orbit, Leonov began training for a long-term Earth orbital mission on a space station. His lunar training diminished, finally ebbing away for good in May 1970. "The goal I had treasured for so long," he wrote, was now out of reach forever. He seriously considered writing another letter to Brezhnev complaining about the management of the space program. He discussed the possibility with Kamanin, who wisely suggested that the backlash would be worse than the current situation, dissuading him.

Within the year, Leonov was chosen to command the second crew scheduled to visit the first Soviet space station, known as Salyut. Pyotr Kolodin and Valery Kubasov were selected as his crew members. Kolodin had been second backup to Leonov for *Voskhod 2*. It had now been five years since Leonov had flown in space, four of which had been spent on a dead-end lunar program. Now he had to change gears and quickly learn the intricacies of an entirely new program, a challenge he admits caused him a great deal of stress. He could only hope his luck would now change.

The first crew to fly to Salyut was not able to correctly dock with the station, and so the mission of Leonov's follow-up crew became all the more critical to the future success and stature of the Soviet space program. The launch date for *Soyuz 11* was set for 6 June 1971, but then fate struck a totally unexpected blow. In the final preflight medical examination, just three days before the launch date, Kubasov underwent a routine chest x-ray. The doctors noticed a spot on his lung and suspected tuberculosis. Kubasov was immediately replaced by his backup, Vladislav Volkov. Leonov was not pleased; Volkov did not seem as well prepared as Kubasov. But even more devastating news was to come. A mere two days before launch, Leonov was told that his entire crew was being replaced by the backup crew. He flew into a rage and promised to contest the decision. He kept that promise, arguing forcefully in the short time remaining before launch that only Kubasov should be replaced on his crew. He had most of the Cosmonaut Training Center on his side, including Kamanin, and he nearly won the argument. Only Mishin opposed him, but that was enough. Leonov was told he had lost the battle. He would sit out this mission, and the backup crew would fly to Salyut in his crew's place.

Twenty-four days after launch, cosmonauts Dobrovolsky, Patsayev, and Volkov, following an incredibly successful mission on board the space station, undocked *Soyuz 11* from Salyut for their return to Earth—and died on reentry when their spacecraft depressurized. Kolodin, Kubasov, and Leonov grieved for their three colleagues, but they also realized that they had been lucky to escape the same fate. Kubasov's supposed "lung ailment," which had saved their lives, turned out to be nothing more dramatic than an allergic reaction to a pesticide.

Although he was assigned to other crews, Kolodin never had the opportunity to fly in space. Once Kubasov had recovered, he and Leonov trained for further Salyut missions—but now as a two-person crew. The space station they had trained so hard to fly to was never visited again. In July 1972, Leonov and Kubasov were ready to fly to a new space station, but it failed to reach orbit. They were ready once again in May 1973 to launch to yet another space station, but it lost attitude control, and the flight was cancelled. Eight years had now passed since Leonov had flown in space. Since that time he had seen all of his planned destinations—lunar orbit, the lunar surface, and three different space stations—taken from him, in some instances just days before a planned launch. His only piece of good fortune had resulted in the death of three colleagues rather than himself—a dubious piece of luck at best. "I was really beginning to lose faith in the program," a wistful Leonov wrote in his memoirs. He must have wondered if he would ever fly in space again.

Then, finally, Leonov's luck changed. Just days after the last space station failure, he and Kubasov were given a prestigious and high-profile assignment that would revitalize both of their careers. As one of the longest-training crews, they were a natural choice to be the Soviet crew for a joint mission with the Americans. "The discussions between Russia and America began in 1972," Leonov told the authors. "President Nixon and Premier Kosygin, and President of Russian Academy of Sciences Keldysh. The decision was, let's do it, with our space programs." Leonov's first concern was that he didn't speak English, but this was no real obstacle—he would soon learn and remain proficient in it for life. He was also a decade older than when he had commanded *Voskhod 2* and had been diagnosed with a minor heart condition that required medication in space. He later told the authors that, to prepare for the mission, he put himself on a strict program to lose weight and regain his former fitness. "In the last four months before the flight, I never drank whisky, never drank vodka. Maybe one week before, I drank one glass of Georgian wine!"

On 15 July 1975, over ten years after *Voskhod 2*, Leonov and Kubasov were launched into orbit aboard *Soyuz 19*. Two days later they docked with the last manned Apollo spacecraft to fly, and Leonov symbolically shook hands with American commander Tom Stafford through the opened hatch. It had been a frustrating decade for Leonov, full of missed opportunities that he was not yet allowed to reveal to his American fellow travelers. But now, in a mission full of symbolism, he'd regained his place as one of the most respected and well-known symbols of his country's space achievements. His personality was perfect for that of international ambassador, and he charmed the Americans when he visited the United States as part of the joint training sessions. He brought a jovial, sunny, engaging face to the Soviet space program that had not been seen since the heady days of Gagarin and Titov, and he knew how to create the perfect sound bite for the media. Leonov even charmed the actor John Wayne, who harbored a deep hatred of Communists. As Leonov told the authors, it was "a new spirit between our countries, very important at that time." Deke Slayton found him "jolly and friendly, always joking, a lot like Wally Schirra." For the first cooperative mission between the superpowers, there could not have been a better choice. In contrast to his previous, dangerous mission, Leonov later said his *Soyuz 19* mission went "as smoothly as a peeled egg."

Reflecting on the frustrating decade between his flights, Leonov claimed that in retrospect he would probably have concentrated his energies differently. "The thing that worries me most is whether I have done all I could. I have a constant desire to redo, to improve something. Ten years have passed, after all. There have been happy moments, but there have been disappointments as well. . . . I should like to rise to altitudes higher than those we fly at today. The Earth must look quite different from there." After the successful completion of *Soyuz 19*, Leonov was promoted to major general and finally achieved his ambition of becoming the commander of the cosmonaut team. Indeed, he was also elevated to deputy director of the Cosmonaut Training Center. He became in many respects the Deke Slayton of the Soviet space program—respected by his colleagues and deeply involved in many of the vital mission-crew choices. And just like their American astronaut counterparts, the cosmonauts knew that an invitation to Leonov's office might provide the first hint of a crew assignment.

Leonov remained in charge of crew training at Star City until 1991, and it was a job he took extremely seriously. In social circles Leonov was smiling and congenial, but when it came to supervising cosmonauts he was all business, sober and professional. The mildly irresponsible Alexei Leonov of the past was now a memory. In his elevated position of responsibility he firmly believed in the principle of leading by example and maintained his own skills to the same high levels he expected of his crews. His easy compatibility with foreign guests made him a natural to personally supervise the training of spacefarers from other countries. In those years he also advanced his own education, obtaining a degree in technical sciences in 1981. Like Slayton, Leonov would accompany each crew to the launch facilities on the day of liftoff. He would usually only soften his authoritative mood on the day of launch itself, often helping the cosmonauts smuggle aboard small items they wanted to take into space with them. He even thought to give them samples of sweet-smelling *polin* grass, which would give the spacefarers something more pleasant to smell in the sterile spacecraft. Once, unofficially, he even tried to sneak a crew away to go fishing in the days before a launch, but the doctors found out and vetoed the plan, worried that the cosmonauts might catch a cold. It may have been against the rules, but Leonov still remembered what it was like to be on the other side of the spacesuit visor.

In early 1982, Leonov officially stood down from an active cosmonaut career. He remained at Star City as deputy director for another nine years and continued to oversee cosmonaut training until he was relieved of the position in September 1991. He would have stayed in the military longer, but a new rule had been passed requiring any officer aged fifty-five or over to retire. The attempted coup to oust Gorbachev from power had also caused a military shake-up, and Leonov was an innocent casualty. He would probably have become the director of Star City had he been allowed to stay. He was bitterly disappointed at his enforced retirement, which formally took place in March 1992. "It was a stab in the back," he would write in his memoirs. "I felt betrayed . . . this training had been my life's work."

An influential national hero, Leonov was unlikely to brood for long. Almost immediately, he began a lucrative career administering a bank and

The irrepressible Alexei Leonov in 2004. Courtesy Francis French.

overseeing international investment funds. At first, this was a risky venture in the administrative chaos that followed the collapse of Communism, but it seems to have paid off for him. Always open to a new challenge, he found the work a refreshing and invigorating change of direction. These days, when Leonov opens his wallet the gleam of many gold credit cards suggests that he is doing very well indeed. He is also pleased to see that the international cooperation of the Apollo-Soyuz mission is continuing. As he told the authors: "I would have liked more missions together at the time. Now, we have the orbital international station. It is a very good base for understanding each other during the next twenty-five years. It is very important." He is still as engaging a character as he ever was. Arthur C. Clarke describes Leonov as "generous and mischievous. . . . When Alexei's up to his tricks, it's not easy to remember what courage it must have taken to be the very first man to step outside a spacecraft and float in the empty void."

Throughout the many years of his cosmonaut career, Leonov had always kept up his artistic talents. He took some crayons with him on *Voskhod 2* and drew sketches in the ship's log. One particularly striking image caught the many-colored layers of an orbital sunrise—something cameras of the time had yet to capture in all its nuances. During the Apollo-Soyuz training and mission he also sketched his crew mates and anticipated the *Soyuz 19* lift-off by capturing it in oils before he flew. "It is not too difficult to write and draw in weightlessness," he explains. "Only the pencil has to be placed with greater power on the paper, and this takes practice." Back on Earth, he created vivid and evocative oil paintings of his space experiences, images that were also reproduced on Soviet stamps. One memorable series of sketches Leonov created chronicles the history of the Baikonur Cosmodrome back to the days of nomads in tents. Leonov finds his artwork a useful form of self-expression and a way of better understanding his colleagues and himself. He was never able to walk on the moon, but says that through his paintings he has explored its mysteries, traveled to the surface of Mars, and journeyed beyond. "I feel obligated to share my vision of space with people by means of the art I have at my disposal," Leonov comments.

In 2001, astronaut Gene Cernan and Alexei Leonov were sharing a drink in a San Diego hotel bar. Cernan pondered the situation for a moment, then announced to the room: "Hey, do you realize that the two of us are the earliest living spacewalkers?" His dangerous, historic spacewalk from *Voskhod 2* has forever enshrined Alexei Leonov in the history books, while his participation in the Apollo-Soyuz flight only added to his legendary stature as a true pioneer of spaceflight. As someone who also survived an incredibly risky spacewalk, Cernan is well qualified to pass judgment on Alexei Leonov. He characterizes him as "one of the gutsiest men alive."

With the Voskhod and Mercury programs now completed, the United States and Soviet Union had each brought to an end their pioneering thrust into the cosmos. The silent sea of space was no longer quite the mystery it had once been.

Surprisingly few of the spacefarers from those years would ever ride a rocket again. For some, such as Shepard, Glenn, and Leonov, many years and space achievements separated their first and second missions. Meanwhile, new groups of astronauts and cosmonauts, coming from even more diverse backgrounds, had been selected to join the great venture. Through their experience and skills they would eventually push the boundaries even further on increasingly complex and challenging Gemini, Soyuz, and Apollo missions.

Just four years had elapsed between Yuri Gagarin's history-making flight into orbit and Alexei Leonov's spacewalk outside of *Voskhod 2*. In the four years that followed those first tentative steps, a whole new breed of spacefarers would travel ever faster, higher, and farther, until humans would lose sight of our Earth altogether, as they gently slipped into the shadows of the moon.

References

Entries in this bibliography are organized in four sections: books, periodicals and online articles, interviews and personal communications, and other sources (films, personal papers, public documents, presentations, CD-ROMs, and reference sources).

Books

Abbas, Khwaja Ahmad. *Till We Reach the Stars: The Story of Yuri Gagarin.* New York: Asia Publishing House, 1961.

Ackmann, Martha. *Mercury 13: The Untold Story of Thirteen American Women and the Dream of Space Flight.* New York: Random House, 2003.

Aldrin, Buzz, and Malcolm McConnell. *Men from Earth.* New York: Bantam Books, 1989.

Always Aim to Be the First: The 30th Anniversary of the Flight of Yuri Gagarin, the First Cosmonaut (booklet). Moscow: Novosti Publishers, 1991.

Baker, David. *The History of Manned Spaceflight.* New York: Crown Publishers, 1981.

Bartos, Adam, and Svetlana Boym. *Kosmos: A Portrait of the Russian Space Age.* New York: Princeton Architectural + PHS, 2001.

Bell, Joseph N. *Seven into Space: The Story of the Mercury Astronauts.* Chicago: Popular Mechanics, 1960.

Boomhower, Ray E. *Gus Grissom: The Lost Astronaut.* Indianapolis: Indiana Historical Society Press, 2004.

Borisenko, I., and A. Romanov. *Where All Roads into Space Begin.* Moscow: Progress Publishers, 1982.

Burchett, Wilfred, and Anthony Purdy. *Gagarin,* London: Panther Books, 1961.

———. *Gherman Titov's Flight into Space.* London: Panther Books, 1962.

Burgess, Colin, Kate Doolan, and Bert Vis. *Fallen Astronauts: Heroes Who*

Died Reaching for the Moon. Lincoln: University of Nebraska Press, 2003.

Caidin, Martin. *Man into Space.* New York: Pyramid Books, 1961.

———. *The Astronauts: The Story of Project Mercury.* New York: E.P. Dutton, 1961.

———. *Rendezvous in Space.* New York: E.P. Dutton, 1962.

Carpenter, Scott, Gordon Cooper, John Glenn, Virgil Grissom, Walter Schirra, Alan Shepard, and Donald Slayton. *We Seven: By the Astronauts Themselves.* New York: Simon & Schuster, 1962.

———, and Kristen Stoever. *For Spacious Skies.* New York: Harcourt, 2003.

Carroll, Peter N. *Famous in America: The Passion to Succeed.* New York: E.P. Dutton, 1986.

Cassutt, Michael. *Who's Who in Space.* International Space Year edition. New York: Macmillan Publishing, 1993.

———. *Who's Who in Space.* 3rd ed. New York: Macmillan, 1998.

Chaikin, Andrew. *A Man on the Moon.* New York: Viking Penguin Group, 1994.

Chappell, Carl L. *Seven Minus One.* Madison IN: New Frontier, 1968.

Clark, Phillip. *The Soviet Manned Space Program.* New York: Salamander Books, 1988.

Cobb, Jerrie. *Solo Pilot.* Sun City Center FL: Jerrie Cobb Foundation, 1997.

———, and Jane Rieker. *Woman into Space: The Jerrie Cobb Story.* Englewood Cliffs NJ: Prentice Hall, 1963.

Cochran, Jacqueline. *The Stars at Noon.* Boston: Little, Brown, 1954.

Cochran, Jacqueline, and Maryann Bucknum Brinley. *Jackie Cochran: The Story of the Greatest Woman Pilot in Aviation History.* New York: Bantam, 1987.

Collins, Michael. *Carrying the Fire: An Astronaut's Journeys.* New York: Farrar, Straus and Giroux, 1974.

———. *Liftoff: The Story of America's Adventure in Space.* New York: Grove Press, 1988.

Cooper, Gordon, with Bruce Henderson. *Leap of Faith: An Astronaut's Journey into the Unknown.* New York: Harper Collins, 2000.

Cormier, Zeke, Barrett Tillman, Wally Schirra, and Phil Wood. *Wildcats to Tomcats: The Tailhook Navy*. St. Paul MN: Phalanx, 1995.

Cunningham, Walter. *The All-American Boys*. New York: Macmillan, 1977.

Doran, Jamie, and Piers Bizony. *Starman: The Truth behind the Legend of Yuri Gagarin*. London: Bloomsbury, 1998.

Fallaci, Oriana. *If the Sun Dies*. Translated by Pamela Swinglehurst. New York: Atheneum, 1966.

Freni, Pamela. *Space for Women: A History of Women with the Right Stuff*. Santa Ana CA: Seven Locks Press, 2002.

Furniss, Tim. *Manned Spaceflight Log*. London: Jane's, 1983.

Gagarin, Valentin. *My Brother Yuri*. Moscow: Progress Publishers, 1973.

Gagarin, Yuri. *Road to the Stars*. Moscow: Foreign Language Publishing, n.d.

———, and Vladimir Lebedev. *Survival in Space*. New York: Frederick A. Praeger, 1969.

Gibson, Edward G., ed. *The Greatest Adventure*. Sydney: C. Pierson Publishers, 1994.

Glenn, John, and Nick Taylor. *John Glenn: A Memoir*. New York: Bantam Books, 1999.

Golovanov, Yaroslav. *Korolev: Fakty I Mify* [Korolev: Facts and myths]. Moscow: Nauka, 1994.

———. *Our Gagarin*. Moscow: Progress Publishers, 1978.

Grissom, Betty, and Henry Still. *Starfall*. New York: Thomas Y. Crowell, 1974.

Grissom, Virgil. *Gemini! A Personal Account of Man's Venture into Space*. London: Macmillan, 1968.

Gubarev, Vladimir. *The Man from a Legend*. Moscow: Progress Publishers, 1988.

Gurney, Gene, and Clare Gurney. *The Soviet Manned Space Programme: Cosmonauts in Orbit*. New York: Franklin Watts, 1972.

Hall, Rex, and David J. Shayler. *The Rocket Men: Vostok & Voskhod, the First Soviet Manned Spaceflights*. Chichester UK: Praxis, 2001.

Harford, James. *Korolev: How One Man Masterminded the Soviet Drive to Beat America to the Moon*. New York: John Wiley & Sons, 1997.

Hartmann, William, Ron Miller, Andrei Sokolov, and Vitaly Myagkov, eds.

In the Stream of Stars: The Soviet/American Space Art Book. New York: Workman, 1990.

Hawthorne, Douglas. *Men and Women of Space.* San Diego: Univelt, 1992.

Haynsworth, Leslie, and David Toomey. *Amelia Earhart's Daughters.* Scranton PA: William Morrow, 1998.

Holden, Henry M., and Captain Lori Griffith. *Ladybirds II: The Continuing Story of American Women in Aviation.* Mt. Freedom NJ: Black Hawk, 1993.

Hooper, Gordon. *The Soviet Cosmonaut Team.* Vol. 2: *Cosmonaut Biographies.* Suffolk UK: GRH Publications, 1990.

Kamanin, Nikolai. *I Feel Sorry for Our Guys: General N. Kamanin's Space Diaries.* NASA Publication No. TT-21658. Washington DC: NASA, 1993.

———. *The Hidden Cosmos.* Moscow: Infortekst, 1995.

Kelley, Kevin W., ed. *The Home Planet.* Reading MA: Addison-Wesley; Moscow: Mir Publishers, 1988.

Kelly, Dr. Fred. *America's Astronauts and Their Indestructible Spirit.* Blue Ridge Summit PA: Aero (Division of TAB Books), 1986.

Kevles, Bettyann Holtzmann. *Almost Heaven: The Story of Women in Space.* New York: Basic Books, 2003.

Kosmos Moskau Berlin: Ein Bilbrand vom Besuch German Titows in der DDR. Berlin: Verlag Kultur und Fortschritt, 1961.

Kraft, Chris. *Flight: My Life in Mission Control.* New York: Penguin / Dutton Books, 2001.

Kranz, Gene. *Failure Is Not an Option.* New York: Simon & Schuster, 2000.

Lambright, W. Henry. *Powering Apollo: James E. Webb of* NASA. Baltimore: Johns Hopkins University Press, 1995.

Lebedev, L., B. Lukyanov, and A. Romanov. *Sons of the Blue Planet.* New Delhi: NASA publication, produced by New Delhi Amerind Publishing, 1973.

Leonov, Alexei. *The Sun's Wind.* Moscow: Progress Publishers, 1977.

———. *I Walk in Space.* Moscow: Malysh Publishers, 1980.

———, and Andrei Sokolov. *The Stars Are Awaiting Us.* Moscow, 1967.

———, and Vladimir Lebedev. *Space and Time Perception by the Cosmonaut.* Moscow: Mir Publishers, 1971.

———, and Andrei Sokolov. *Life among Stars.* Moscow, 1981.

Lewis, Richard S. *Appointment on the Moon.* New York: Viking Press, 1968.

Lothian, Lady Antonella. *Valentina, First Woman in Space: Conversations with A. Lothian.* Edinburgh, Scotland: Pentland Press, 1993.

Maranin, I. A., S. Shamsutdinov, and A. Glushko. *Soviet and Russian Cosmonauts 1960-2000.* Moscow: Novosti Kosmonautika Publishers, 2001.

McDonnell, Virginia B. *Dee O'Hara: Astronauts' Nurse.* New York: Thomas Nelson & Sons, 1965.

Murray, Charles, and Catherine Bly Cox. *Apollo: The Race to the Moon.* New York: Simon & Schuster, 1989.

Newport, Curt. *Lost Spacecraft: The Search for Liberty Bell 7.* Burlington, Ontario: Apogee Books, 2002.

Nolen, Stephanie. *Promised the Moon: The Untold Story of the First Women in the Space Race.* New York: Four Walls Eight Windows, 2003.

Oberg, James. *Red Star in Orbit.* New York: Random House, 1981.

O'Leary, Brian. *The Making of an Ex-Astronaut.* Boston: Houghton Mifflin Company, 1970.

Outer Space and Man. Moscow: Mir Publishers, 1967.

Pogue, William. *Space Trivia.* Burlington, Ontario: Apogee Books, 2003.

Ponomareva, Valentina. *The Female Face of the Cosmos.* Moscow: Gelios Publications, 2002.

Project Mercury. Editors of Time Life. New York: Time Life, 1964.

Pushkin, Alexander. *Eugene Onegin.* London: Penguin Classics, 1977.

———. *The Queen of Spades and Other Stories.* London: Penguin Classics, 1962.

Results of the Second U.S. Manned Suborbital Flight. Washington DC: NASA HQ History Office, 1961.

Riabchikov, Evgeny. *Russians in Space.* New York: Doubleday, 1971.

Romanov, A. *Spacecraft Designer: The Story of Sergei Korolev.* Moscow: Novosti Press, 1976.

Sacknoff, Scott, ed. *In Their Own Words: Conversations with the Astronauts and Men Who Led America's Journey into Space.* Bethesda MD: Space Publications, 2003.

Schefter, James. *The Race: The Uncensored Story of How America Beat Russia to the Moon.* New York: Random House, 1999.

Schirra, Wally, and Richard Billings. *Schirra's Space.* Boston: Quinlan Press, 1988.

Scott, David, and Alexei Leonov. *Two Sides of the Moon.* New York: Simon & Schuster, 2004.

Shatalov, Vladimir, and Mikhail Rebrov. *Cosmonauts of the USSR.* Moscow: Prosveshcheniye, 1980.

Sharpe, Mitchell R. *It Is I, Seagull.* New York: Thomas Y. Crowell, 1975.

———. *Yuri Gagarin, First Man in Space.* Huntsville AL: The Strode Publishers, 1969.

Shelton, William. *Man's Conquest of Space.* Washington DC: National Geographic Society, 1968.

———. *Soviet Space Exploration: The First Decade.* London: Arthur Barker Ltd., 1968.

Shepard, Alan, and Deke Slayton. *Moon Shot: The Inside Story of America's Race to the Moon.* Atlanta: Turner Publishing, 1994.

Schick, Ron, and Julia Van Haaften. *The View from Space: American Astronaut Photography 1962-1972.* New York: Clarkson N. Potter, Inc., 1988.

Siddiqi, Asif. *Challenge to Apollo: The Soviet Union and the Space Race, 1945-1974.* Washington DC: NASA History Office Publication, 2000.

———. *Sputnik and the Soviet Space Challenge.* Gainesville: University Press of Florida, 2003.

———. *The Soviet Space Race with Apollo.* Gainesville: University Press of Florida, 2003.

Silverberg, Robert. *First American in Space.* Derby CT: Monarch Books, 1961.

Slayton, Donald K., and Michael Cassutt. *Deke! U.S. Manned Space: From Mercury to the Shuttle.* New York: Forge Books, 1994.

Smaus, Jewel Spangler, and Charles B. Spangler. *America's First Spaceman.* New York: Doubleday, 1962.

Sobel, Lester A., ed. *Space: From Sputnik to Gemini.* New York: Facts On File, 1965.

Soviet Man in Space. Moscow: Foreign Languages Publishing House, 1961.

Steadman, Bernice, and Jody M. Clark. *Tethered Mercury.* Traverse City MI: Aviation Press, 2001.

Suvorov, Vladimir, and Aleksandr Sabelnikov. *The First Manned Spaceflight: Russia's Quest for Space.* New York: Nova Science Publishing, 1999.

Swenson, Loyd S., Jr., James Grimwood, and Charles C. Alexander. *This New Ocean: A History of Project Mercury.* Washington DC: NASA Publications, 1998.

Syme, Anthony. *The Astronauts.* Sydney: Horwitz Publications, 1965.

Tereshkova, Valentina. *Stars Are Calling.* Moscow, 1963.

The Astronauts Book. Norwich UK: Panther Books, 1966.

Thompson, Neal. *Light This Candle: The Life & Times of Alan Shepard, America's First Spaceman.* New York: Crown Publishers, 2004.

Titov, Gherman. "An Amazing Art." In *The Soviet Circus: A Collection of Articles.* Moscow: Progress Publishers, 1967.

———. *700,000 Kilometers through Space.* Moscow: Foreign Languages Publishing House, n.d.

———, and Martin Caidin. *I Am Eagle!* New York: Bobbs-Merrill, 1962.

———, with Pavel Barashev and Yuri Dokuchayev. *Gherman Titov: First Man to Spend a Day in Space.* New York: Crosscurrents Press, 1962.

Trux, John. *The Space Race: From Sputnik to Shuttle.* London: New English Library, 1986.

Tsymbal, Nikolai (editor). *First Man in Space: The Life and Achievement of Yuri Gagarin.* Moscow: Progress Publishing, 1984.

Vladimirov, Leonid. *The Russian Space Bluff.* London: Tom Stacey Ltd., 1971.

Walk into Space. Moscow: Novosti Press Agency Publishing House, 1965.

Weiss, Otis L., ed. *The United States Astronauts and Their Families.* New York: World Book Encyclopedia Science Service, 1965.

Wendt, Guenter, and Russell Still. *The Unbroken Chain.* Burlington, Ontario: Apogee Books, 2001.

White, Frank. *The Overview Effect: Space Exploration and Human Evolution.* Wilmington MA: Houghton Mifflin, 1987.

Wilford, John Noble. *We Reach the Moon.* New York: W.W. Norton, 1969.

Wolfe, Tom. *The Right Stuff.* New York: Farrar, Straus and Giroux, 1979.

Periodicals and Online Articles

Ackmann, Martha. "For a Shot at Space." *Houston Chronicle*, 27 February 2000.

Agle, D. C. "Flying the Gusmobile." *Air & Space*, August/September 1998. http://www.airspacemag.com/ASM/Mag/Index/1998/AS/ftgm.html.

"Ask an Astronaut, Q&A with John Glenn." National Space Society, November 1996. http://www.ari.net/nss/askastro/Glenn/answers.html.

Baker, Tobie. "Pilot Plans to Continue Space Travel Quest." *(Grenada) Daily Sentinel-Star*, 19 May 2000.

"Bid to Kill Russian Spacemen." *Daily Telegraph* (Sydney), 23 January 1969.

Burgess, Colin. "Lost Mercury: Alan Shepard's Unflown Mission." *Spaceflight*, vol. 45 (May 2003), 208-11.

———, and Francis French. "Only Males Need Apply: The Lovelace Women and the Not-So-Right Stuff." *Spaceflight*, vol. 41, no. 1 (January 1999), 28-32.

Dalavai, Leona. "Wally Funk: Born to Fly." *Distinctive Lifestyles*, September/October 1996.

Dokuchayev, Yuri. "Yuri Gagarin." *Soviet Life*, April 1981, 9-16.

"Ex-cosmonaut Savors Thrill but Rues Waste," *Russian Weekly*, 23 October 1998.

Farley, Tracy. "Gordon Cooper Reaches for the Stars." *Shawnee News-Star*, 17 October 1999.

"The 'First Lady of Space' Remembers." *Quest*, vol. 10, no. 2 (2003), 6-21. (Translated excerpt from Tereshkova, *Stars Are Calling*, Moscow, 1963.)

"40 Years of First Woman Cosmonaut." *Kathmandu Post* (*Sunday Post*), June 2003.

"45 Days Deep Study." *The Australian*, 30 August 1962.

French, Francis. "Yuri Gagarin's Visit to Manchester." *Spaceflight*, vol. 40, no. 7 (July 1998), 261-63.

———. "Lost Faith: A Lone Rebel and Space Bureaucracy." *Spaceflight*, vol. 43, no. 9 (September 2001), 374-80.

———. "I Worked with NASA, Not for NASA: An Interview with Wally Schirra." *Spaceflight*, vol. 43, no. 11 (November 2001), 471-75.

———. "A Vital Link: Guenter Wendt and the Meaning of Responsibility." *Collect Space: The Source for Space History and Artifacts*. http://www.collectspace.com/news/news-052102a.html.

———, and Wally Funk. "Another Giant Leap." *Spaceflight*, vol. 41, no. 2 (December 1999), 517-20.

Funk, Wally. "Training with the Cosmonauts." *International Women Pilots*, July/August 2000.

———, and Francis French. "Wally Funk Trains with the Cosmonauts." *Aviation for Women* January/February 2001.

Gray, Tara. "40th Anniversary of the Mercury 7." NASA HQ History Office (Washington DC). http://www.nasa.gov/office/pao/History/40th merc7.htm.

Halvorson, Todd. "Cosmonauts Saddened by Titov's Death." Space.com. http://www.space./com/peopleinterviews/cosmonauts titov 000921.html.

"Gherman Titov" (obituary). *The Guardian* (London), 22 September 2000.

"Gherman Titov" (obituary). *The Times* (London), 22 September 2000.

"Gherman Titov" (obituary). *Weekly Telegraph* (London), 4 October 2000.

"Gherman Titov, Second Person in Orbit, Buried in Moscow." *Florida Today*, 26 September 2000. http://www.flatoday.com/space/explore/stories/2000b/092600b.htm.

"Gordon Cooper, the Man." Gordon Cooper Technology Center (Shawnee OK). http://www.getech.org/about/geman.htm.

"Gunman in Moscow." *Daily Telegraph* (Sydney), 23 January 1969.

Hehs, Eric. "Jackie Parker: Fighter Pilot." *Code One*, April 1995.

Ikeda, Daisaku. "Valentina Tereshkova: First Woman in Space." *Living Buddhism*, September 1999.

Jessen, Gene Nora. Letter to Editor. *International Women Pilots,* March/April 1999.

Kinnunen, Helena. "The First Lady of Space: Valentina Tereshkova." *Helsingin Sanomat,* 17 September 2002.

Krum, Sharon. "Space Cowgirl." *The Guardian,* 2 April 2002.

"Man Shot into Orbit by Russia." *Sydney Morning Herald,* 15 August 1962.

Moore, Roger E. "No Go: Project Mercury Missions and Spacecraft That Never Left Earth." *AOL Hometown.* http://members.aol.com/toger70129/nogo.html.

Nikolayeva, Anna, and Roman Popovich. "Mother of Cosmonaut-3." *Soviet Weekly,* October 1997, 22.

———. "Father of Cosmonaut-4." *Soviet Weekly,* October 1997, 22.

Noble, Marilyn. "Wally Funk: From Cowgirl to Spacegirl." *Dallas/Fort Worth Aviation & Business,* February 2003.

O'Meara, Kelly Patricia. "The First Women Astronauts." *Insight on the News,* 13 March 2000.

"Pair Might Swap Ships," *Daily Mirror* (Sydney), 17 August 1962.

Pugh, Eddie. "Interview with Scott Carpenter." *Space Flight News,* no. 54 (June 1990), 20-25.

Reese, Jack. "Cooper: A Glimpse at an Astronaut." *Shawnee (Kansas) News-Star,* 29 June 1963.

Rimmer, Terri. "Pilot to Realize Dream of Traveling in Space." *Keller (Texas) Citizen,* 1 February 2000.

"Russian Up for Long Flight." *Daily Telegraph* (Sydney), 15 August 1962.

"Safe Splashdown." *The Australian,* 30 August 1965.

"Second American in Orbit." *Flight International,* 31 May 1962, 871.

"Space Flight by Russian Woman Near." *Daily Mirror* (Sydney), 14 August 1962.

"Spaceman Whirls to Record." *Daily Telegraph* (Sydney), 19 August 1962.

"Undersea Aquanauts Get Space Phone Call." *Sydney Morning Herald,* 29 August 1965.

Uvarova, Marina. "On the Heels of Gagarin." Interview with Valentina Tereshkova. *Moscow Times,* 6 April 2001.

"Valentina Tereshkova—First Woman in Space." *Russian Center for Science and Culture* (newsletter), April 2003.

Wilbur, Ted (Commander). "Once a Fighter Pilot." *Naval Aviation News*, November 1970, 4-7.

Wilson, Keith T. "Mercury Atlas 10: A Mission Not Flown." *Quest*, vol. 2, no. 4 (1993), 22-25.

"Woman in Orbit." *Daily Telegraph* (Sydney), 17 August 1962.

Interviews and Personal Communications

Ackmann, Martha. E-mail correspondence with Francis French, 5/6 January 2004.

Ary, Max. Interview by Lawrence McGlynn (telephone). 28 March 2003.

Berry, Charles. Interview by Carol Butler. 29 April 1999. NASA Oral History. http://www.jsc.nasa.gov/history/oral_histories/oral_histories.htm.

Bykovsky, Valery. Interview by Colin Burgess. Coventry UK. 11 October 2004.

Carpenter, Scott. Interview by Francis French. Los Angeles CA. 27 September 2003.

———. Interview by Colin Burgess (telephone). 16 December 2002.

Cooper, Gordon. Interview by Roy Neal. 21 May 1998. NASA Oral History. http://www.jsc.nasa.gov/history/oral_histories/oral_histories.htm.

———. Interviews by Francis French. Ventura CA, 25 April 2001, and Santa Monica CA, 26 October 2002.

Frankman, Betty Skelton. Interview by Carol Butler. 19 July 1999. NASA Oral History. http://www.jsc.nasa.gov/history/oral_histories/oral_histories.htm.

Frasketi, Joseph, Jr. E-mail correspondence with Colin Burgess. 26 December 2002.

Funk, Wally. Interview by Carol Butler. 18 July 1999. NASA Oral History. http://www.jsc.nasa.gov/history/oral_histories/oral_histories.htm.

———. E-mail correspondence with Francis French. January 1999 to January 2005.

Grissom, Lowell. E-mail correspondence with Colin Burgess. 8 July 2003.

Haney, Paul. Interview by Sandra Johnson. 20 January 2000. NASA Oral

History. http://www.jsc.nasa.gov/history/oral_histories/oral_histories.htm.

———. E-mail correspondence with Colin Burgess. 18 September 2003 to February 2005.

Kranz, Gene. Interviews by Roy Neal. 19 March 1998 and 28 April 1999. NASA Oral History. http://www.jsc.nasa.gov/history/oral_histories/oral_histories.htm.

———. Interview by Rebecca Wright. 8 January 1999. NASA Oral History. http://www.jsc.nasa.gov/history/oral_histories/oral_histories.htm.

Leonov, Alexei. Interviews by Francis French. San Diego CA, 23/24 March 2001, and Burbank CA, 2 September 2004.

Lewis, Jim. E-mail correspondence with Colin Burgess. November 2002 to August 2003.

Melroy, Pamela. E-mail correspondence with Colin Burgess. 2/10 December 2003.

Nikolayev, Andrian. Interview by Rex Hall. Coventry UK. 22 November 2003.

Northcutt, Robert. E-mail correspondence with Colin Burgess. December 2004/January 2005.

O'Hara, Dee. Interview by Rebecca Wright. 23 April 2002. NASA Oral History. http://www.jsc.nasa.gov/history/oral_histories/oral_histories.htm.

———. Interview by Colin Burgess and Francis French. San Diego CA. 18 January 2003.

Ponomareva, Valentina. Interview by Bert Vis and Rex Hall. Moscow. 14 April 2000. Used by permission of B. Vis and R. Hall.

Popovich, Pavel. Interview by Rex Hall. Coventry UK. 22 November 2003.

———. interview conducted by Bert Vis, Ottawa, Canada, 30 September 1996. Used by permission of B. Vis.

Schirra, Wally. Interview with Roy Neal. 1 December 1998. NASA Oral History. http://www.jsc.nasa.gov/history/oral_histories/oral_histories.htm.

———. Interviews by Francis French. San Diego CA, 19 February 2001 and 2 May 2001, and Los Angeles CA, 26/27 September 2003.

Shepard, Alan B. Jr. Interview for Academy of Achievement. http://www.achievement.org/autodoc/page/sheoint-1.

———. Interview by Pam Platt. *Florida Today*, 22 July 1998. http://www.flatoday.com/space/today/072298g.htm.

———. Interview by Roy Neal. 20 February 1998. NASA Oral History. http://www.jsc.nasa.gov/history/oral_histories/oral_histories.htm.

Solovyova, Irina. Interview by Bert Vis. Moscow. 5 March 2001. Used by permission of B. Vis.

Stoever, Kristen. E-mail correspondence with Colin Burgess and Francis French. October 2002/January 2004.

Tereshkova, Valentina. Interview by Kerrie Dougherty. Sydney, Australia. 13 March 1996. Used by permission of K. Dougherty.

———. Interview by Slava Gerovitch. Moscow. 17 May 2002.

———. Interviews by Colin Burgess. Coventry UK. 8/11 October 2004.

Titov, Gherman. Interview by CNN Interactive Cold War Experience. http://www.cnn.com/SPECIALS/cold.war/episodes/08/interviews/titov/.

Wendt, Guenter. NASA Oral History interview with Doyle McDonald, 16 January 1998.

———. NASA Oral History interview with Catherine Harwood, 25 February 1999.

Yazdovsky, Valery. interview with Bert Vis, Moscow, 14 April 2000. Used by permission of B. Vis.

Other Sources

Carpenter, Scott. Presentation at Reuben H. Fleet Science Center, San Diego CA, 20 January 2003, hosted by Francis French.

Encyclopedia Astronautica/astronautix.com. http://www.astronautix.com.

Funk, Wally. Personal papers and archives. Roanoke TX.

NASA Mission Transcripts. CD-Rom number SP-2000-4602. Washington DC:NASA HQ.

The Hunt for Liberty Bell 7 (documentary film). Discovery Channel, 1999.

U.S. Congress. House. Committee on Science and Astronautics. *Qualifications for Astronauts: Hearings before the Special Subcommittee on the Selection of Astronauts.* 87th Cong., 2d sess, July 17/18 1962. Washington DC: USGPO, 1962.

University of Nebraska Press
Also of Interest:

Fallen Astronauts

Heroes Who Died Reaching for the Moon
By Colin Burgess and Kate Doolan, with Bert Vis
Foreword by Captain Eugene A. Cernan U.S. Navy (Ret.), Commander, Apollo 17

This book enriches the saga of mankind's greatest scientific undertaking, Project Apollo, and conveys the human cost of the space race—by telling the stories of those sixteen astronauts and cosmonauts who died reaching for the moon.

ISBN: 0-8032-6212-4; 978-0-8032-6212-6 (paper)

Teacher in Space

Christa McAuliffe and the Challenger Legacy
By Colin Burgess
Foreword by Grace George Corrigan

Christa McAuliffe's name is deeply entrenched in American history as the teacher who died when the *Challenger* exploded in January 1986. *Teacher in Space* explores and celebrates Christa's life and legacy and suggests that her goals of involving and educating children are being fulfilled even today.

ISBN 0-8032-6182-9; 978-0-8032-6182-2 (paper)

A Journal for Christa

Christa McAuliffe, Teacher in Space
By Grace George Corrigan

"A straightforward account of one woman's 'ordinary' life: a life made extraordinary nevertheless by a passion for teaching coupled with a rare and unexpected chance to chase [her] dreams toward the stars."—*Boston Globe*

ISBN 0-8032-6411-9; 978-0-8032-6411-3 (paper)